Case Studies in Neural Data Analysis

Computational Neuroscience

Terrence J. Sejnowski and Tomaso A. Poggio, editors

For a complete list of books in this series, see the back of the book and http://mitpress.mit.edu/Computational_Neuroscience

Case Studies in Neural Data Analysis
A Guide for the Practicing Neuroscientist

Mark A. Kramer and
Uri T. Eden

The MIT Press
Cambridge, Massachusetts
London, England

This book was set in Times Roman by diacriTech, Chennai.

Printed and bound in the United States of America.

Library of Congress Cataloging-in-Publication Data

Names: Kramer, Mark A., author. | Eden, Uri T., author.
Title: Case studies in neural data analysis : a guide for the practicing neuroscientist / Mark A. Kramer and Uri T. Eden.
Description: Cambridge, MA : The MIT Press, 2016. | Series: Computational neuroscience series | Includes bibliographical references and index.
Identifiers: LCCN 2016014656 | ISBN 9780262529372 (pbk. : alk. paper)
Subjects: LCSH: Neural analyzers. | Neuropsychological tests.
Classification: LCC QP357.5 .K69 2016 | DDC 616.8/0475—dc23 LC record available at https://lccn.loc.gov/2016014656

10 9 8 7 6 5 4 3 2 1

It is these boundary regions of science which offer the richest opportunities to the qualified investigator. They are at the same time the most refractory to the accepted techniques of mass attack and the division of labor.
—Norbert Wiener, *Cybernetics*

I am writing this book for students, dressmakers, secretaries, artists, lazy people, poets, [people] of action, dreamers, scientists, and everyone else who has only an hour for lunch or dinner but still wants thirty minutes of peace to enjoy a cup of coffee.
—Edouard de Pomiane, *French Cooking in Ten Minutes*

Contents

Preface

Before beginning in earnest, some quick notes.

The aim of this book is to teach skills in practical data analysis through a case study approach. Each chapter begins with a motivating dataset and then proceeds with hands-on methods to analyze these data. In this way, we hope to motivate exploration of data analysis methods. Although we do provide some mathematical discussions, this book does not provide a deep exploration of mathematical or statistical theory; for that, we have included references throughout.

There are multiple paths through this book. If you've never used MATLAB before and are new to data analysis, we recommend beginning with chapter 1. If you have some MATLAB experience, you might narrow your focus by data type, either neural field data (chapters 2–7) or spike train data (chapters 8–10). Or, you might narrow your focus by method type, namely, spectral analysis (chapters 3, 4, 10), generalized linear models (chapters 9, 10), coherence (chapters 5 and 11), or cross-frequency coupling (chapter 7). Each chapter consists of a separate case study and may be selected à la carte for targeted investigation.

In the chapters that follow, we present the data as case studies and provide a narrative describing how each dataset was collected. However, all the data presented here were simulated by the authors. We include only synthetic data to emphasize particular points that commonly appear in actual recordings. Readers should not base any scientific conclusions on these synthetic data.

We chose to use MATLAB in this book because, in our opinion, MATLAB currently serves as the programming language of neuroscience. This may change. If so, all the ideas and methods presented here can be translated to a different programming language.

We welcome corrections from readers and will try to maintain a running list of errors online. The website `http://github.com/Mark-Kramer/Case-Studies-Kramer-Eden` will contain these corrections. All of the data files and MATLAB code implemented in this book are also available on this website.

The authors thankfully acknowledge support from the Burroughs Wellcome Fund (Career Award at the Scientific Interface), the Simons Foundation (320136), the National Science Foundation (#1451384), the National Institutes of Health (R01 NS072023, R01 MH105174, R01 NS073118, R25 GM114827), and their students and colleagues.

1 Introduction to MATLAB

Synopsis

Data Simple numerical examples.

Goal Introduce basic operations in MATLAB.

Tools Create vectors and arrays, load data, make plots.

1.1 Introduction

In this book, we use the software package MATLAB. The best way to learn new software (and probably most things) is when you are motivated by a particular problem. In subsequent chapters, we pursue specific questions driven by neural data, and use our desire to understand these data to motivate the development and application of computational methods.

This chapter focuses on basic coding techniques and principles in MATLAB. In examples, you are asked to execute code in MATLAB. If your MATLAB experience is limited, you should actually *do* this, not just read the text.

This chapter follows in spirit and sometimes in detail chapter 2 of *MATLAB for Neuroscientists*, an excellent reference for learning to use MATLAB with many additional examples [1]. If you have never used MATLAB before, you might examine the *MATLAB Primer* (available on the MathWorks website).

1.2 Starting MATLAB

To begin, gain access to MATLAB. To do so, visit the MathWorks website, http://www.mathworks.com/, and check for licensing options from your local institution, if relevant. We assume in this book that you have access to MATLAB, the ability to launch the software, and a basic familiarity with the programming environment.

1.3 MATLAB Is a Calculator

Enter the following command at the MATLAB prompt:

```
4+3
```

Q: What does MATLAB return? Does it make sense?

You may notice that the font for `4+3` differs from the rest of the text. This font indicates MATLAB code. In this case, you are being asked to enter the command `4+3` at the MATLAB prompt, execute the command, and evaluate the output.

1.4 MATLAB Can Compute Complicated Quantities

Enter the following command at the MATLAB prompt:

```
4/10^2
```

Q: What does MATLAB return? Does it make sense? You may use parentheses to alter the result. How does

```
(4/10)^2
```

compare to your previous answer?

1.5 Built-in Functions

A *function* is a program that operates on inputs. In the following,

```
sin(2*pi)
```

`sin` is a function that computes the sine of its input; in this case, the input is `2*pi`. Here are three more examples of functions that operate on inputs:

```
cos(2*pi + 1/10)
exp(-2)
atan(2*pi)
```

Q: What is the function `atan`?

A: Try using MATLAB Help. For example, at the MATLAB prompt, type

```
help atan
```

to read a brief description of the function `atan`.

The MATLAB `help` function is extremely useful. It's always a good place to look when you have questions about a function.

1.6 Vectors

A *vector* is a list of numbers. Let's define a vector that consists of four numbers:

```
[1 2 3 4]
```

Q: What happens when you execute this above code in MATLAB?

1.7 Manipulating Vectors with Scalars

A *scalar* is a single number. Consider

```
[1 2 3 4] * 3
```

which consists of the product of a vector (`[1 2 3 4]`) and a scalar (`3`).

Q: Upon executing this statement, what do you find? Does it make sense?

1.8 Manipulating Vectors with Vectors

Consider

```
[1 2 3 4] .* [1 2 3 4]
```

Q: Does this operation return a vector or a scalar? Does the result make sense?

The notation `.*` is used to multiply. This causes the vectors to be multiplied element by element. Namely, the first element of the first vector is multiplied by the first element of the second vector (i.e., `1*1`); the second element of the first vector is multiplied by the second element of the second vector (i.e., `2*2`); the third element of the first vector is multiplied by the third element of the second vector (i.e., `3*3`); and so on. To perform this calculation, the vectors must have the same number of elements. Consider

```
[1 2 3 4] .* [1 2 3 4 5]
```

Q: What happens when you execute this expression?

A: The expression should fail to execute. To understand why, consider the number of elements in each vector. The second vector has one additional element (the scalar `5`) compared to the first vector. So, if we try to multiply the two vectors element by element, we cannot; there is no value in the first vector to multiply the `5` in the second vector.

The operator `.*` differs from the operator `*`. The latter operator performs matrix multiplication, which is very different from the element by element multiplication we just performed. If you've taken a class in linear algebra, then you've seen matrix multiplication. We do not use matrix multiplication extensively here, but it is an essential tool in more advanced data analysis. The important point is the following.

Alert! The operators `.*` and `*` are very different.

1.9 Defining Variables

In the previous examples, we typed `[1 2 3 4]` over and over again. A better approach is to assign a *variable* to this vector and then simply reference this variable to access the vector. Consider

```
a = 2
b = [1 2 3 4]
```

Here we have defined two variables. The variable `a` is assigned the number `2` and is therefore a scalar. The variable `b` is assigned the vector `[1 2 3 4]` and is therefore a vector. To work with this scalar and vector, we can now refer to the variables. Consider

```
c = a*b
d = b.*b
```

Q: What are the results in the new variables c and d?

1.10 Probing the Defined Variables

Once we define variables, we often want to check the size of these variables. A good place to start is the *workspace window* of MATLAB. Look for this window on your desktop.

Q: What variables do you see? What information do you learn about the variables by examining the workspace?

To see a list of the variables you've defined, type

```
who
```

To determine the size of a variable, for example, the variable c defined in the previous section, type

```
size(c)
```

1.11 Summing Elements in a Vector

Sometimes we need to add up all the elements in a vector. In simple cases, we can perform this operation by hand. For example, consider the vector c used in the previous example. To recall the values in c, type c at the MATLAB command prompt and evaluate it:

```
c
```

To sum the elements of c, we may examine the output of the previous command, which lists all the values in c, and add these values by hand. Doing so is not difficult, but it is tedious. You might imagine the difficulty of this approach when the vector to sum contains many elements. For example, manually computing the sum of a vector containing 100 elements is not especially difficult but highly error prone. Fortunately, there's an easier way: we can use the MATLAB command sum:

```
sum(c)
```

Q: Notice that we've used a new function: `sum`. What is the input to this function? What is the output of this function? You might consider `help sum`.

Q: Compare the result of the MATLAB command `sum(c)` and your by-hand sum. Are the two the same? *Hint*: They should be.

1.12 Clearing All Variables

To clear all variables from the workspace, enter

```
clear
```

Q: Are all the variables gone? How can you check? *Hint*: Consider the command `who`.

1.13 Matrices

A more complicated list of numbers contains multiple rows or columns. This is called a *matrix*. We can think of a matrix as a group of vectors. Consider the following:

```
p = [1 2 3; 4 5 6]
```

This operation creates a matrix with two rows and three columns.

Q: How can you verify the size of `p` in MATLAB? *Hint*: Consider the command `size`.

We can manipulate matrices just as we manipulate vectors. Consider adding the scalar `2` to the matrix:

```
p + 2
```

Q: What happens?

Q: How would you create a matrix with three rows and four columns? Create this matrix, and verify the size using the command `size`.

1.14 Indexing Matrices and Vectors

Matrices and vectors are lists of numbers, and sometimes we want to access individual elements or small subsets of these lists. That's easy to do in MATLAB. Consider

```
a = [1 2 3 4 5]
b = [6 7 8 9 10]
```

To access the second element of `a` or `b`, enter

```
a(2)
b(2)
```

Q: Do the results make sense? How would you access the fourth element of each vector?

We can combine `a` and `b` to form a matrix.

```
c = [a; b]
```

To learn about `c`, enter

```
size(c)
```

The size of `c` is [2 5]. It has two rows and five columns. To access the individual element in the first row and fourth column of `c`, type

```
c(1,4)
```

We access matrices using (row, column) notation. So `c(1,4)` indicates the element in row 1, column 4, of `c`.

To access all columns in the first row of `c`, enter

```
c(1,:)
```

The notation `:` means "all indices". Here we use this notation to indicate all columns of `c`. To access the second through fourth columns of row 1 of `c`, type

```
c(1,2:4)
```

The notation `2:4` means "all integers from 2 to 4," which in this case is 2, 3, 4.

Q: How could you access all rows in the second column of `c`?

1.15 Finding Subsets of Elements in Matrices and Vectors

Sometimes we're interested in locating particular values within a matrix or vector. For example, let's first define a vector:

```
b = [1,2,3,4,5,6,7,8,9,10]*2
```

Q: What is the size of the variable `b`? What are the values in the variable `b`?

There's an easier way to define the same list of numbers. Consider the command

```
a = (1:1:10)*2
```

Q: What is the size of `a`? What are the values in `a`? How do the values in `a` compare to the values in `b`?

Now let's find the indices for all values in `a` that exceed 10. We could, of course, do so manually.

Q: Determine by inspection the indices of `a` that exceed 10. What are they?

There's a simple way to perform this same operation in MATLAB,

```
indices = find(a > 10)
```

Q: `find` is a function built into MATLAB. What does this function do? *Hint*: Consider `help find`.

Q: What is now stored in the variable `indices`?

A: The variable `indices` indicates all indices of the vector `a` with values greater than `10`. To verify this, examine the value of `a` at each index in `indices`; it should be greater than `10` for each index.

We can examine the values of a at the indices:

```
a(indices)
```

Q: What do you find? Do all the values exceed 10?

Beyond simple examinations, we can also manipulate the values of a at the indices. For example,

```
a(indices) = 0
```

Q: After performing this operation, what do you find is the largest value stored in the vector a? What has happened to all values of a that exceed 10?

1.16 Plotting Data

It's usually not easy to look at lists of numbers and gain an intuitive feeling for their behavior, especially when the lists contain many elements. In these cases, it's much better to visualize the lists of numbers in a plot. Consider

```
x = (0:1:10)
```

This above line constructs a vector that starts at 0, ends at 10, and takes steps of size 1 from 0 to 10. Then define

```
y = sin(x)
```

We have now applied the function sin to the list of numbers in x.

Q: Look at the values of y printed as output upon evaluating the line of code defining y. Can you deduce the behavior of y?

A: A list of the values in y is not particularly informative. Let's examine a printout of y:

```
0 0.8415 0.9093 0.1411 -0.7568 -0.9589 -0.2794 0.6570
0.9894 0.4121 -0.5440
```

Examining this list, we get a sense for the values of y. We find an interval of some positive values, followed by an interval of negative values, and so on. However, it's difficult to get a deep intuitive sense for the behavior of y through inspection of this printout alone.

To visualize y versus x, execute

```
plot(x,y)
```

This command produces a plot of x versus y (see figure 1.1a). Visual inspection reveals the curve is a bit jagged, not smooth like a sinusoid. To make the curve smoother, let's redefine x as

```
x = (0:0.1:10);
```

Q: Compare this definition of x to the previous definition. How do these two definitions of x differ?

Q: What is the size of x now? Does this make sense?

You might notice that we've added a semicolon (;) to the end of the definition of the vector x. This semicolon tells MATLAB not to print out the results of this command in the command window. We don't want to write out x because it contains lots of elements, and it's not useful for us to look at them right now. Now recompute y as

```
y = sin(x);
```

We again include the semicolon at the end to prevent a printout of many not currently informative values. Now, let's plot the two new variables:

```
plot(x,y)
```

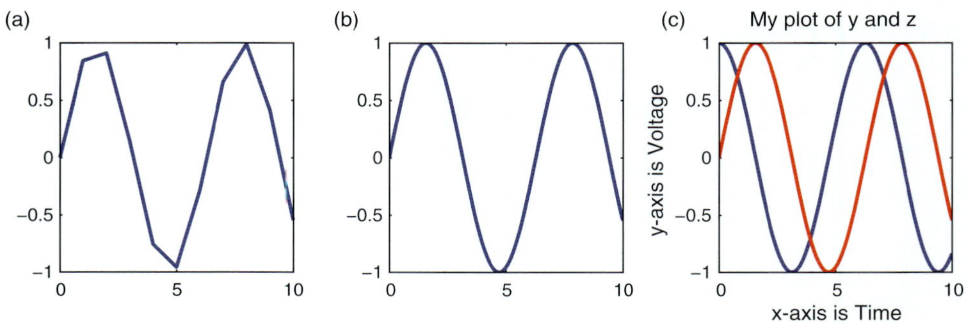

Figure 1.1
Plots of x versus y for (*a*) low density of x, and (*b*) high density of x. (*c*) a Plot of x, y, and z.

Q: Compare the newly created plot (figure 1.1b) with the original plot (figure 1.1a). How do the two differ?

1.17 Multiple Plots, One atop the Other

Continuing the example in the previous section, let's define a second vector

```
z = cos(x);
```

and plot it:

```
plot(x,z)
```

We'd now like to compare the two variables `y` and `z`. To do this, let's plot both vectors on the same figure. First, open a new figure:

```
figure
```

and plot `x` versus `z`:

```
plot(x,z)
```

Now, tell MATLAB to freeze, or "hold," the graphics on this figure:

```
hold on
```

With the graphics now frozen, let's also plot on this figure `x` versus `y`:

```
plot(x,y, 'r')
```

Notice that we've included a third input to the function `plot`. Here the third input tells MATLAB to draw the curve in a particular color: `'r'` for red. There are many options we can use to plot; to see more, check out `help plot`. When we're done *holding* the figure, it's polite to turn "hold" off:

```
hold off
```

We can also label the axes and give the figure a title:

```
xlabel('x-axis is Time')
ylabel('y-axis is Voltage')
title('My plot of y and z')
```

1.18 Random Numbers

To generate a single random number in MATLAB, type

```
randn(1)
```

Q: Execute this command a few times. What values do you get?

Sometimes, we would like to generate a list of random numbers. To do so, we can again use the `randn` command but this time with different inputs. Consider the following line of code:

```
r = randn(10,1)
```

In this line, we again call the function `randn`. But this time we call the function with two inputs, `(10,1)`, and store the results in the variable `r`.

Q: What is the size of the variable `r`? What numbers do you find in `r`?

Q: What happens to the size of `r` as you change the values of the inputs? Consider, for example,

```
r = randn(100,1);
r = randn(10,10);
```

What is the size of `r` in each case? What are the types of values you observe in `r` in each case?

We sometimes need to generate long lists of random numbers. We can do so using the `randn` command as follows:

```
r = randn(1000,1);
```

Q: How many elements are in the vector `r`?

To see how these random numbers are distributed, we can visualize the results. Let's plot the variable `r`:

```
plot(r)
```

Q: Examine this plot. What values does `r` take on? What is the biggest value of `r`? the smallest? the most common?

1.19 Histograms

There are many ways to display the data in the variable `r`. Another way is to construct a *histogram*. A histogram displays the number of times we observe a value in a list. Consider the following list of numbers:

```
a = [0,0,0, 1,1,1,1, 3,3];
```

It's easy to construct a histogram of the values in `a` using the built-in MATLAB function `hist`:

```
hist(a, (0:1:5))
```

Q: The result of this command is to produce the plot in figure 1.2. Examine this plot. What do you find?

The histogram (figure 1.2) indicates the number of times we observe each value in the variable `a`. In this case, we observe the values

0, observed three times;

1, observed four times;

3, observed two times.

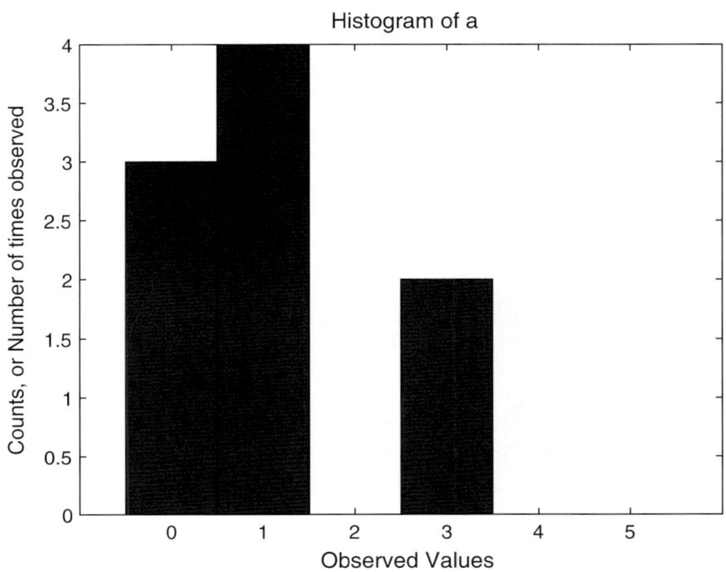

Figure 1.2
Histogram of variable `a`. Horizontal axis indicates values in `a`; vertical axis, number of times each value appears.

In the histogram plot, there are three bars. The first bar has an x-coordinate of 0 and a y-coordinate of 3. This tells us that in the variable `a`, we observe the value 0 three times. In the `hist` command, we specify two inputs:

first input: `a`, the variable of interest;

second input: the histogram *bins*, which specify the center value of each bin.

In this example, for the second input, we specify six bins: the bins start at a center value of 0, and step forward in values of 1, up to a center value of 5. For completeness, let's label the histogram:

```
xlabel('Observed Values')
ylabel('Counts, or Number of times observed')
title('Histogram of a')
```

Q: Consider the x-coordinate of 1 in the plot. Do you observe a bar at this x-coordinate? If so, what is the height? How does this height correspond to the number of 1s you observe in your list `a`? What about the x-coordinate of 2? the x-coordinate of 3?

Q: Consider how the histogram plot changes as the second input to `hist` (the bins) is varied. More specifically, consider `hist(a, (0:2:5))` and `hist(a, (0:0.1:5))`. Compare the histograms you create in each case. Do they make sense?

Finally, let's return to the list of random numbers (generated in section 1.18) that we stored in the variable `r`, and plot a histogram:

```
hist(r, (-5:0.5:5))
```

Q: Consider the histogram plot of `r`. What is the most common value, that is, which value has the most counts?

Q: How does the histogram of `r` change as you vary the bins (the second input to the `hist` command)?

See `help hist` to learn about the function `hist`.

1.20 Repeating Commands

Sometimes we want to repeat the same command over and over again. For instance, if we want to plot `sin(x + k*pi/4)`, where `k` varies from 1 to 5 in steps of 1, how do we do it? Consider the following (in which the % symbols indicate comments; comments are not evaluated by MATLAB, but help explain the code):

```
x = (0:0.1:10);          %Define x from 0 to 10 with step 0.1.
k=1;                     %Fix k=1,
y = sin(x + k*pi/4);     %... and define y at this k.

figure                   %Now, make a new figure window,
plot(x,y)                %... and plot y versus x.

k=2;                     %Let's repeat this, for k=2,
y = sin(x + k*pi/4);     %... and redefine y at this k,
hold on                  %... and plot it.
plot(x,y)
hold off

k=3;                     %Let's repeat this, for k=3,
y = sin(x + k*pi/4);     %... and redefine y at this k,
hold on                  %... and plot it.
plot(x,y)
hold off

k=4;                     %Let's repeat this, for k=4,
y = sin(x + k*pi/4);     %... and redefine y at this k,
hold on                  %... and plot it.
plot(x,y)
hold off

k=5;                     %Let's repeat this, for k=5,
y = sin(x + k*pi/4);     %... and redefine y at this k,
hold on                  %... and plot it.
plot(x,y)
hold off
```

This code is correct but horrible. To execute our objective, we cut and paste the same lines of code over and over again. As a general rule, if you're repeatedly cutting and pasting code, you're doing something inefficient and typically error prone. There's a much more elegant way to achieve the objective, and it involves making a *for-loop*:

```
x = (0:0.1:10);              %First, define the vector x,
figure                       %... and open a new figure.
for k=1:1:5                  %For k from 1 to 5 in steps of 1,
    y = sin(x + k*pi/4);     %... define y (note 'k' in sin),
    hold on                  %... hold the figure,
    plot(x,y)                %... plot x versus y,
    hold off                 %... and release the figure.
end
```

This small section of code replaces the code involving repeated cutting and pasting. Here, we update the definition of `y` with different values of `k` and plot it within a for-loop.

> **Q:** Spend some time studying this for-loop. Does it make sense?

Some of the subsequent code we develop includes for-loops. So, it's useful to be familiar with this type of operation.

1.21 Defining a New Function: The m-File

So far, we haven't explicitly saved any code. Instead, we entered all commands directly into the MATLAB command line. Sometimes, especially when developing a series of complex commands, we want to save the code so as not to have to reenter it. In other cases, we want to define and save new MATLAB functions for specific purposes.

A powerful feature of MATLAB allows us to create our own functions to achieve our goals. As an example, let's write a new MATLAB function: the function will take as input a vector and scalar, and return as output each vector element squared plus the scalar. Ideally, this function would be called in MATLAB as follows:

```
v = (0:1:10);                   %Define a vector.
b = 2.5;                        %... and a scalar,
v2 = my_square_function(v,b);   %... as inputs to the function.
```

However, these lines of code do not yet work. We first need to define the new function `my_square_function`. To do so, we create a new file in the MATLAB editor. Let's begin by entering at the MATLAB command prompt,

```
edit
```

MATLAB will open a new file called `Untitled` in the editor. This file is initially blank (i.e., it contains no text). We use this blank template to create the new function. Type into the `Untitled` file the following three lines of code:

```
function output = my_square_function(input1, input2)
  output = input1.^2 + input2;
end
```

In the first line, we define labels for the input and output variables to the our function. The keyword `function` lets MATLAB know we're defining a new function. We define the function name (`my_square_function`) and specify this function to have two inputs (`input1` and `input2`). If we desired more inputs, we would simply add additional variables in the parentheses that follow the function name. We also specify a single output for this function (`output`). The syntax of this first line of code makes the relation of inputs and outputs clear; the output is equal to the function evaluated at the inputs.

Q: Consider the second line in this code. It defines the body of the new function and performs the actual work of the desired computation. Does this line achieve the stated goal of this function—to return as output the vector elements squared plus the scalar?

The last line of code, `end`, serves to end the function definition.

Q: Consider changing the variable names of the inputs from `input1` and `input2` to `a` and `b`, respectively. How must we change the function to perform the same computation?

A: The input variables are called dummy variables. We're free to chose these names to be (almost) anything. It's good practice to choose informative names, not confusing names. For example, a confusing name for an input variable would be `output`. That choice of name is not wrong but aesthetically unpleasant. Such a choice may make it difficult for a colleague (or your future self) to understand the function's behavior. Changing the variable names requires us to change the function as follows:

```
function output = my_square_function(a, b)
  output = a.^2 + b;
end
```

Notice that we've simply replaced each occurrence of `input1` with `a`, and each occurrence of `input2` with `b`. The resulting function behaves exactly the same as the original function. We have only changed some of the internal labels to the function, which has no impact on the desired computation.

Having input these three lines of code, we save this m-file with the name `my_square_function.m`. Notice the `.m` file extension, which indicates a MATLAB function or script. You might perform this save through key strokes (e.g., Command-S or Control-S, depending on your operating system) or by using the mouse.

> The file name and function name must match. In this case, the file name (`my_square_function.m`) matches the function name (`my_square_function`) that we call at the MATLAB command line prompt.

> To call this function, we first need to navigate MATLAB to the directory that holds the function file. Otherwise, when we attempt to execute
>
> ```
> v2 = my_square_function(v,b);
> ```
>
> we get an error because MATLAB can't locate the file `my_square_function.m`. So, first navigate MATLAB to the directory where this file lives.

We have created a new function. This function takes two inputs, and returns one output. Having created the function and navigated MATLAB to the appropriate directory, we may finally execute this function:

```
v = (0:1:10);
b = 2.5;
v2 = my_square_function(v,b);
```

Notice that when the function is called, the variables `v` and `b` act as the input arguments. Within the function, variable `v` is assigned a different label (e.g., `input1`) and the variable `b` is assigned a different label (e.g., `input2`). However, this relabeling doesn't matter when we call the function; it happens behind the scenes, within the function. In the same way, within the function, the output variable is labeled `output`. But in our call to the function, we relabel the output `v2`.

> **Q:** Examine the output variable `v2`. Does it match your expectations?

> **Q:** Consider this call to the new function,
>
> ```
> v2 = my_square_function(b,v);
> ```
>
> We have swapped the order of the inputs. Does the function still perform the desired computation? If not, how could we fix the function? *Hint:* This question is rather advanced.

1.22 Saving Your Work

The file `my_square_function.m` defines a MATLAB function that we can execute. In addition, we can also create MATLAB files called scripts. Scripts are useful for executing a series of commands and saving the work when calling functions. To create a new MATLAB script, enter at the MATLAB command prompt

```
edit
```

This creates a blank file named `Untitled` in the MATLAB editor, into which we may enter any commands we like. For example, type into this blank file the following commands:

```
x = (0:0.1:10);             %First, define the vector x,
figure                      %... and open a figure.
for k=1:1:5                 %For k from 1 to 5 in steps of 1,
    y = sin(x + k*pi/4);    %... define y (note 'k' in sin),
    hold on                 %... hold the figure,
    plot(x,y)               %... and plot x versus y,
    hold off                %... and release the figure.
end
```

It is then easy to execute the code in this script. You may do so by copying the code from the script and pasting it into the command line, or (more elegantly) by using the mouse to select Run on the MATLAB editor tab. This will execute all the commands within the script. You can then save this script (with the `.m` file extension) and return to it later. For example, you might redefine `x = (0:0.1:25);` and reevaluate your script. To do so requires only changing one line of the script (the first line) and reexecuting all the code in the script. By saving the list of commands as a script, you do not need to retype all the other commands when you make a change and reevaluate the results. Saving a list of commands in a script is convenient and less prone to error.

1.23 Loading Data

Sometimes we need to load data (collected in an experiment, say) into MATLAB. Fortunately, that's easy to do. For an example of this procedure, visit,

`http://github.com/Mark-Kramer/Case-Studies-Kramer-Eden`

and download the file `Ch1-example-data.mat`. We now would like to open this dataset. The first step is to direct MATLAB to the same directory that contains the downloaded file. Once in the correct directory, type

```
clear                       %First clear the workspace,
load Ch1-example-data.mat   %... then load the data.
```

The first command, `clear`, clears the MATLAB workspace. This command is not necessary, but we perform it here so that any new variables we subsequently load are obvious. The second line of code, `lcad Ch1-example-data.mat`, reads in the data stored in the file `Ch1-example-data.mat`.

> **Q:** What variables now exist in your workspace? What is the size of variables `t1` and `v1`? Plot `v1` and `t1` to get a sense for how the two variables behave.

Note that the file extension `.mat` denotes a MATLAB save file. Typically, a `.mat` file is created in MATLAB (using the `save` command; see MATLAB Help) and then loaded into MATLAB with the `load` command. Other file types can also be loaded into MATLAB; see MATLAB Help.

1.24 Loading Additional Functionality

A powerful feature of MATLAB (or any programming language) allows us to use sophisticated functions developed by others. Throughout this book, we make use of the open source software package *Chronux*, developed for the analysis of neural data [2]. The Chronux software is a MATLAB library, and to acquire its functionality we must download and install it. To do so, visit the Chronux website,

```
http://chronux.org/
```

On this website, you will find detailed information about the Chronux software package and its functionality. By navigating through this website, you will find the latest version of the Chronux software for MATLAB. Download and install this software following the setup instructions included in the downloaded software manual. The software consists of a directory with many MATLAB m-files that contain (sophisticated) functions.

To utilize these files, you must add the Chronux directory and all its subdirectories to the MATLAB path. To do so, check out the functions `genpath` and `addpath` in MATLAB Help. Once you have successfully added the Chronux directory and all subdirectories to the path, then Chronux should be accessible in MATLAB.

> **Q:** To check that Chronux has been successfully installed, type
>
> ```
> help chronux
> ```
>
> This should display on the MATLAB command line detailed information about Chronux. If it does, you have successfully installed Chronux.

We perform many interesting tasks with Chronux in subsequent chapters.

2 The Event-Related Potential from a Scalp Electroencephalogram

Synopsis

Data 1 s of scalp EEG data sampled at 500 Hz during 1,000 trials in two conditions.

Goal Characterize the response of the EEG in the two conditions.

Tools Visualization, event-related potential, confidence intervals, bootstrapping.

2.1 Introduction

2.1.1 Background

Voltage recordings from the scalp surface—the electroencephalogram (EEG)—provide a powerful window into brain voltage activity. Some of the earliest human EEG recording occurred in 1924, when Dr. Hans Berger made a remarkable discovery: the EEG of a human subject at rest with eyes closed exhibits rhythmic activity, an approximately 10 Hz oscillation that he labeled the *alpha rhythm* [3]. Although now studied for nearly 100 years, the definitive functional role (if any) of the alpha rhythm remains unknown. Since then, many other EEG rhythms have been detected and labeled (typically with Greek letters), and the analysis of EEG rhythms remains an active area of research [4].

Compared to other modalities for measuring brain activity, the EEG possesses both advantages and disadvantages. Perhaps the most important advantages are that the EEG is noninvasive and permits a high temporal resolution (on the order of milliseconds). But the EEG measure also suffers from significant disadvantages, most notably, poor spatial resolution: a single scalp electrode detects the summed activity from approximately 10 cm^2 of cortex [5].

In this chapter, we consider EEG data recorded from a single scalp electrode. We analyze these data to determine what (if any) activity is evoked following two different types of stimuli presented to a human subject. In doing so, we continue demonstrating MATLAB features and how this powerful tool can help us understand these time series data. We begin with a brief description of the EEG data.

2.1.2 Case Study Data

An undergraduate student volunteers to participate in a psychology study at his university. In this study, EEG electrodes (sampling rate 500 Hz, i.e., 500 samples per second) are placed on the student's scalp, and he is set in a comfortable chair in a dark, electrically isolated room. Upon entering the room, the student is instructed to place headphones over his ears and listen to a series of repeated sounds. The sounds consist of two tones: either a high-pitch tone or a low-pitch tone. A single tone is presented once every few seconds, and the student responds with a button press to the low-pitch tone. The tone presentation is repeated to collect the EEG response to numerous presentations of the two tones (figure 2.1). Our collaborator leading this research study has agreed to provide us with EEG data recorded at a single electrode for 1,000 presentations of the high-pitch tone and 1,000 presentations of the low-pitch tone. In each presentation, or trial, she provides us with 1 s of EEG data, such that the tone occurs at 0.25 s into the trial (gray-shaded regions in figure 2.1). She asks us to analyze these data to determine whether the EEG signal differs following the two tone presentations.

2.1.3 Goal

Our goal is to analyze the collection of EEG data recorded in the two conditions. We would like to understand the EEG features evoked by the stimulus, and how these features differ in the two conditions.

2.1.4 Tools

We develop two primary tools: visualization techniques and the *event-related potential*. In computing the latter, we consider two techniques to create confidence intervals for the ERP, one relying on the central limit theorem, the other a bootstrap technique. We also discuss how to implement a bootstrap test to assess the difference in response between the two conditions.

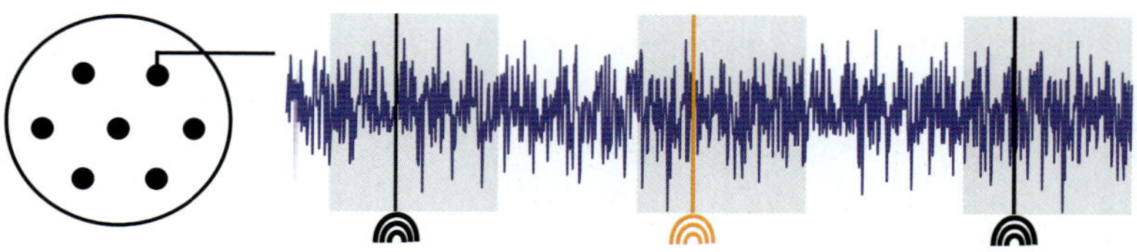

Figure 2.1

Cartoon illustration of EEG experiment. *Left*, EEG electrodes are placed on scalp surface of a human subject. *Right*, EEG activity (*blue*) recorded as a function of time during presentation of high-pitch tones (black) and low-pitch tones (*orange*).

2.2 Data Analysis

To access the EEG data for this chapter, visit

http://github.com/Mark-Kramer/Case-Studies-Kramer-Eden

and download the file Ch2-EEG-1.mat.

2.2.1 Visual Inspection

The best place to begin the data analysis is with visual inspection of the time series. The instructions to perform this initial analysis are simple: plot the data and look at them. Of course, before we can visualize the data, we must load them into MATLAB. To do so, we use the function load:

```
load('Ch2-EEG-1.mat')      %Load the EEG data.
```

where the file Ch2-EEG-1.mat resides in the current MATLAB directory.[1] To understand the outcome of issuing this command, let's examine the variables now present in the MATLAB workspace by asking who:

```
who                        %List variables in the workspace.
```

The who command lists all the variables that currently reside in the MATLAB workspace. We find there are three: EEGa, EEGb, and t. These correspond to the EEG data recorded in the two conditions (EEGa to condition A and EEGb to condition B) as well as a time axis (t). Let's further investigate the structure of the EEG variables by determining their size:

```
size(EEGa)                 %Determine the dimensions of EEGa.
```

Upon executing this command, MATLAB returns the value [1000, 500]. The variable EEGa is a *matrix* with 1,000 rows and 500 columns. Our collaborator tells us that each row corresponds to a separate trial, and each column to a point in time. So, there are 1,000 total trials, each consisting of 500 time points. As a matter of convenience, we define a new variable to record the number of trials:

```
ntrials = size(EEGa,1); %Define variable to record no. of trials.
```

Notice that in using the size function, we supply two inputs; the second input requests the size of the first dimension (i.e., the number of trials). This variable will be useful later.

Q: Determine the size of the variable EEGb. How many rows and columns does it possess? Which dimension of the matrix corresponds to trials, which dimension to time?

---

1. An alternative approach to loading the data: the file Ch2-EEG-1.mat may be dragged from a folder on your computer and dropped into the MATLAB command window.

Both EEGb and EEGa are complicated variables that contain many elements. To understand these data, we might attempt to read the values contained in each element. For example, we can print to the MATLAB screen the EEG data for the first trial of condition A.

```
EEGa(1,:)          %Print to screen data from condition A, trial 1.
```

In this command, we index the first row of the matrix EEGa and print out all columns (corresponding to all moments of time).

Q: Upon issuing this command, what do you find? Does the printout help you understand these data?

A: You should observe a list of 500 numbers that begins

```
-0.1859 0.4499 1.0607 -0.4713 1.6867 0.9382 0.2212...
```

We might conclude that these numbers exhibit variability (i.e., the values are both positive and negative), but examining the data in this way is not particularly useful. For example, determining trends in the behavior (such as intervals of repeated activity) through inspection of these printed numbers alone is extremely difficult.

Printing out the data to the screen is not useful in this case. How else can we deepen our understanding of these data? Let's make a plot. Fortunately, plotting the data is easily done in MATLAB:

```
plot(EEGa(1,:))        %Plot the data from condition A, trial 1.
```

The results of this command are shown in figure 2.2a. Visualizing the data in this way, we immediately notice many features. First, let's consider the axes. The horizontal axis extends from 1 to 500. This corresponds to the 500 columns in the variable EEGa. It would be more informative to plot the EEG data as a function of time rather than indices. Fortunately, we possess a variable t in the workspace that corresponds to the time axis. Determining the size of the variable t, we find it is a vector with 1 row and 500 columns. Each column corresponds to a point in time. Plotting the variable t (figure 2.2b), we see it ranges from 0 to 1. This corresponds to the 1 s of EEG data recorded in each trial. We can also use the variable t to determine the *sampling interval*,

```
dt = t(2) - t(1);         %Determine the sampling interval.
```

The new variable dt corresponds to the time between samples.

Q: What is the value of dt? We were told by our collaborator that the sampling frequency is 500 Hz. Is the value of dt consistent with this sampling frequency?

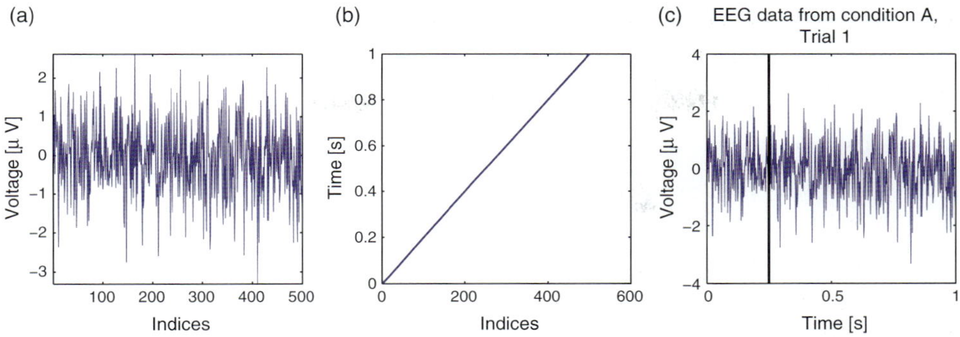

Figure 2.2

Initial exploration of the variables received. (*a*) Plot of EEG data from condition A in first trial. (*b*) Plot of variable `t`. (*c*) Plot of EEG data from condition A in first trial with appropriate axes, labels, and a vertical line indicating time of stimulus presentation.

A: Yes, it is consistent. We find that `dt` is 0.002 s, or 2 ms. The sampling frequency of 500 Hz corresponds to one sample of the EEG data every 1/(500 Hz) = 2 ms. If the two were not consistent, we would return to our collaborator and figure out what has gone wrong. In general, it's useful to ask such questions along the way to make sure we understand the formatting of the data and catch any potentially serious misunderstandings early in the analysis.

We can now combine the time axis with the EEG data to make a more complete plot. Let's also label the axes and give the plot a title.

```
plot(t,EEGa(1,:))            %Plot condition A, trial 1 data vs t.
xlabel('Time [s]')           %Label the x-axis as time.
ylabel('Voltage [\mu V]')    %Label the y-axis as voltage.
title('EEG data from condition A, Trial 1')     %Add a title.
```

This plot (figure 2.2c) provides a nice summary of the data in the first trial of condition A. A visual inspection of the plot suggests that these data exhibit complicated activity. We know from our collaborator that the stimulus occurs at time 0.25 s in each trial. This time is indicated in figure 2.2c as a vertical line. To add this vertical line to the figure, we first instruct MATLAB to freeze the plot (to not erase the current plot contents) using the `hold on` command:

```
hold on      %Freeze plot & indicate stimulus delivery time ...
plot([0.25, 0.25], [-4,4], 'k', 'LineWidth', 2)
hold off     %Release the current plot.
```

This `plot` command includes additional options that make the line black (`'k'`) and a bit wider (`'LineWidth', 2`). We end this code with `hold off` to release the plot; the next issuance of the `plot` command will erase the current plot contents. Releasing a plot is not always necessary but always polite.

> **Q:** What else, if anything, can you say about the single trial of EEG data plotted in figure 2.2c? Does visual inspection reveal any particular change in the EEG activity following the stimulus presentation?

So far we have visualized only the data from condition A. Because we are interested in whether the EEG behaves differently in the two conditions, visualizing both conditions simultaneously would be of use. Using the `hold on` command, we can do this in MATLAB:

```
plot(t,EEGa(1,:))          %Plot condition A, trial 1, data vs t,
hold on                    %... freeze the plot, and then plot
plot(t,EEGb(1,:), 'r')     %... data from condition B, trial 1,
hold off                   %... and release the plot.
xlabel('Time [s]')             %Label the x-axis as time.
ylabel('Voltage [\mu V]')      %Label the y-axis as voltage.
title('EEG data from conditions A (blue) and B (red), Trial 1')
```

> **Q:** Compare the voltage traces from the first trial of conditions A and B as plotted in figure 2.3. What similarities and differences do you observe?

> **Q:** The analysis has so far focused only on the first trial. Repeat this visual inspection of the data for different trials. What do you find? What similarities and differences exist between the two conditions across trials?

These techniques allow us to visualize the data one trial at a time. That is useful but can be time consuming, especially for a large number of trials. For the EEG data of interest here, each condition contains 1,000 trials, and to visualize each trial separately could require 2,000 plots. We can certainly create 2,000 plots, but the subsequent visual inspection would be time consuming and difficult. Fortunately, a more efficient visualization approach exists: we can display the entire structure of the data across both time and trials as an *image*. Doing so is straightforward in MATLAB:

```
imagesc(t,(1:ntrials),EEGa);     %Image the data from condition A.
xlabel('Time [s]')               %Label the x-axis.
```

```
ylabel('Trial #')                  %Label the y-axis.
hold on                            %Indicate time of stimulus onset.
plot([0.25, 0.25], [0 1000], 'k', 'LineWidth', 2)
hold off
```

The `imagesc` command allows us to visualize the entire matrix `EEGa` as a function of trial number and time (figure 2.4). Each row corresponds to a single trial of duration 1 s, and the color indicates the voltage, with warm (cool) colors indicating higher (lower) voltages.[2] This plot also indicates the time of stimulus presentation with a vertical black line as a cue to assist visual inspection.

> **Q:** Upon close inspection of figure 2.4, what response, if any, do you observe following the stimulus presentation? (Look *really* carefully.) Repeat this visualization and analysis for `EEGb`. How do the two conditions compare?

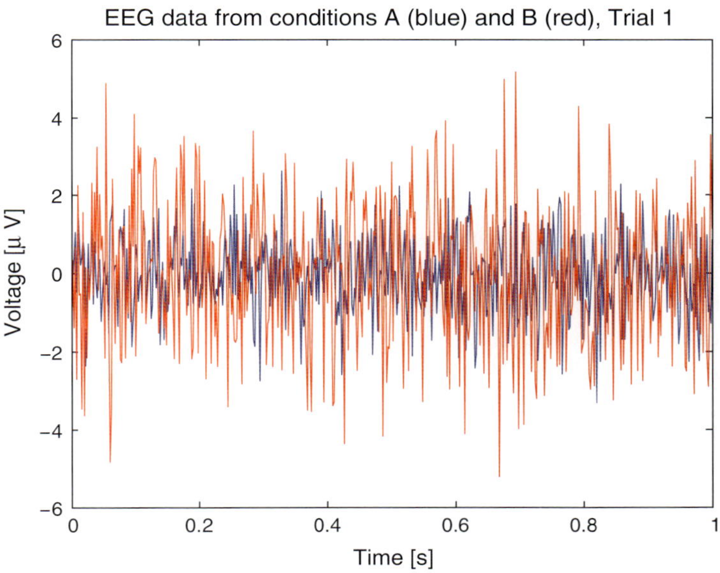

Figure 2.3
EEG data from first trial of condition A (*blue*) and condition B (*red*).

---

2. We have used the default color settings. There are many other options; check out `help colormap` for details.

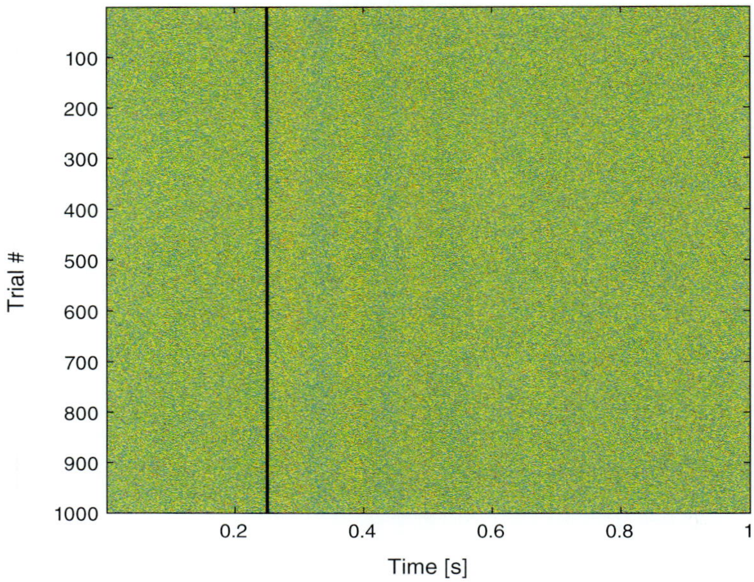

Figure 2.4

Image of EEG data from condition A for all trials. Time of stimulus presentation is indicated with a vertical black line.

2.2.2 Plotting the ERP

Visual inspection of the EEG data has so far come up empty. The EEG traces appear noisy or perhaps rhythmic, but from visual inspection of the individual trials it's difficult to make a decisive conclusion of underlying structure (e.g., figure 2.3). To further investigate the activity in these data, we compute the *event-related potential* (*ERP*). To compute the ERP, we first assume that each trial evokes an instantiation of the same underlying brain process. So, in this case, we assume that the same brain response is evoked 1,000 times (once for each trial) for each condition. However, the evoked response due to the stimulus is small and hidden in the EEG signal by other ongoing activity unrelated to the stimulus (e.g., daydreaming, thoughts of dinner, thoughts of homework). Therefore, to tease out the weak evoked effect, we average the EEG responses across trials. Ideally, EEG activity unrelated to the stimulus will cancel out in the average, while EEG activity evoked by the stimulus will sum constructively. The procedure to perform and display this averaging can be done in MATLAB:

```
load('Ch2-EEG-1.mat')          %Load the EEG data.
plot(t, mean(EEGa,1))          %Plot the ERP of condition A.
xlabel('Time [s]')             %Label the x-axis as time,
ylabel('Voltage [\mu V]')      %...and the y-axis as voltage,
title('ERP of condition A')    %...and provide a useful title.
```

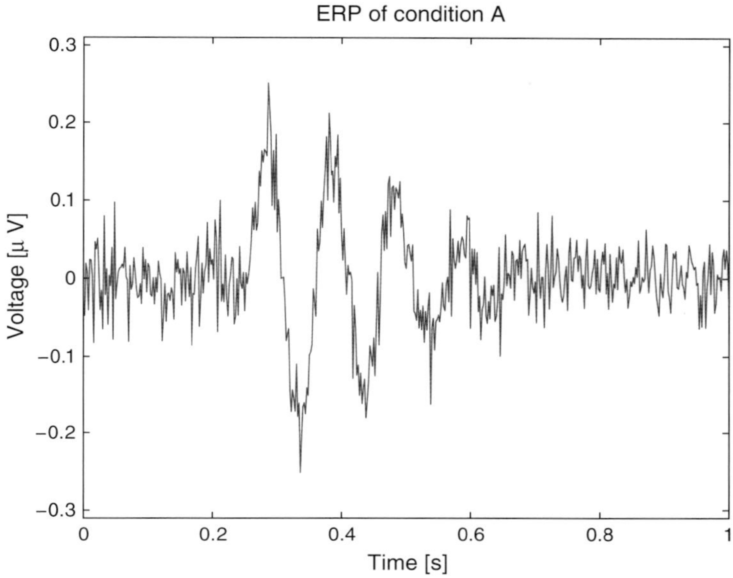

Figure 2.5
ERP for condition A computed by averaging EEG response over all trials.

Notice that in the second line, we use the MATLAB function `mean` to compute the mean of `EEGa` over the first dimension (the number of trials). The result is the ERP for condition A (figure 2.5).

> **Q:** Consider the ERP for condition A plotted in figure 2.5. Update this figure to include a vertical line at the location of the stimulus, and the ERP for condition B. How, if at all, do the ERPs for Conditions A and B differ?

The ERP in figure 2.5 shows the mean voltage across trials at each moment in time. Visual inspection suggests that before stimulus presentation (i.e., times 0 s to 0.25 s) the EEG fluctuates around zero. Then, after stimulus presentation, the ERP increases and decreases substantially above and below zero. Which, if any, of these deviations following stimulation are significant? To address this, we make use of the trial structure of the EEG data to compute *confidence bounds* for the ERP. We do so in two ways.

2.2.3 Confidence Intervals for the ERP (Method 1)

To compute the ERP we average the EEG data across many trials. Because of this, we may make use of a powerful theorem in statistics—the *central limit theorem* (CLT)—to include approximate confidence bounds in the ERP figure. Briefly, this theorem states that the mean of a sufficiently large number of independent random variables, each with finite mean and variance, will be approximately normally distributed. Remember that the ERP at each moment in time is the sum of EEG activity across trials (then scaled by a constant, the number of trials). Let's assume that the trials are independent (i.e., one trial does not depend on any other trial). Let's also assume that the EEG data at each moment in time have finite mean and variance. With those assumptions, we have satisfied the CLT and may therefore conclude that the ERP at each moment in time is approximately normally distributed.

Q: To use the CLT, we make two assumptions about the EEG data. Are these assumptions reasonable?

A: We assume that the EEG data are independent across trials. This assumption may fail if, for example, the activity in one trial influences the activity in the next trial. We also assume that the EEG data are "well-behaved" (i.e., have finite mean and variance). That is a reasonable assumption for physical data we observe from the brain; we expect the EEG data to always remain finite and not diverge to plus or minus infinity.

This conclusion—that the ERP at each moment in time is approximately normally distributed—is useful because the normal distribution (also known as the Gaussian distribution or bell curve) possesses many convenient properties. First, a normal distribution is relatively simple; it can be completely specified with two parameters: the mean value and the standard deviation (figure 2.6). Second, 95% of the values drawn from a normal distribution lie within approximately two standard deviations of the mean (figure 2.6).

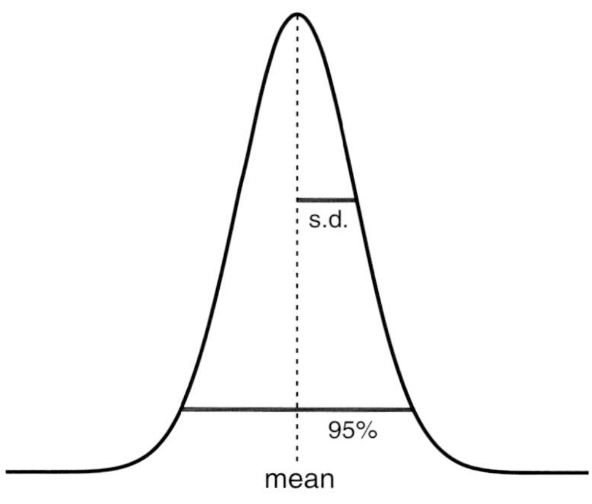

Figure 2.6

Canonical normal distribution showing mean (*dotted vertical line*) and standard deviation (*blue*). Ninety-five percent of values lie within the interval indicated by the red bar.

Therefore, to construct a 95% confidence interval for the ERP, we need to determine the mean and standard deviation of the mean across trials at each point in time.[3] To compute the mean in MATLAB is easy:

```
mn = mean(EEGa,1);        %Compute mean across trials (the ERP).
```

We again note that the second input to the function `mean` specifies the dimension in which we compute the mean, in this case, across the first dimension of the variable `EEGa` corresponding to the trials. To compute the standard deviation of the mean, we start by computing the standard deviation of the data:

```
sd = std(EEGa,1);         %Compute std across trials.
```

But we're not interested in the standard deviation of the EEG data across trials; instead, we're interested in the standard deviation of the estimate of the mean. To calculate the standard deviation of the mean, we divide the standard deviation of the data by the square root of the number of trials (i.e., the number of terms used to compute the mean; see the appendix at the end of this chapter). In MATLAB,

```
sdmn=sd/sqrt(ntrials);  %Compute the std of the mean.
```

Now, having found the mean and the standard deviation of the mean, we can compute a 95% confidence interval for the ERP. We again exploit the observation, based on the central limit theorem, that the ERP is normally distributed at each instant of time. With these calculations, the following MATLAB code plots the ERP and the 95% confidence interval:

```
plot(t, mn, 'LineWidth', 3)      %Plot the ERP of condition A.
hold on                          %Freeze the plot,
plot(t, mn+2*sdmn);              %... and include the upper CI,
plot(t, mn-2*sdmn);              %... and the lower CI.
hold off                         %Release the plot.
xlabel('Time [s]')               %Label the x-axis as time.
ylabel('Voltage [\mu V]')        %Label the y-axis as voltage.
title('ERP of condition A')      %Provide a useful title.
```

The ERP computed with confidence intervals allows us to ask specific questions about the data. For example, does the ERP ever differ significantly from zero? To answer this, we

3. We could instead write the *sample* mean because we use the observed data to estimate the theoretical mean that we would see if we kept repeating this experiment. This distinction is not essential to our goals here, but it is important when talking to your statistics-minded colleagues. Throughout the book, we omit the term *sample* when referring to sample means, variances, covariances, and so forth, unless this distinction is essential to the discussion.

look for intervals of the ERP for which the confidence intervals do not include zero. To aid visual inspection, we add to the ERP plot a horizontal line at 0:

```
hold on                            %Freeze the plot,
plot(t, zeros(size(mn)), 'k')      %... add horizontal line at 0,
hold off                           %... and release the plot.
```

Q: What is the role of the MATLAB function `zeros` in this code? *Hint*: If you have not encountered `zeros` before, look it up in MATLAB Help.

We find three time intervals at which the confidence intervals of the ERP do not include zero: near 0.27 s, near 0.37 s, and near 0.47 s (see figure 2.7). These results suggest that for an interval of time following the stimulus presentation in condition A, the observed ERP is not a random fluctuation about zero but instead contains consistent structure across trials.

Q: Construct the ERP with confidence intervals for condition B. As for condition A, you should find that before stimulus presentation the ERP fluctuates around zero. What intervals of time, if any, differ significantly from zero?

2.2.4 Comparing ERPs

In the previous section, we implemented a procedure to compute confidence intervals for the ERPs in conditions A and B. To investigate *differences* between the ERPs in the two

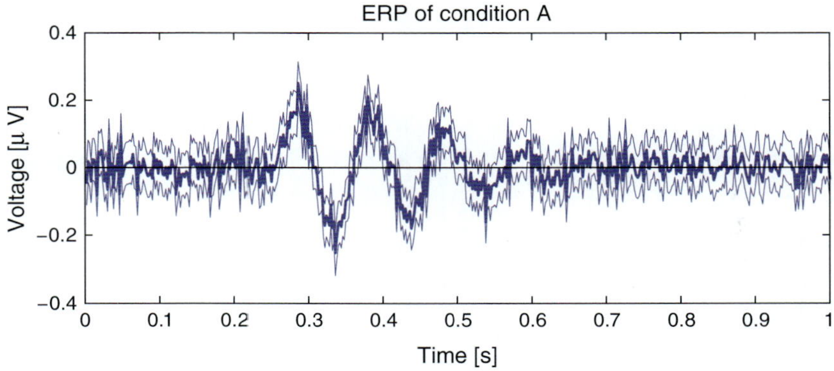

Figure 2.7

The ERP for condition A with 95% confidence intervals. The thick line indicates the ERP (i.e., the mean of the EEG across trials) while the thin lines indicate the 95% confidence intervals.

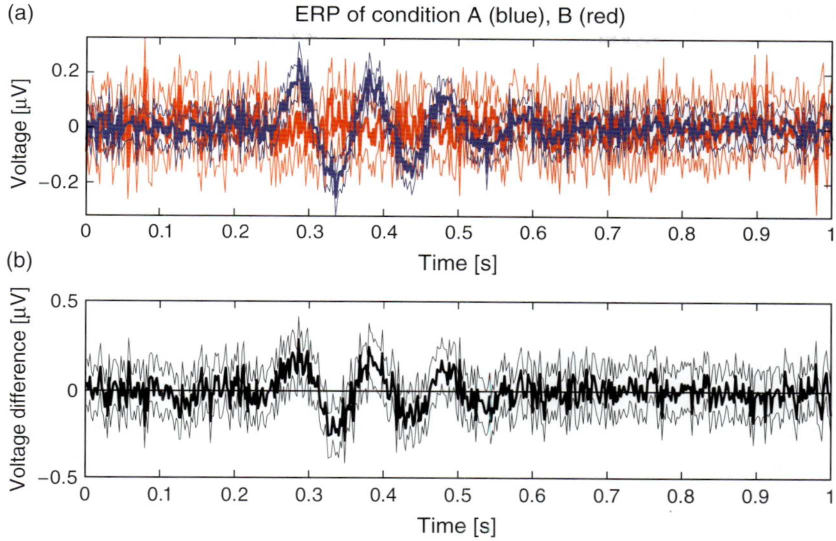

Figure 2.8
(*a*) ERPs with confidence intervals for condition A (*blue*) and condition B (*red*). (*b*) The differenced ERP. Thick lines indicate the mean; 95% of values lie between the thin lines.

conditions, we can use a similar approach. To start, let's plot the ERPs with confidence intervals for both conditions and attempt to identify periods for which the confidence intervals do not overlap (such intervals would correspond to significant differences between the responses of the two conditions). The plot of both ERPs (figure 2.8a) is rather messy; it's difficult to determine through visual inspection alone in which intervals the ERPs exhibit significant separation.

To facilitate further inspection of the data, we compute the *difference* between the ERPs in the two conditions. In the differenced signal, large deviations between the two conditions will appear as large differences from zero. To determine whether a deviation is significantly different from zero, we need to determine the confidence interval for the differenced ERP. This requires we propagate the standard deviation of the mean for both ERPs to the new differenced ERP. The propagated standard deviation of the mean at a fixed moment in time is computed as

$$\sigma = \sqrt{\frac{\sigma_A^2}{K} + \frac{\sigma_B^2}{K}}, \tag{2.1}$$

where σ_A is the standard deviation of condition A, σ_B is the standard deviation of condition B, and K is the number of trials. In MATLAB we compute the differenced ERP and standard deviation of the mean of this difference as follows:

```
mnA=mean(EEGa,1);                      %ERP of condition A,
sdmnA=std(EEGa,1)/sqrt(ntrials);%...and standard dev of mean.
mnB=mean(EEGb,1);                      %ERP of condition B,
sdmnB=std(EEGb,1)/sqrt(ntrials);%...and standard dev of mean.
mnD=mnA-mnB;                           %The differenced ERP,
sdmnD=sqrt(sdmnA.^2 + sdmnB.^2);%...and its standard dev.
```

In this code we first compute the ERP and standard deviation of the mean for each condition. We then compute the differenced ERP (`mnD`) and the standard deviation of the mean of this difference (`sdmnD`) using equation (2.1). We note that `sdmnA` $= \sqrt{\sigma_A^2/K}$ and therefore `sdmnA.^2` $= \sigma_A^2/K$, with similar expressions for condition B. The resulting differenced ERP with 95% confidence intervals is shown in figure 2.8b. The hope is that from this figure we can more easily identify significant differences between the two conditions.

> **Q:** Examine the differenced ERP in figure 2.8b. In what intervals of time do the EEG responses in the two conditions significantly differ?

2.2.5 Confidence Intervals for the ERP (Method 2)

So far we have computed confidence intervals for the ERPs by relying on the central limit theorem and approximating the average voltage values at each point in time as normally distributed. That's a completely reasonable approach. And because the normal distribution is so well-behaved, it's easy to compute the 95% confidence intervals. An alternative approach to generate confidence intervals is through a *bootstrap* procedure. Bootstrapping is a resampling method that allows us to estimate the sampling distribution of many different statistics. In this chapter, we implement a *nonparametric bootstrap*. To do so, we generate new *pseudodata* from the observed EEG data.[4] We begin by using a bootstrapping procedure to create confidence intervals for the ERPs observed in each condition.

We implement the bootstrapping procedure to compute pointwise confidence intervals. By pointwise we mean that the confidence intervals are computed separately for each point in time, and interactions across time are not considered. The prescription for the bootstrapping procedure follows four steps:

4. Briefly, there is strong theoretical justification for the nonparametric bootstrap. The fundamental idea is that resampling the data with replacement is equivalent to sampling new pseudodata from the empirical cumulative distribution function (eCDF) of the observed data. For a large sample of independent, identically distributed random variables, the distribution of the pseudodata generated from the eCDF will be close to the true distribution of the data. Note the important caveat that the variables are independent, identically distributed; this assumption fails in many cases, such as for time series. Here, we assume that each trial is drawn independently from the same distribution (i.e., the trials are independent, identically distributed variables). For more details, see [6].

1. Sample with replacement 1,000 trials of the EEG data from condition A.
2. Average these 1,000 trials to create a resampled ERP.
3. Repeat these two steps 3,000 times to create a distribution of ERPs.
4. For each time point, identify the values greater than 2.5% and greater than 97.5% of all 3,000 values. This range determines the 95% confidence interval of the ERP for that time point.

Let's now implement each step in MATLAB. In step 1 we must sample with replacement from the EEG data. To visualize this procedure, imagine placing 1,000 marbles in an opaque bag. Each marble is assigned a unique integer value from 1 to 1,000. Now, reach your hand into the bag, grab a marble, record its number, and replace the marble in the bag. We assume that each marble is equally likely to be selected at each draw (i.e., there are no special features that allow some marbles to be drawn more often). Repeat this procedure 1,000 times to create a list of 1,000 integers. Notice that after recording the drawn marble's number, we replace it in the bag. So, we could potentially draw the same marble 1,000 times, although that's extremely unlikely. Performing this sampling with replacement procedure by hand would, of course, be extremely time consuming (e.g., who will paint integers on each marble?). Fortunately, MATLAB provides a function to perform sampling with replacement:

```
%Draw 1000 integers with replacement from (1,1000);
i=randsample(ntrials,ntrials,1);
```

The first and second inputs to `randsample` specify the maximum integer to draw and the number of integers to draw, respectively. In this case, both inputs equal the number of trials (1,000). The last input is a flag that tells `randsample` to sample with replacement.

> **Q:** Execute this command and examine the values of `i` returned. What values do you find?

The result `i` provides a list of integers between 1 and 1,000. These values specify the trials to use in creating the resampled EEG. This resampled EEG will contain the same number of trials as the original EEG (i.e., 1,000 trials) but in a different order and with possibly repeated trials. For example, if the sampling with replacement procedure returns

```
i = [10, 941, 3, 400, 10, . . .
```

then the first and fifth trials of the resampled EEG will equal the tenth trial of the original EEG. We create the resampled EEG in MATLAB as follows:

```
EEG0 = EEGa(i,:);               %Create the resampled EEG.
```

In this code we use the variable `i` as the index to the rows of `EEGa`.

Q: What is the `size` of the new variable `EEG0`? Is this size consistent with the original EEG datasets?

That completes step 1 of the resampling procedure. Step 2 is easy: we create a resampled ERP from the resampled EEG data. Computing the resampled ERP requires only one line of code in MATLAB:

```
ERP0 = mean(EEG0,1);          %Create the resampled ERP.
```

Q: What is the difference between the resampled EEG and the resampled ERP? Explain your answer in words.

Q: Plot the resampled ERP that we created. What does it look like?

In the first two steps of the resampling procedure we created a single resampled ERP. In step 3 we are instructed to repeat this procedure 3,000 times and create a distribution of ERPs. How can we do so? One potential solution is to cut and paste the code we developed over and over again, for example:

```
i=randsample(ntrials,ntrials,1);  %Draw integers,
EEG1 = EEGa(i,:);                 %... create resampled EEG,
ERP1 = mean(EEG1,1);              %... create resampled ERP.

i=randsample(ntrials,ntrials,1);  %Draw integers,
EEG2 = EEGa(i,:);                 %... create resampled EEG,
ERP2 = mean(EEG2,1);              %... create resampled ERP.

i=randsample(ntrials,ntrials,1);  %Draw integers,
EEG3 = EEGa(i,:);                 %... create resampled EEG,
ERP3 = mean(EEG3,1);              %... create resampled ERP.
```

In these lines, we have created three resampled ERPs, each with its own variable name. We could, of course, repeat this procedure and eventually define the variable `ERP3000`.

Q: Is defining the resampled ERPs in this way a good idea?

A: No! We should let the computer execute this repeated procedure for us.

> If you find yourself cutting and pasting the same code over, you're probably doing something inefficient, inelegant, and error-prone.

A better approach to create the 3,000 resampled ERPs is with a *for-loop*. We do so in MATLAB with the `for` command:

```
ERP0 = zeros(3000,size(EEGa,2));       %Create empty ERP variable.
for k=1:3000                           %For each resampling,
    i=randsample(ntrials,ntrials,1);%... choose the trials,
    EEG0 = EEGa(i,:);                  %... create resampled EEG,
    ERP0(k,:) = mean(EEG0,1);          %... save resampled ERP.
end
```

In the first line, we define an empty matrix with 3,000 rows and 500 columns; each row will contain a resampled ERP of length 500 elements, corresponding to the 1 s of EEG data (sampled at 500 Hz) considered here. In the second line, the for-loop begins. For each value of k from 1 to 3,000, we draw the trial indices with replacement (third line), create the resampled EEG (fourth line), and compute and save the resampled ERP (fifth line). This completes step 3 of the bootstrapping procedure.

A note about this segment of MATLAB code: We did not need to include the first line. The purpose of this line is to create an empty variable, in this case a matrix filled with zeros. Then, within the for-loop, we add information to this variable (we fill each row with a resampled ERP). By defining the variable `ERP0` before executing the for-loop, we make MATLAB more efficient, and the code will execute faster. In addition, defining the variable `ERP0` in advance forces us to consider the dimensions of the result, which can be useful in verifying our understanding of the variables that already exist (i.e., `EEGa`) and that we newly create (i.e., here `ERP0`).

Step 4 of the bootstrapping procedure is to determine for each time point the values greater than 2.5% and greater than 97.5% of all values. There are many ways to perform this operation in MATLAB, perhaps the easiest being to *sort* from smallest to largest the 3,000 resampled ERP values at each time point. With the resampled values sorted in this way, we then find the resampled ERP value at index $0.025 \times 3000 = 75$ and $0.975 \times 3000 = 2925$. These indices correspond to the resampled ERP values greater than 2.5% of all values and greater than 97.5% of all values, respectively, and therefore define the lower and upper confidence intervals at each moment in time. We can compute both confidence intervals in MATLAB, and (at last) plot the ERP for condition A with confidence intervals computed using the bootstrapping procedure:

```
sERP0=sort(ERP0);           %Sort each column of resampled ERP.
ciL  =sERP0(0.025*size(ERP0,1),:);  %Determine lower CI.
ciU  =sERP0(0.975*size(ERP0,1),:);  %Determine upper CI.
```

```
mnA = mean(EEGa,1);          %Determine ERP for condition A,
plot(t, mnA, 'LineWidth', 3)         %... and plot it.
hold on                      %Freeze the current plot,
plot(t,ciL)                  %... and plot lower CI,
plot(t,ciU)                  %... and upper CI.
hold off                     %Release the current plot.

ylabel('Voltage [\mu V]')    %Label the y-axis as voltage.
xlabel('Time [s]')           %Label the x-axis as time.
title('ERP of condition A with bootstrap confidence intervals')
```

The results are plotted in figure 2.9. We can use these results to identify, for example, intervals in which the ERP differs significantly from zero by finding periods in which the confidence intervals do not include zero. The advantage of the bootstrapping procedure over other approaches is that this procedure requires few assumptions about the distribution of the statistic of interest, and that we use the observed data to probe the distribution of the statistic. The disadvantage of the bootstrapping procedure is that it is computationally intensive. Here we considered 3,000 resamplings, but we could easily consider more.

Q: Compare the confidence intervals plotted in figures 2.7 and 2.9. How are the two results similar or different? What happens to the confidence intervals if you change the number of resamplings in step 3 from 3,000 to 10,000?

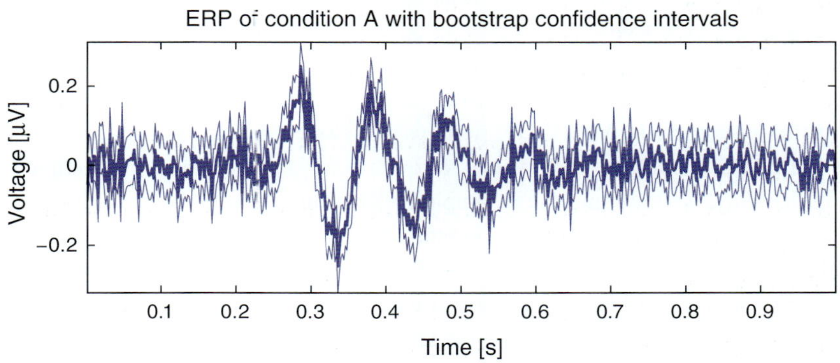

Figure 2.9

ERP with confidence intervals computed using the bootstrapping procedure for condition A. Thick line indicates the mean; 95% of values lie between the thin lines.

Q: Compute the confidence intervals using the bootstrapping procedure for the ERP of condition B. What do you find?

2.2.6 A Bootstrap Test to Compare ERPs

The bootstrapping procedure provides a powerful technique to construct confidence intervals for the ERPs using only the observed EEG measurements. We can apply a similar technique to search for significant differences between the ERPs in conditions A and B. To do so, we first choose a *statistic*, a measure of some attribute of the difference between the two ERPs. There are many choices, some informative and some not. Let's choose as our statistic the maximum absolute value of the difference in the ERPs across time. Computing this statistic is straightforward in MATLAB:

```
mnA = mean(EEGa,1);          %Determine ERP for condition A.
mnB = mean(EEGb,1);          %Determine ERP for condition B.
mnD = mnA-mnB;               %Compute the differenced ERP.
stat = max(abs(mnD));        %Compute the statistic.
```

The variable `stat` is a number, in this case 0.2613.

Q: Given the value we determined for `stat`, are the ERPs for the two conditions different?

In isolation, the numerical value for `stat` is not very useful or interesting. Is the value for `stat` consistent with noisy scalp EEG data lacking an evoked response? Or is the value for `stat` large and unexpected to occur unless the ERPs in the two conditions are different? To make the statistic useful, we need `stat` to be interpretable, which we pursue here through a bootstrapping procedure. We assume that no difference exists between the two conditions; in the language of statistics, this is called the null hypothesis. If the null hypothesis holds, then we can pool all the EEG signals together from both conditions (for a total of 2,000 trials) and draw from this combined distribution to create resampled ERPs representative of either condition.

It may seem odd to create pseudodata by selecting trials across *both* conditions; intuitively, we may expect the data to differ in these two conditions and feel uncomfortable making a pseudodata set that includes trials from both conditions. But under the null hypothesis, we assume no difference between the EEG responses in conditions A and B, and we are therefore free to create pseudodata drawing from trials in both conditions. We do so with the goal of creating a distribution of values for `stat` under the null hypothesis that conditions A and B exhibit no difference. We then compare the observed value of `stat` with this distribution of `stat` values. If there *is* a difference between the two conditions, we expect to find the observed value of `stat` to be very different from the distribution of `stat` values generated from the pseudodata under the null hypothesis.

To create the distribution of `stat` values under the null hypothesis of no difference between the two conditions, we perform a bootstrap test. The idea is similar to the bootstrapping procedure used to construct the confidence intervals for the ERP (see figure 2.9). We proceed as follows:

1. Merge the 1,000 trials each of EEG data from conditions A and B to form a combined distribution of 2,000 trials.
2. Sample with replacement 1,000 trials of EEG data from the combined distribution, and compute the resampled ERP.
3. Repeat step 2 and compute a second resampled ERP.
4. Compute the statistic, the maximum absolute value of the difference between the two resampled ERPs.
5. Repeat steps 2–4 3,000 times to create a distribution of statistic values.
6. Compare the observed statistic to this distribution of statistic values. If the observed statistic is greater than 95% of the bootstrapped values, then reject the null hypothesis that the two conditions are the same.

The MATLAB code to implement this procedure is similar to the bootstrapping procedure we have already implemented to compute the confidence intervals for the ERP:

```
EEG = [EEGa; EEGb];              %Merge EEG data from all trials.
statD = zeros(3000,1);           %Empty variable to hold results.
for k=1:3000                     %For each resampling,
    i=randsample(2*ntrials,ntrials,1);  %... choose trials,
    EEG0 = EEG(i,:);             %... create resampled EEG,
    mnA = mean(EEG0,1);          %... create resampled ERP.

    i=randsample(2*ntrials,ntrials,1);  %Re-choose trials,
    EEG0 = EEG(i,:);             %... create resampled EEG,
    mnB = mean(EEG0,1);          %... create resampled ERP.

    mnD = mnA-mnB;               %Compute differenced ERP,
    statD(k)= max(abs(mnD));%... and the statistic.
end
```

In this code, we first combine `EEGa` and `EEGb` in a new variable `EEG`. We then resample from `EEG` two times and each time compute a resampled ERP. We use the resampled ERPs to compute the statistic, which is saved in the variable `statD`. We perform this operation

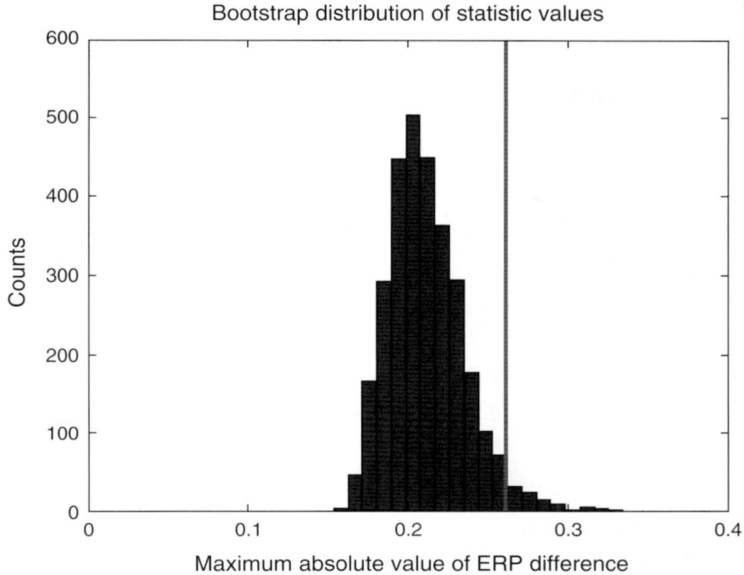

Figure 2.10

Distribution of values for the test statistic under the null hypothesis of no difference between the two conditions. Red line indicates observed statistic from EEG data.

in a for-loop to repeat the procedure 3,000 times. A histogram of the values for `statD` is shown in figure 2.10.

Q: Given the distribution of `statD` values shown in figure 2.10, and the value of `stat` computed from the original data, can we conclude that the difference between the two conditions is significant with this statistic?

A: Yes. Under the null hypothesis, the distribution of the statistic ranges from approximately 0.15 to 0.33 (figure 2.10). The observed statistic `stat` = 0.2613 exceeds most values in this distribution. Computing `length(find(statD > stat))` we find in this example that only 88 of the 3,000 values in the distribution exceed the observed statistic. This corresponds to a proportion of $88/3000 = 0.029$. We therefore reject the null hypothesis of no difference between the ERPs of conditions A and B. This result may be surprising, given how similar the two ERPs appear and the large variability in their differences (figure 2.8).

This result illustrates the power of the bootstrapping procedure. We proposed a complicated statistic (the maximum absolute value of the difference between the two resampled ERPs).

For this statistic, we do not possess an obvious formula to decide whether the resulting statistic is significant (we cannot rely on the CLT, for example). To determine significance, we employ a bootstrapping procedure (also known as a permutation test), which we can perform even for the relatively complicated statistic. In this way, we may devise complicated measures of data and construct error bars or compute statistical significance, provided our computational resources are sufficient.

Summary

In this chapter, we considered scalp EEG data recorded from a single electrode during an auditory task. The task consisted of two conditions, and we sought to uncover the difference in the EEG responses between the two conditions. We began with a visual inspection of the EEG recordings from individual trials and from all trials, and concluded that the data were quite noisy; any evoked response due to the stimulus was not obvious in the single-trial data. To emphasize the evoked signal, we computed the ERP, which involved averaging the EEG signal across trials. By doing so, we uncovered interesting structure in condition A, but not much in condition B. We then developed two techniques to add error bars to an ERP. One technique relied on the central limit theorem, and the other technique involved a computationally expensive bootstrapping procedure. Both techniques suggested that the ERP in condition A differed significantly from zero following the stimulus at time 0.25 s. Finally, we assessed whether the two ERPs differed. We did so through visual inspection, by comparing the differences in the ERPs, and by computing a statistic and assessing its significance through a bootstrapping procedure. Using the last procedure, we concluded that the ERP in the two conditions significantly differed.

Problems

2.1. Consider an alternative statistic to assess the difference in the ERPs between conditions A and B: the integrated area of the squared difference between the two ERPs. Compute this statistic for the data, and use a bootstrapping procedure to test the null hypothesis of no difference in this statistic between the two conditions.

2.2. Load the file `Ch2-EEG-2.mat`, available at

    `http://github.com/Mark-Kramer/Case-Studies-Kramer-Eden`

    into MATLAB. You'll find two variables:
    `EEG` = the EEG data, consisting of 1,000 trials, each observed for 1 s;
    `t` = the time axis, which ranges from 0 s to 1 s.
    These data have a similar structure to the data studied in this chapter. To collect these data, a stimulus was presented to the subject at 0.25 s. Analyze these data for the presence of an evoked response. To do so, answer the following questions.

a. What is the time between samples (`dt`) in seconds?

b. Examine some individual trials of these data. Explain what you observe in pictures and words. From your visual inspection, do you expect to find an ERP in these data?

c. Compute the ERP for these data, and plot the results. Do you observe an ERP (i.e., times at which the 95% confidence intervals do not include zero)? Include 95% confidence intervals in your ERP plot, and label the axes. Explain in a few sentences the results of your analysis, as you would to a collaborator who collected these data.

2.3. Load the file `Ch2-EEG-3.mat`, available at

`http://github.com/Mark-Kramer/Case-Studies-Kramer-Eden`

into MATLAB. You'll find two variables:
`EEG` = the EEG data, consisting of 1,000 trials, each observed for 1 s;
`t` = the time axis, which ranges from 0 s to 1 s.
These data have a similar structure to the data studied in this chapter. To collect these data, a stimulus was presented to the subject at 0.25 s. Analyze these data for the presence of an evoked response. To do so, answer the following questions.

a. What is the time between samples (`dt`) in seconds?

b. Examine some individual trials of these data. Explain what you observe in pictures and words. From your visual inspection, do you expect to find an ERP in these data?

c. Compute the ERP for these data, and plot the results. Do you observe an ERP (i.e., times at which the 95% confidence intervals do not include zero)? Include 95% confidence intervals in your ERP plot, and label the axes. Explain in a few sentences the results of your analysis, as you would to a collaborator who collected these data.

2.4. Load the file `Ch2-EEG-4.mat`, available at

`http://github.com/Mark-Kramer/Case-Studies-Kramer-Eden`

into MATLAB. You'll find two variables:
`EEG` = the EEG data, consisting of 1,000 trials, each observed for 1 s;
`t` = the time axis, which ranges from 0 s to 1 s.
These data have a similar structure to the data studied in this chapter. To collect these data, a stimulus was presented to the subject at 0.25 s. Analyze these data for the presence of an evoked response. To do so, answer the following questions.

a. What is the time between samples (`dt`) in seconds?

b. Examine some individual trials of these data. Explain what you observe in pictures and words. From your visual inspection, do you expect to find an ERP in these data?

c. Compute the ERP for these data, and plot the results. Do you observe an ERP (i.e., times at which the 95% confidence intervals do not include zero)? Include 95% confidence intervals in your ERP plot, and label the axes. Explain in a few sentences the results of your analysis, as you would to a collaborator who collected these data.

2.5. In the previous problem, you considered the dataset `Ch2-EEG-4.mat` and analyzed these data for the presence of an ERP. To do so, you averaged the EEG data across trials. The results may have surprised you. Modify your analysis of these data in some way to better illustrate the appearance (or lack thereof) of an evoked response. Explain what's happening in these data, as you would to a colleague or experimental collaborator.

2.6. Load the file `Ch2-EEG-5.mat`, available at

`http://github.com/Mark-Kramer/Case-Studies-Kramer-Eden`

into MATLAB. You'll find two variables:
`EEG` = the EEG data, consisting of 1,000 trials, each observed for 1 s.
`t` = the time axis, which ranges from 0 s to 1 s.
These data have a similar structure to the data studied in this chapter. To collect these data, a stimulus was presented to the subject at 0.25 s. Analyze these data for the presence of an evoked response. To do so, answer the following questions.

a. What is the time between samples (`dt`) in seconds?

b. Examine some individual trials of these data. Explain what you observe in pictures and words. From your visual inspection, do you expect to find an ERP in these data?

c. Compute the ERP for these data, and plot the results. Do you observe an ERP (i.e., times at which the 95% confidence intervals do not include zero)? Include 95% confidence intervals in your ERP plot, and label the axes. Explain in a few sentences the results of your analysis, as you would to a collaborator who collected these data.

2.7. Compare the datasets `Ch2-EEG-3.mat` and `Ch2-EEG-4.mat` studied in the previous problems. Use a bootstrap procedure to test the hypothesis that the evoked response is significantly different in the two datasets.

2.8. The goal of this problem is to explore the central limit theorem. We informally defined the CLT in this chapter. Let's perform a numerical experiment to test it.

a. Generate a large set of independent random numbers. These are many options to choose from. For example, use MATLAB to produce 500 uniformly distributed random numbers:

```
R = rand(500,1);
```

Describe this list of random numbers. Include a histogram (with axes labeled).

b. Compute the mean of this list of numbers. What value do you find for the mean? Does the mean value make sense when compared to the histogram you created?

c. Generate another instance of the independent random numbers you chose in part (a) and compute its mean, Repeat 5,000 times, saving the result each time. In the end, you should have 5,000 values, each corresponding to a mean of a different realization of the variable `R`. Describe the distribution of these mean values of `R`. What do you find? Is your result consistent with the CLT? (*Hint*: Make a histogram and consider its shape ...)

d. Repeat your analysis with a different choice for the independent random numbers. Note that MATLAB provides many different options. For your new choice of random numbers, describe a single instance of 500 values, as in part (a), and then describe the distribution of the mean of these random numbers, as in part (c).

Appendix: Standard Error of the Mean

Assume the values,

$$x = \{x_1, x_2, x_3, \ldots, x_K\},$$

are independent data drawn from a distribution with a theoretical mean μ and theoretical variance σ^2. The sample mean of x is

$$\bar{x} = \frac{1}{K}\sum_{k=1}^{K} x_k .$$

Since each of the values x_k is a sample of a random variable with a probability distribution, so is the sample mean. We are interested in computing the properties of the distribution of the sample mean of the original data.

First, we compute the expected value, or theoretical mean, of the sample mean,

$$E[\bar{x}] = \frac{1}{K}\sum_{k=1}^{K} E[x_k] = \frac{K\mu}{K} = \mu .$$

In other words, the sample mean has the same expected value, or theoretical mean, as the original data. This is why we often use the sample mean to estimate the theoretical mean of the data.

Next, we compute the theoretical variance of the sample mean of the data. The variance is the squared deviation of the sample mean of the data from the theoretical mean,

$$\begin{aligned}\mathrm{Var}\left(\bar{x}\right) &= E\left[\left(\frac{1}{K}\sum_{k=1}^{K} x_k - \mu\right)^2\right] \\ &= E\left[\left(\frac{1}{K}\sum_{k=1}^{K} (x_k - \mu)\right)^2\right] \\ &= \frac{1}{K^2} E\left[\left(\sum_{k=1}^{K} (x_k - \mu)\right)^2\right].\end{aligned}$$

The expected value in the last line of this expression is the theoretical variance of the sum of zero-mean random variables $(x_k - \mu)$. Assuming that the variables x_k are independent, the variance of the sum becomes the sum of the variances, so

$$\begin{aligned}\mathrm{Var}\left(\bar{x}\right) &= \frac{1}{K^2}\sum_{k=1}^{K} E\left[(x_k - \mu)^2\right] \\ &= \frac{1}{K^2}\left(\mathrm{Var}\left[x_1\right] + \mathrm{Var}\left[x_2\right] + \mathrm{Var}\left[x_3\right] + \cdots + \mathrm{Var}\left[x_K\right]\right) \\ &= \frac{1}{K^2}\left(\sigma^2 + \sigma^2 + \sigma^2 + \cdots + \sigma^2\right) \\ &= \frac{1}{K^2}\left(K\sigma^2\right) \\ &= \frac{1}{K}\sigma^2,\end{aligned}$$

where Var with a capital V refers to the theoretical variance of a random variable (we use lowercase var to refer to the sample variance computed from the data). The last expression defines the variance of the sample mean of x in terms of the variance of each observation of x. To know σ^2, we would need information about the distribution from which each x_k was drawn. We usually do not have this information, so instead we assume that each x_k is drawn from the same distribution, and approximate σ^2 as the sample variance of x. The advantage of this approach is that we can then estimate σ^2 from the observed data, that is, $\hat{\sigma}^2 = \mathrm{var}\left[x\right]$. Then

$$\mathrm{Var}\left(\bar{x}\right) \approx \frac{1}{K}\mathrm{var}\left[x\right].$$

In words, to find the theoretical variance of the sample mean of x we first compute the sample variance of x and then divide this sample variance by the number of observations K.

The sample standard deviation of the sample mean of x, which is also called the *standard error of the mean* (sem), is then

$$\begin{aligned}\text{sem}[x] &= \sqrt{\text{Var}[\bar{x}]} \\ &\approx \sqrt{\frac{1}{K}\,\text{var}[x]} \\ &= \frac{\text{std}[x]}{\sqrt{K}},\end{aligned}$$

where $\text{std}[x]$ is the sample standard deviation of x.

3 Analysis of Rhythmic Activity in the Scalp Electroencephalogram

Synopsis

Data Field data: 2 s of scalp EEG data sampled at 1000 Hz.

Goal Characterize the observed rhythms in these data.

Tools Fourier transform, power spectral density, spectrogram.

3.1 Introduction

3.1.1 Background

In this chapter, we again consider data recorded in the scalp EEG. The EEG provides a measure of brain voltage activity with high temporal resolution (typically on the order of milliseconds) but poor spatial resolution (on the order of 10 cm^2 of cortex). Here we consider EEG activity recorded from a single scalp electrode. We analyze these data to determine what (if any) rhythmic activity is present. We explain an important technique to characterize rhythms in data—the Fourier transform and power spectral density, or spectrum—and the many subtleties associated with this technique.

3.1.2 Case Study Data

A patient enters the Massachusetts General Hospital (MGH) emergency room unconscious. As part of his clinical workup, electrodes are placed on the scalp surface and the EEG recorded. We assume that the skilled technicians at MGH record the EEG data with no artifacts (i.e., correctly placed electrodes in good electrical contact with the scalp). Twenty-one electrodes simultaneously record the EEG data for 10 minutes sampled at 1000 Hz (i.e., 1,000 samples per second). To start, we receive from our clinical collaborator a 2 s segment of EEG data recorded from a single electrode. If we find anything interesting in this 2 s segment, our clinical collaborator has promised to provide additional EEG data from this patient and others.

3.1.3 Goal

Our goal is to analyze the 2 s of EEG data by characterizing the observed activity. There are many ways to do so, and we focus on developing techniques to characterize the observed rhythms. By the end of this chapter, you should be familiar with the principles of the Fourier transform, how to compute the spectrum in MATLAB, and the time-windowed spectrum.

3.1.4 Tools

The primary tool we develop here is the Fourier transform. We demonstrate how to compute the Fourier transform and the associated spectrum in MATLAB. The spectrum provides a powerful technique to assess rhythmic structure in time series data.

3.2 Data Analysis

To access the EEG data for this chapter, visit

`http://github.com/Mark-Kramer/Case-Studies-Kramer-Eden`

and download the file `Ch3-EEG-1.mat`.

3.2.1 Visual Inspection

A good place—often the best place—to begin data analysis is visual inspection of the time series. Let's plot the data in MATLAB:

```
load('Ch3-EEG-1.mat')          %Load the EEG data,
plot(t,EEG)                    %... plot it vs time,
xlabel('Time [s]')             %Label time axis.
ylabel('Voltage [ \muV]')      %Label voltage axis.
```

Q: The EEG data are plotted in figure 3.1. What do you see?

You might notice by visual inspection a dominant rhythm. We can approximate the *frequency* of this rhythm by counting the number of oscillations that occur in a 1 s interval. To do so, we might count the total number of maxima in figure 3.1 and divide by 2 (because we observe 2 s of data). However, counting so many maxima over an extended time interval is quite error prone. Instead, let's count the number of maxima in the first 0.2 s and then multiply by 5; that will approximate the total number of peaks in a 1 s interval. We count about 12 peaks in the first 0.2 s, which corresponds to approximately 60 peaks in 1 s, or approximately 60 cycles per second, or 60 Hertz (Hz).

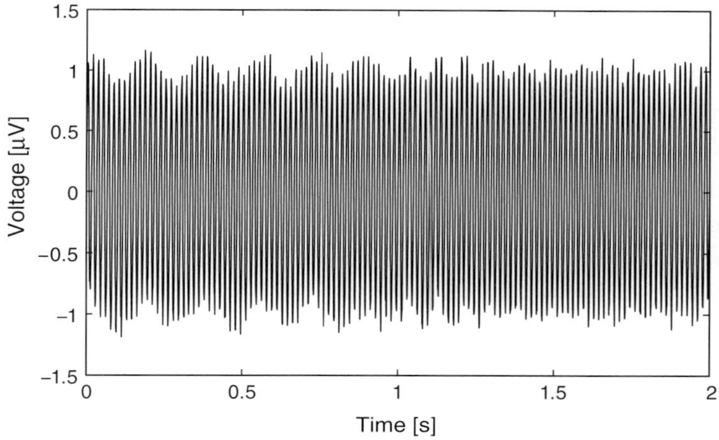

Figure 3.1
EEG data provided by our collaborator.

> **Q:** What if you counted the minima instead of the maxima? Do you get the same answer? What if you counted the zero crossings?

Visual inspection suggests a dominant rhythmic activity at a frequency of 60 Hz. High-frequency oscillations in the 40–80 Hz band (the gamma band) are thought to be important for cognitive processing in the brain [7]. But there's a reason for the label gamma *band*: the rhythmic activity observed *in vivo* is typically diffuse, spread over a range of rhythms at neighboring frequencies. The rhythmic activity observed here is concentrated and remarkably regular for EEG data. This perhaps makes us doubtful, and rightly so.

> The alternating current in any North American electrical socket alternates at 60 Hz.

We therefore propose the most likely conclusion: the data are dominated by electrical noise. We have at least three options: (1) Try to reduce the impact of this noise during the clinical recording procedure, (2) abandon this analysis and pursue other research projects, or (3) continue with additional analysis beyond visual inspection of the time series data. The first option is too expensive and time consuming; the clinical recording system cannot be changed without encountering significant regulations in the monitoring of human subjects. The second option is not viable. So, we'll pursue the third. Why? Visual inspection suggests a dominant 60 Hz signal, but looks can be deceiving. Perhaps something else is there, lurking in the signal background.

> The 60 Hz noise is not all bad. Because we expect it (in North America), we can use this information to our advantage. As we build measures to characterize rhythms, we can always check these measures by confirming that the 60 Hz rhythm appears.

If only visual inspection were enough! Sometimes it is, especially when something has gone wrong (e.g., if the EEG trace were zero for all time, we should be suspicious). But looks can be deceiving. The voltage trace in figure 3.1 looks like a continuous line drawn on the page. That's incorrect. If we look more closely, we find that the data consist of discrete points (figure 3.2). Although the true brain signal may evolve as a continuous voltage trace in time, we do not observe this true signal. Instead, we observe a discrete sampling of this signal in time. The spacing between these samples is determined by the recording device collecting the EEG data. In this case, the spacing between samples is small; our collaborator has told us that the data were sampled at 1000 Hz, which corresponds to a sample of data every 1 ms. So, we observe not the (presumably) continuous true voltage signal but instead discrete samples of this signal in time. Note that in figure 3.1 the discrete points are so closely spaced that the discrete sampling of the signal appears to be a continuous curve evolving in time.

To understand the impact of this discrete sampling, we first require some definitions. Let's define Δ as the time between samples, in this case $\Delta = 1$ ms. We also define N as the

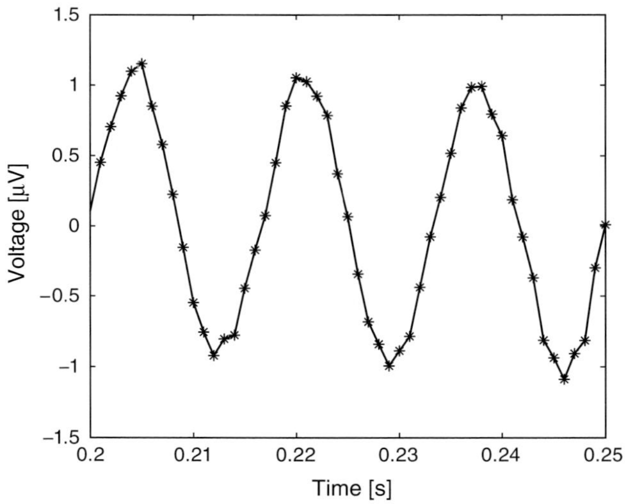

Figure 3.2

Inspection of a small window of data to investigate discrete sampling of the EEG. Asterisks indicate individual sample points. Time interval between points is the sampling interval, Δ.

total number of points observed, and T as the total time of the recording. These three terms are related: $T = N\Delta$. For $T = 2$ s of EEG data, there are $N = T/dt = 2/0.001 = 2000$ points. From this, we can also define the sampling frequency $f_0 = 1/\Delta$, which in this case is 1000 Hz. Finally, we define a symbol for the data, x, which we also write as x_n to explicitly indicate the index $n \in \{1, 2, 3, \ldots, N\}$ corresponding to the sample number. Let's also define all of these variables in MATLAB:

```
x  = EEG;            %Relabel the data variable.
dt = t(2)-t(1);      %Define the sampling interval.
N  = length(x);      %Define the total number of data points.
T  = N*dt;           %Define the total duration of data.
```

We need to keep the sampling interval Δ and the total recording duration T in mind; both play fundamental roles in the characterization of the rhythmic activity.

Q: In the second line of this code we define the sampling interval as `dt=t(2)-t(1)`. How else could we have defined `dt`? Would `t(10)-t(9)` be appropriate?

3.2.2 Mean, Variance, and Standard Deviation

As a first step in analyzing the EEG data, let's define two of the simplest measures we can use to characterize data x: the *mean* and the *variance*.[1] To estimate the mean $\bar{x}$, or average value of x, we compute

$$\bar{x} = \frac{1}{N}\sum_{n=1}^{N} x_n. \tag{3.1}$$

In words, we sum the values of x for all n time indices, then divide by the total number of points summed (N). To estimate the variance σ^2 of x we compute,

$$\sigma^2 = \frac{1}{N}\sum_{n=1}^{N} (x_n - \bar{x})^2, \tag{3.2}$$

which characterizes the extent of fluctuations about the mean. The *standard deviation* is simply the square root of the variance, σ. It's straightforward to compute all three quantities using built-in MATLAB functions:

```
mn = mean(x);        %Compute the mean of the data.
vr = var(x);         %Compute the variance of the data.
sd = std(x);         %Compute the standard deviation of the data.
```

1. We could instead write the *sample* mean because we use the observed data to estimate the theoretical mean that we would see if we kept repeating this experiment. However, this distinction is not essential to our goals here.

Q: Compare the mean computed in MATLAB with the EEG data plotted in figure 3.1. Are the two consistent? How does the standard deviation computed in MATLAB compare with the EEG fluctuations in figure 3.1?

A: The computed mean is approximately 0. Visual inspection of figure 3.1 suggests that the EEG data fluctuates around a center value of 0, so the computed mean is consistent with the visual inspection of the data. The computed standard deviation is approximately 0.71. We expect that most of the signal fluctuations lie within two standard deviations (i.e., $\pm 2\sigma$) of the mean. We therefore expect to observe EEG values mostly between $0 \pm 1.4 = (-1.4, 1.4)$, which is in fact what we observe.

The mean, variance, and standard deviation provide single numbers that summarize the EEG trace. In this case, these numbers are not particularly useful. They may depend on many factors, including the electrical contact between the electrode and scalp surface, the noise in the signal, and the cognitive state of the subject. Here, we're more interested in how the EEG activity is distributed across rhythms. We've already begun to assess rhythms in the EEG data through visual inspection of the time series. To further characterize these rhythms, we employ another powerful tool, the Fourier transform. Before introducing the Fourier transform, we consider a related measure, the autocovariance.

3.2.3 Autocovariance

Visual inspection strongly suggested a prominent feature in the data: rhythmic activity. Rhythmic activity represents a type of dependent structure in the data. For example, if we know the data tend to oscillate near 60 Hz, then given the value of the EEG data now, we can accurately predict the value of the EEG data 1/60 s in the future (i.e., one cycle of the 60 Hz activity); it should be similar. One technique to assess the dependent structure in the data is the *autocovariance*. The formula for the autocovariance, $r_{xx}[L]$, evaluated at lag L, is

$$r_{xx}[L] = \frac{1}{N} \sum_{n=1}^{N-L} (x_{n+L} - \bar{x})(x_n - \bar{x}). \tag{3.3}$$

In words, the autocovariance multiplies the data x at index $n+L$ by the data x at index n, and sums these products over all indices n. Notice that in both terms the mean value $\bar{x}$ is subtracted from x before computing the product, and we divide the resulting sum by N, the total number of data points in x. Note that (3.3) is a *biased* estimate of the autocovariance; we compare this to an unbiased estimate of the autocovariance in the next section.

To gain some intuition for the autocovariance, let's represent x graphically as a one-dimensional row vector (figure 3.3a). For the case $L = 0$, the autocovariance is simply

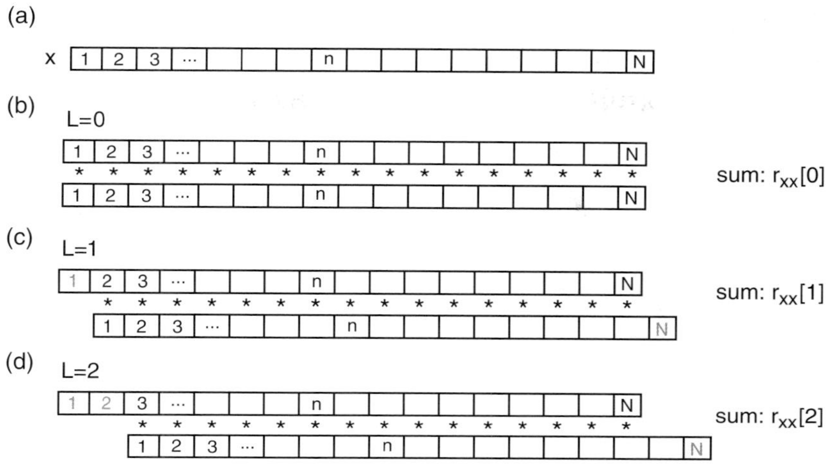

Figure 3.3

Cartoon representation of autocovariance. (*a*) Data x are represented as one-dimensional vector; with indices $n = \{1, 2, 3, \ldots, N\}$. Autocovariance at (*b*) lag 0, (*c*) lag 1, and (*d*) lag 2. Asterisks indicate multiplication between elements of the two vectors. Gray index labels at beginning and end of vectors indicate data points not involved in computing autocovariance at the chosen lag L.

the element by element product of x with itself, summed over all indices (figure 3.3b). For the case $L = 1$, we shift x by one index, multiply element by element the original (unshifted) x by the shifted version, and sum over all indices (figure 3.3c). This process of shifting, multiplying element by element, and summing can be repeated for both positive and negative values of the lag L. Notice that for larger values of L, we lose values at the beginning and ends of the autocovariance.

> **Q:** What is the largest reasonable value of L to consider? For example, does a value of L greater than N make sense?

The autocovariance will be largest at the lag L for which the values of x match. For most functions the autocovariance is largest at $L = 0$ (of course, x matches itself with zero shift) and tends to decrease as the magnitude of L increases. Physically, the decrease in autocovariance with lag is consistent with the notion that data become less similar as time progresses. For example, in an EEG recording, we expect the activity now to be similar to the activity in the immediate future but different from the activity in the more distant future. As the brain responds to different internal and external cues, we expect different EEG activities to emerge and associations between the EEG activity now and later to decay. Functions x that exhibit dependent structure possess informative features in the autocovariance.

Q: Compare the autocovariance at $L = 0$ and the variance (3.2). Notice anything similar?

To compute the autocovariance of the EEG data, we execute the following commands in MATLAB:

```
%Compute the autocov for L +/- 100 indices.
[ac,lags] = xcorr(x-mean(x),100,'biased');
%... and plot the autocov vs lags in units of time
plot(lags*dt,ac)
%... with axes labeled.
xlabel('Lag [s]')
ylabel('Autocovariance');
```

Q: Examine the autocovariance of the EEG data plotted in figure 3.4. What do you observe?

The first input to the function `xcorr` is the EEG data with the mean subtracted (`x - mean(x)`). One striking feature of the autocovariance is the periodicity. A careful

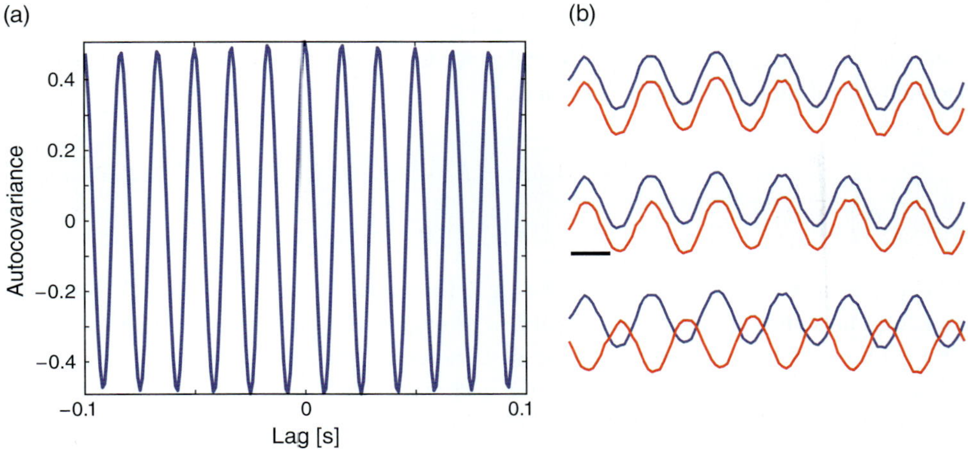

Figure 3.4

(*a*) Autocovariance of the EEG. (*b*) Examples of EEG at different shifts. With no shift (*top*) traces exactly overlap. With shift of one 60 Hz cycle (*middle*) traces appear quite similar. With shift of one-half the 60 Hz cycle (*bottom*) traces move in opposite directions. Scale bar indicates 10 ms.

inspection shows that the autocovariance exhibits repeated peaks and troughs approximately every 0.0166 s.

Q: Why does the autocovariance exhibit repeated peaks and troughs approximately every 0.0166 s?

A: The autocovariance reflects the dominant rhythmic activity in the data. Remember that the EEG data are dominated by a 60 Hz rhythm (figure 3.1). To gain intuition for how this rhythmic activity affects the autocovariance, we plot in figure 3.4 examples of the EEG data aligned with different lags L. At zero lag ($L = 0$), the two time series are identical. Therefore, the product

$$(x_{n+0} - \bar{x})(x_n - \bar{x}) = (x_n - \bar{x})(x_n - \bar{x}) = (x_n - \bar{x})^2$$

is non-negative for all indices n (the product may sometimes be zero, but it's never negative). To compute the autocovariance, we sum this product over all indices n, and divide by N, as defined in (3.3). Because we sum many positive terms, we expect to find a large positive value for $r_{xx}[0]$. Shifting the EEG data by an integer multiple of the 60 Hz cycle, we observe a similar voltage trace (figure 3.4b). Therefore, at this lag L, we again expect the summed product

$$(x_{n+L} - \bar{x})(x_n - \bar{x})$$

over all indices n to be large, and to find a large positive value for $r_{xx}[L]$. In fact, we expect the autocovariance to be large and positive whenever the lag L is an integer multiple of the 60 Hz cycle (i.e., an integer multiple of $1/60 \approx 0.0166$ s); this is exactly what we find in figure 3.4a.

Shifting the EEG data by half of the 60 Hz cycle, we observe a different type of relation; at this lag, call it L^*, positive values in the unshifted EEG correspond to negative values in the shifted EEG (figure 3.4b). Therefore, most terms in the product

$$(x_{n+L^*} - \bar{x})(x_n - \bar{x})$$

are negative, and summing up these terms to compute the autocovariance (3.3), we find a large negative value for $r_{xx}[L^*]$.

The autocovariance is a useful tool for assessing the dependent structure in the EEG data. Visual inspection of the EEG reveals a specific type of dependent structure—a strong rhythmic component—in the data. This dependent structure is further characterized in the autocovariance, in which the dominant 60 Hz activity manifests as periodic peaks and troughs. In the next section, we consider a second tool, the spectrum, for assessing dependent structure in time series data. The autocovariance and spectrum are related in a remarkable way.

Biased Versus Unbiased Autocovariance The equation for the autocovariance (3.3) is a biased estimate of the true autocovariance. To compute an *unbiased* measure of the autocovariance, we replace the $1/N$ term in (3.3) with $1/(N-L)$,

$$r^*_{xx}[L] = \frac{1}{N-L}\sum_{n=1}^{N-L}(x_{n+L}-\bar{x})(x_n-\bar{x}). \tag{3.4}$$

To examine the difference in the biased versus the unbiased autocovariance, let's compute both for the EEG data over a broad interval of lags:

```
[ac_u,lags]=xcorr(x-mean(x),2000,'unbiased');%Unbiased autocov,
plot(lags*dt,ac_u)                            %... plot it.
[ac_b,lags]=xcorr(x-mean(x),2000,'biased');  %Biased autocov,
hold on                                       %... plot it.
plot(lags*dt,ac_b, 'r')
hold off
xlabel('Lag [s]')                             %Label the axes.
ylabel('Autocovariance');
```

Here we compute the autocovariance for ±2000 lags (second input to the `xcorr` function) and choose either an `'unbiased'` or a `'biased'` autocovariance (third input to the `xcorr` function). We see in figure 3.5 the similarities and differences between these two estimates. At small lags (near 0 s), the biased and unbiased estimates of the autocovariance are similar. In this case, L is small, so the factor $1/(N-L)$ in the unbiased estimate is similar to the factor $1/N$ in the biased estimate. At large lags (away from 0 s), the biased and unbiased estimates of the autocovariance are quite different. As L approaches ±2 s, the biased estimate approaches zero.

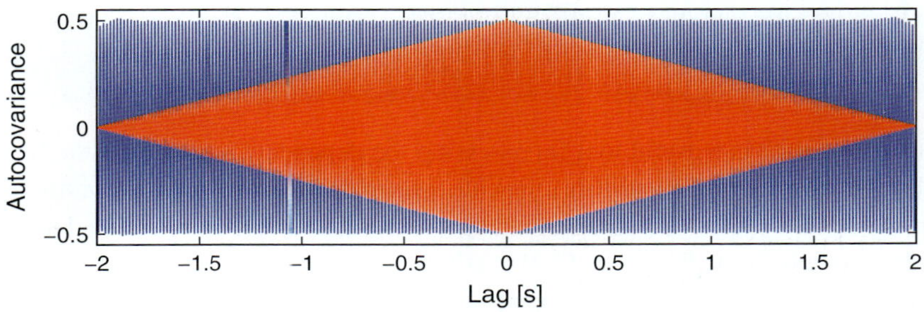

Figure 3.5
Autocovariance of the EEG for biased (*red*) and unbiased (*blue*) estimates over a wide range of lags.

Q: Why does the biased estimate of the autocovariance approach zero at large lags?

A: For concreteness, let's consider the case when $L = N - 2$. In this case L is large, and nearly equal to the number of points in the data (N). When L is large, we shift the time series x so that only a subset of indices overlap. Consider the $L = 2$ case in figure 3.3 and the extension to $L = N - 2$. Because we only compute the product

$$(x_{n+L} - \bar{x})(x_n - \bar{x})$$

for the overlapping indices of x_n and x_{n+L}, we only include two terms in the summation (3.3). The sum of these two terms is then divided by N, which results in a small number that approaches zero as L approaches N.

Compare this observation to the *unbiased* estimate of the autocovariance. In this example, we see that the unbiased estimate of the autocovariance remains large even as L approaches ± 2 s (see figure 3.5).

Q: Why does the unbiased estimate of the autocovariance remain large at large lags?

A: As in the biased case, consider $L = N - 2$. In this case, L is large, and we shift the time series x so that only a subset of indices overlap. Therefore, the product

$$(x_{n+L} - \bar{x})(x_n - \bar{x})$$

again only includes two terms in the summation (3.4). However, in the unbiased case, the sum of these terms is divided by $N - L = N - (N - 2) = 2$; as L approaches N, the term $N - L$ approaches zero. In this case, we find a balance between the summation of two terms and then a division by the number of terms in the sum (in this example, division by 2). This balance allows the unbiased estimate of the autocovariance to remain large as L approaches N.

Careful inspection of the blue curve in figure 3.5 reveals another feature of the unbiased estimate; the estimated values at large lags become more variable (look carefully at lags near ± 1.75 s and beyond). Increased variability at large lags occurs because as L approaches N, we have less data to compare in the assessment of the autocovariance. Notice that when $L = N - 1$, the estimate of the autocovariance utilizes only two data points from x (i.e., the product consists only of one term: $(x_N - \bar{x})(x_1 - \bar{x})$). We do not expect a reliable assessment of associations in the data with so few data points to compare.

With those observations, should we use the biased or unbiased estimator of the autocovariance? Statisticians typically prefer the biased estimator for a variety of reasons [8]. First, for many stationary processes, the mean squared error of the biased estimator is smaller than that of the unbiased estimator. The mean squared error depends on both the variance and the bias of the estimator:

$$\text{Mean squared error} = \text{Variance} + (\text{Bias})^2. \tag{3.5}$$

Although the biased estimator is "biased," the variability of the unbiased estimator is more harmful. We saw a hint of this increased variability in the unbiased estimator at large lags in figure 3.5. To make this observation more explicit, let's consider the lag $L = N - 1$, and compute the expression for the biased estimator (3.3),

$$\begin{aligned} r_{xx}[N-1] &= \frac{1}{N}\sum_{n=1}^{N-(N-1)} (x_{n+(N-1)} - \bar{x})(x_n - \bar{x}) \\ &= \frac{1}{N}\sum_{n=1}^{1}(x_{n+(N-1)} - \bar{x})(x_n - \bar{x}) \\ &= \frac{1}{N}(x_N - \bar{x})(x_1 - \bar{x}). \end{aligned} \tag{3.6}$$

The expression for the unbiased estimator (3.4) becomes

$$\begin{aligned} r^*_{xx}[N-1] &= \frac{1}{N-(N-1)}\sum_{n=1}^{N-(N-1)} (x_{n+N-1} - \bar{x})(x_n - \bar{x}) \\ &= \sum_{n=1}^{1}(x_{n+N-1} - \bar{x})(x_n - \bar{x}) \\ &= (x_N - \bar{x})(x_1 - \bar{x}). \end{aligned} \tag{3.7}$$

These two expressions reveal that at a large lag $L = N - 1$, the variance of the unbiased estimator (3.7) is N^2 times the variance of the biased estimator (3.6). The dramatic increase in variance of the unbiased estimator leads to unreliable estimates of the autocovariance at large lags. Also, we note that the biased estimator behaves nicely as L increases to N; we see from (3.6) that $r_{xx}[N-1]$ approaches zero when N is large. This is arguably the behavior we want. We have few data points to compare at large lags, and therefore an unreliable estimate of the autocovariance, so we're better off disregarding these values. For these reasons, we use the biased estimator; in this estimate, autocovariance values at large lags, which utilize less data and are typically noisier, are reduced.

For the EEG data of interest here, the unbiased estimator outperforms the biased estimator. For these data, which are dominated by a 60 Hz rhythm, there is significant autocovariance even at long lags. In this case, the biased estimator leads to an interpretation of decreasing autocovariance, even though that is not true. However, for most brain signals (not saturated by 60 Hz line noise), we expect the autocovariance to decrease in time.

3.2.4 Power Spectral Density, or Spectrum

There are many techniques to assess rhythmic activity in the EEG data. Here, we compute the *power spectral density*, or simply the spectrum, of x using a well-established technique, the Fourier transform.[2]

> The *spectrum* of the data x is the magnitude squared of the Fourier transform of x. The spectrum indicates the amplitude of rhythmic activity in x as a function of frequency.

There are many subtleties associated with computing and interpreting the spectrum. We explore some of them here; in doing so, we build our intuition for spectral analysis and our ability to deal with future, unforeseen circumstances in other data we encounter in research.

Computing the Spectrum. We start by presenting all the formulas and MATLAB code necessary to compute the spectrum of the data. Then throughout the chapter, we circle back and consider each step of the computation in detail.

We first need a formula for the discrete-time Fourier transform of the data x:

$$X_j = \sum_{n=1}^{N} x_n \exp(-2\pi i f_j t_n). \tag{3.8}$$

The Fourier transform computes the sum over all time indices $t_n = \Delta\{1,2,3,\ldots,N\}$ of the data x_n multiplied by sinusoids oscillating at a given frequency $f_j = j/T$, where $j = \{-N/2+1, -N/2+2, \ldots, N/2-1, N/2\}$. The result is a new quantity X_j, the signal as a function of frequency f_j rather than time t_n. The spectrum is then

$$S_{xx,j} = \frac{2\Delta^2}{T} X_j X_j^*, \tag{3.9}$$

which is the product of the Fourier transform of x with its complex conjugate (indicated by the superscript $*$), scaled by the sampling interval and the total duration of recording. The term $2\Delta^2/T$ is simply a numerical scaling (see appendix C at the end of this chapter). The units of the spectrum are, in this case, $(\mu V)^2/\text{Hz}$. Computing the spectrum in MATLAB requires only a few lines of code:

```
xf  = fft(x-mean(x));              %Compute Fourier transform of x.
Sxx = 2*dt^2/T*(xf.*conj(xf));     %Compute spectrum.
Sxx = Sxx(1:length(x)/2+1);        %Ignore negative frequencies.
```

2. The power spectral density describes the extent to which sinusoids of a single frequency capture the structure of the data. To compute the power over any range of frequencies, we would integrate (or for discrete frequencies, sum) the spectrum over that frequency range.

```
df = 1/max(T);                    %Determine frequency resolution.
fNQ = 1/ dt / 2;                  %Determine Nyquist frequency.
faxis = (0:df:fNQ);               %Construct frequency axis.

plot(faxis, Sxx)                  %Plot spectrum vs frequency.
xlim([0 100])                     %Select frequency range.
xlabel('Frequency [Hz]')          %Label the axes.
ylabel('Power [ \muV^2/Hz]')
```

That's not so bad; the code to compute and display the spectrum fits in 12 lines (with spacing for aesthetics). The results of this computation are plotted in figure 3.6. Notice the large peak at 60 Hz. This peak is consistent with our visual inspection of the EEG data, in which we approximated a dominant rhythm at 60 Hz by counting the number of peaks that appeared in the voltage traces. So, our computation of the spectrum at least matches our initial expectation deduced from visual inspection of the data.

So, we've managed to compute and plot the spectrum in MATLAB, and our analysis results match our expectations. We could choose to stop here. But a danger persists: we've blindly entered MATLAB code and achieved an expected result. What are the frequency resolution and Nyquist frequency mentioned in the comments of the code? Maybe this procedure is fraught with pitfalls, and we simply got lucky in this case? Does the spectrum

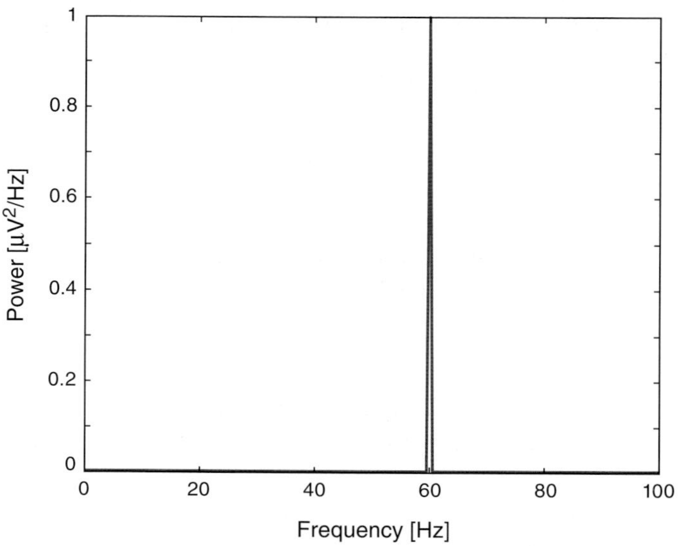

Figure 3.6
Power spectrum of EEG data. A large peak occurs at 60 Hz along the horizontal axis.

provide additional information that was not immediately uncovered in figure 3.6? How will we react and adapt when the spectrum results do not match our intuition? To answer these questions requires developing more intuition for the Fourier transform and spectrum. In the following sections, we examine equations (3.8) and (3.9), and the MATLAB code. In doing so, we explore some subtleties of this measure and strengthen our intuition for this measure's behavior. Building this intuition is perhaps the most important part for dealing with unforeseen circumstances arising in your own data.

What Is the Fourier Transform Actually Doing? The Fourier transform represents the data x as a linear combination of sinusoids with different frequencies. To see this, consider again (3.8):

$$X_j = \sum_{n=1}^{N} x_n \exp(-2\pi \, i f_j \, t_n).$$

What is this expression actually doing? Let's consider this equation piece by piece. The first term,

$$\sum_{n=1}^{N}$$

represents a sum over the indices of x_n. Because each subsequent index represents a step forward in time of Δ, this sum represents a sum over time. In fact, this sum extends from the first index ($n = 1$) to the last index ($n = N$) of the data, so the summation occurs over the entire duration of the recording. Adding the second term,

$$\sum_{n=1}^{N} x_n,$$

the summation now acts on the data x_n, so we're summing up the data over all indices or equivalently over the entire time of recording. The third term,

$$\exp(-2\pi \, i f_j \, t_n),$$

consists of an exponential (exp) operating on the product of five individual terms, each of which is a number. The first two numbers are simple: -2π. The third number is $i \equiv \sqrt{-1}$, a quantity representing an *imaginary unit*. The utility of the imaginary unit is that we can rewrite the exponential as the sum of a sine and cosine function. Remember from calculus the remarkable Euler's formula:

$$\exp(-2\pi \, i f_j \, t_n) = \cos(-2\pi \, f_j \, t_n) + i \sin(-2\pi \, f_j \, t_n).$$

Notice that the i now appears multiplying the sine term, and not inside of the arguments of either sinusoid. Both sinusoids operate on the product of 2π and two terms, the frequency f_j and the time t_n. Rewriting (3.8) using Euler's formula, we find

$$X_j = \left(\sum_{n=1}^{N} x_n \cos(-2\pi f_j t_n)\right) + i\left(\sum_{n=1}^{N} x_n \sin(-2\pi f_j t_n)\right), \tag{3.10}$$

where the summation is distributed over both terms.

Written in this way, the Fourier transform becomes easier to interpret. Let's consider the first term of (3.10). For each index n, we multiply the data x_n by a cosine function evaluated at frequency f_j and time t_n. We then sum the results of this multiplication over all indices from $n = 1$ to $n = N$, or equivalently, from time $t_n = \Delta$ to time $t_n = T$. So, we multiply the data by a cosine function at frequency f_j for each point in time, and sum the product over time. The second term of (3.10) is like the first, except we multiply the data x_n by a sine function.

We therefore think of the Fourier transform as comparing the data x to the sinusoids oscillating at frequency f_j. When the data and sinusoid at frequency f_j align the summation in the Fourier transform is large and the result X_j is a large number. When the data and sinusoid at frequency f_j do not align, the summation in the Fourier transform is small and X_j is a tiny number. To make these ideas more concrete, consider the examples in figure 3.7. In these simple examples, the data x are a perfect cosine with frequency 10 Hz

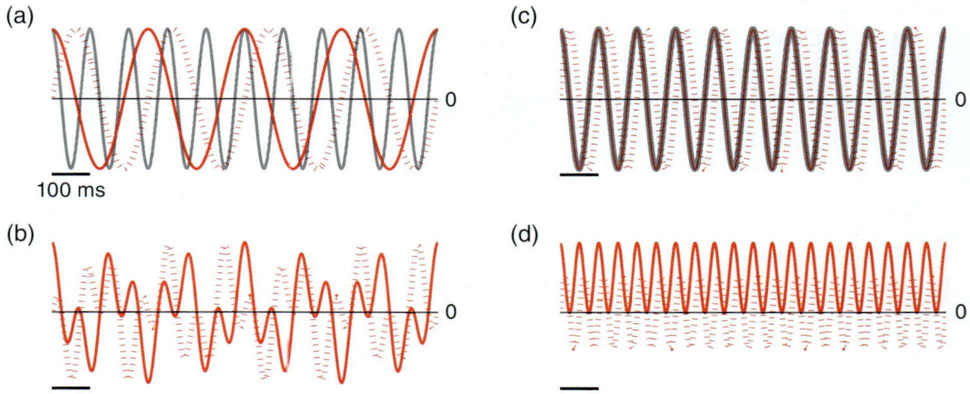

Figure 3.7

Artificial data, a 10 Hz cosine function, compared to sinusoids at different frequencies. Data are depicted by gray curves; cosine functions, by red solid curves; and sine functions, by red dotted curves. Scale bars indicate 100 ms. (*a*) Data compared to cosine and sine functions, each with frequency $f_j = 4$ Hz. (*b*) Product of data and 4 Hz cosine function, and product of data and 4 Hz sine function. (*c*) Data compared to 10 Hz cosine and 10 Hz sine functions. (*d*) Product of data and 10 Hz cosine function, and product of data and 10 Hz sine function.

(figure 3.7a). Choosing $f_j = 4$ Hz, we construct a sine and cosine function each oscillating at 4 Hz (figure 3.7a). Then, to perform the calculation of (3.10) we multiply the data x by the sinusoids at each point in time. The results of this computation are plotted in figure 3.7b. Notice that the products alternate between positive and negative values throughout time.

Q: In this case, what is the approximate value of X_j when $f_j = 4$ Hz?

A: From (3.10) we know that X_j is the summation over time of the product of x and the sinusoids. These products over time are plotted in figure 3.7b. Because each product alternates between positive and negative values roughly equally over time, the summation of the product over time is approximately zero. We therefore conclude that both the real part (i.e., the cosine term in (3.10)) and the imaginary part (i.e., the sine term in (3.10)) are small and $X_j \approx 0 + 0i$ when $f_j = 4$ Hz.

In this case, the sinusoids at frequency $f_j = 4$ Hz do not align with the data x, and X_j is nearly zero in both its real and imaginary parts. Now consider the case in which we choose $f_j = 10$ Hz. With this choice of f_j, the data x and the cosine function align perfectly (figure 3.7c). The product of the cosine function and the data is always non-negative (figure 3.7d), and therefore its summation over time results in a large positive number. In this case, the real part of X_j is large because the cosine function with frequency $f_j =10$ Hz and the data x match. In this sense, the Fourier transform reveals the dominant frequencies of the underlying time series.

Q: What is the approximate value of the *imaginary* part of X_j for $f_j = 10$ Hz? *Hint*: Consider the plot of the product of the sine function and data in figure 3.7d.

Relation of the Spectrum to the Autocovariance. We've now introduced two tools for assessing dependent structure in the EEG data: the autocovariance and the spectrum. Remarkably, these two measures are related in an important way.

The spectrum is the Fourier transform of the autocovariance.

The spectrum and autocovariance both assess dependent structure in the data but in different domains, the spectrum in the frequency domain and the autocovariance in the time domain. Notice that the spectrum $S_{xx,j}$ in (3.9) is a function of frequency index j, while the autocovariance $r_{xx}[L]$ in (3.3) is a function of time lag L. For the EEG data of interest here, the dominant 60 Hz rhythm manifests as periodicity in the autocovariance as a function of lag (figure 3.4a), and a peak in the spectrum as a function of frequency (figure 3.6). Although the two measures are related through the Fourier transform, each provides a different perspective on the dependent structure in the data. In practice, applying and visualizing both measures is often of use. See appendix A at the end of this chapter for a more mathematical discussion of this relation.

Here's an associated mathematical nugget. The spectrum is the Fourier transform of x multiplied by its complex conjugate. That produces a real number (i.e., the imaginary part is 0), which is convenient for plotting and visualization. The autocovariance is necessarily symmetric with respect to lag, that is, $r_{xx}[L] = r_{xx}[-L]$. In other words, the autocovariance is an *even function*. The Fourier transform of an even function is real, so the Fourier transform of the autocovariance is also a real number, consistent with the values of the spectrum.

Relation of the Spectrum to Multiple Linear Regression. As a final perspective, we consider a statistical modeling approach, with the goal of characterizing the rhythms that appear in the EEG data. To develop this approach, we first introduce *linear regression*. The idea of linear regression is to express a response variable at time n (call it x_n) in terms of predictor variables (call them $z_{1n}, z_{2n}, \ldots, z_{pn}$ for p predictor variables) as

$$x_n = \beta_0 + \beta_1 z_{1n} + \beta_2 z_{2n} + \cdots + \beta_p z_{pn} + \epsilon_n, \tag{3.11}$$

where ϵ_n is a random variable. This formulation represents the model we use in *multiple linear regression*. The term *multiple* comes from the multiple predictors used. The term *linear* expresses the fact that each predictor appears linearly in the expression (3.11). The challenge is to identify the unknown coefficients (the β's) given the observed response and predictor variables. We note that the case of $p = 1$ corresponds to simple linear regression; in that case, the goal is to fit the line,

$$x_n = \beta_0 + \beta_1 z_{1n} + \epsilon_n,$$

with slope β_1 and intercept β_0.

Let's consider application of multiple linear regression to the EEG data with a specific purpose: to remove the 60 Hz line noise. Recall that we found that the spectrum was dominated by a 60 Hz peak (figure 3.6). We expect this 60 Hz is due to electrical noise in the system, and this large noise peak may mask other interesting features occurring in the EEG data. Therefore, our analysis of the EEG data may benefit by removing this large 60 Hz signal. To do so, we first fit a multiple linear regression model to the data x_n with the following form,

$$x_n = \beta_0 + \beta_1 \sin(2\pi 60\, t_n) + \beta_2 \cos(2\pi 60\, t_n) + \epsilon_n, \tag{3.12}$$

where x_n is the EEG data at index n, t_n is the corresponding time axis at index n in units of seconds, and ϵ_n is a random variable.

Q: The model consists of three predictors. What are they?

A: The predictors are a constant term, a sine function at 60 Hz, and a cosine function at 60 Hz. Our goal is to solve for the unknown coefficients β_0, β_1, and β_2 given the EEG data.

This multiple linear regression is performed in MATLAB as follows:

```
%Define the model,
model = [ones(size(x)) sin(2*pi*60*t') cos(2*pi*60*t')];
%... and perform regression.
b = regress(x, model);
```

In the first line we define the model, which consists of a constant term (the variable `ones`), the 60 Hz sine function, and the 60 Hz cosine function. We note that the variable `t` is transposed so that `model` is a matrix with dimensions $[2000, 3]$. In the second line we ask MATLAB to perform the regression using the function `regress`, and return the β values in the single output variable `b`.

Q: Examine the output variable `b`. What do you find?

A: We find

```
b = -0.0000 0.9989 -0.0032
```

This result indicates that the constant predictor and the 60 Hz cosine predictor do not contribute much to the data; the values `b[1]` and `b[3]` are both near zero. However, the 60 Hz sine function makes a much large contribution; the value `b[2]` is near 1.

To see how well our multiple linear regression model fits the data, let's evaluate the model, and compare it to the original EEG data:

```
%Evaluate the model.
x_60Hz_modeled = b(1)+b(2)*sin(2*pi*60*t)+b(3)*cos(2*pi*60*t);
plot(t,x)                        %Plot the EEG data.
hold on                          %... freeze the graphics window,
plot(t,x_60Hz_modeled, 'r')      %... and plot the modeled EEG data,
hold off                         %... release the graphics window,
```

```
xlim([0.5 1])                   %... examine 0.5 s of data,
xlabel('Time [s]')              %... and label the axes.
ylabel('EEG and Modeled EEG [\muV]')
```

We plot the EEG data and model fit in figure 3.8. The model, which contains only three predictors, appears to do quite a good job at capturing the 60 Hz activity in the EEG data.

Q: The multiple linear regression model (3.12) is not a *perfect* fit to the data. Why?

A: The multiple linear regression model includes only three predictors. We expect the EEG data to consist of other features, including rhythms at other frequencies. Therefore, this simple model cannot account for all features of the EEG time series. That's okay. The goal of this model was to fit a particular rhythm, the 60 Hz activity, not every aspect of the data.

Now, let's examine a powerful application of the model. We've constructed the model to fit the 60 Hz line noise in the EEG data. Let's now use the model to *remove* the 60 Hz line noise from the original EEG data. To do so, we subtract the model fit from the EEG data and then plot the resulting new signal:

```
%Evaluate the model.
x_60Hz_modeled = b(1)+b(2)*sin(2*pi*60*t)+b(3)*cos(2*pi*60*t);
x_cleaned=x-x_60Hz_modeled';    %... remove it from the EEG data,
plot(t,x_cleaned);              %... and plot the result.
```

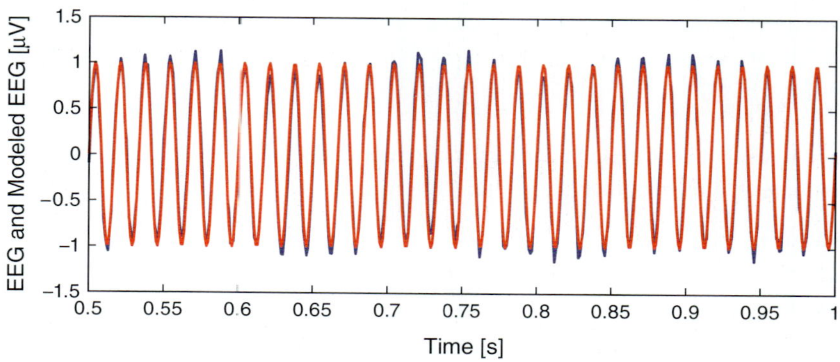

Figure 3.8
Comparison of 0.5 s segment of EEG data (*blue curve*) and multiple linear regression fit (*red curve*) shows good agreement.

In the first line we use the model fit (the β's in variable `b`) to define the modeled 60 Hz EEG activity. Then, in the second line, we subtract this modeled 60 Hz activity from the EEG data. We plot the results of the cleaned EEG data in figure 3.9.

Q: Consider the cleaned EEG data in figure 3.9. What activity do you now notice? Compare the EEG data in this plot to the original EEG data (which includes the 60 Hz noise) in figure 3.1. What's different?

In this example, we used multiple linear regression to model a particular rhythmic component of the EEG data, the 60 Hz activity. We may also use the model result to estimate the *power* at 60 Hz. In MATLAB,

```
Sxx_model_60Hz = b(2)^2+b(3)^2;
```

The power estimate from the model consists of two terms: the squared coefficient of the sine function (`b(2)^2`) plus the squared coefficient of the cosine function (`b(3)^2`). Note that the variable `Sxx_model_60Hz` has units of mV^2.

Q: Compare the power estimate from the model (the variable `Sxx_model_60Hz`) to the power spectral density at 60 Hz computed using the Fourier transform (see the MATLAB code following (3.9)). What do you find?

A: We note that the units of the power spectral density (variable `Sxx`) are mV^2/Hz, while the units of the power estimated in variable `Sxx_model_60Hz` are mV^2. To convert the power spectral density to (integrated) spectral power, we must integrate the variable `Sxx` over a frequency range. Here, we choose a 1 Hz interval centered at 60 Hz, which corresponds to a single index of the variable `faxis`; the frequency

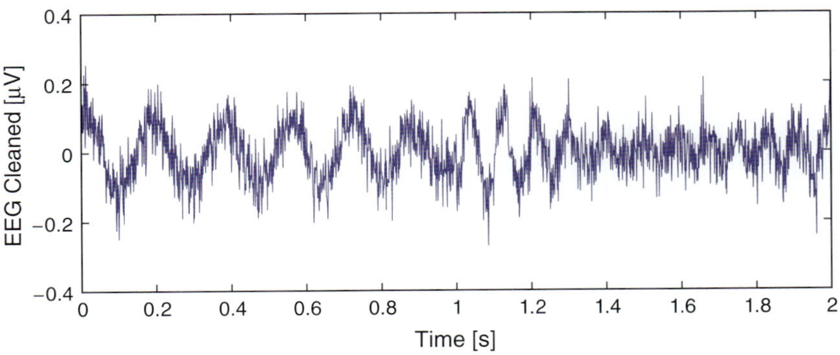

Figure 3.9
EEG data after removal of 60 Hz line noise using multiple linear regression.

resolution for these data is 0.5 Hz (see next section). Then the approximate integrated power over this 1 Hz interval is `Sxx(121)=0.9979`, identical to the value in `Sxx_model_60Hz`, and with the same units.

This example, in which we focused on the 60 Hz activity in the EEG, illustrates how we may use multiple linear regression to estimate the power. We could extend this procedure to include additional rhythms in the model beyond 60 Hz (e.g., sine and cosine functions at 1 Hz, 2 Hz, 3 Hz, etc.). In doing so, we would add more terms to the multiple linear regression model and have more β's to determine from the data. Multiple linear regression provides a way to decompose the EEG data into sine and cosine functions at different frequencies—just as we proposed to do using the Fourier transform—and then compute the power at each frequency. Using either multiple linear regression or the Fourier transform, we aim to decompose the EEG into sine and cosine functions oscillating at different frequencies.

Discrete Fourier Transform in MATLAB. Computing the spectrum of a signal x in MATLAB can be achieved in two simple steps. The first step is to compute the Fourier transform of x:

```
xf = fft(x-mean(x));                    %The Fourier transform of x.
```

We subtract the mean from `x` before computing the Fourier transform. This is not necessary but often useful. For these neural data, we're not interested in the very slow (0 Hz) activity; instead, we're interested in rhythmic activity. By subtracting the mean, we eliminate this low-frequency activity from the subsequent analysis.

The second step is to compute the spectrum, the Fourier transform of x multiplied by its complex conjugate:

```
Sxx = 2*dt^2/T*(xf.*conj(xf));      %Compute the spectrum.
```

In this computation, we use the operation `.*` to multiply x by its complex conjugate. This operation multiplies the two vectors element by element.

Figure 3.10 plots the resulting variable `Sxx`. Upon examining the horizontal axis in this figure, we find it corresponds to the indices of x, beginning at index 1 and ending at index $N = 2000$. Computing the Fourier transform and multiplying by the complex conjugate does not change the length of the data x.

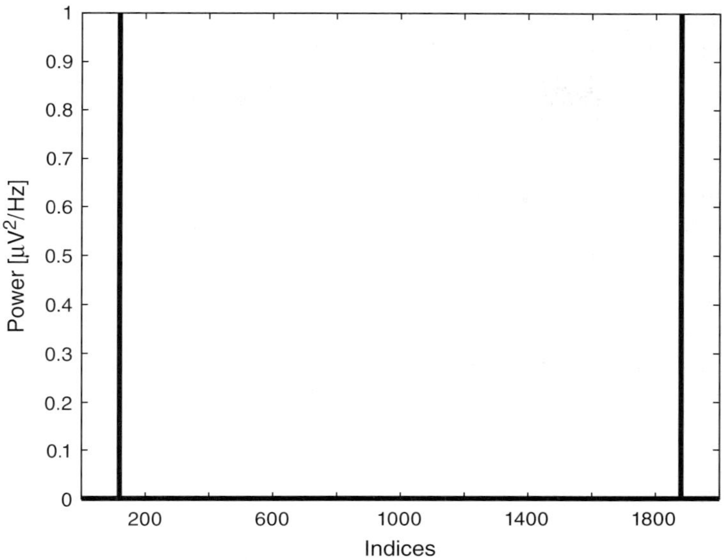

Figure 3.10
Power spectrum of EEG data for all indices. Horizontal axis is in indices rather than frequency.

> **Q:** Confirm in MATLAB that the variables `x`, `xf`, and `Sxx` all have the same size.

Inspection of figure 3.10 reveals a strange characteristic: there are two large peaks, and the plot exhibits a particular symmetry. If we were to cut this plot from the printed page and fold the resulting piece of paper at index 1000, we would find that the two peaks in `Sxx` align; both peaks appear to be the same number of indices away from the center index 1000. This suggests that a redundancy occurs in the variable `Sxx`. In fact, this redundancy is due to the way MATLAB relates the indices and frequencies of `Sxx`. To define this relation requires two new quantities:

- the *frequency resolution*, $df = \frac{1}{T}$, or the reciprocal of the total recording duration;
- the *Nyquist frequency*, $f_{NQ} = \frac{f_0}{2} = \frac{1}{2\Delta}$, or half of the sampling frequency $f_0 = \frac{1}{\Delta}$.

For the clinical EEG data considered here, the total recording duration is 2 s ($T = 2$ s), so the frequency resolution $df = 1/(2 \text{ s}) = 0.5$ Hz. The sampling frequency f_0 is 1000 Hz, so $f_{NQ} = 1000/2$ Hz $= 500$ Hz. There's much more to say about both quantities, but for

Table 3.1
Relation between Vector Sxx Indices and Corresponding Frequencies.

Sxx index j	1	2	3	...	$N/2$
Frequency formula	0	df	$2\,df$	...	$f_{NQ} - df$
Frequency value	0	0.5 Hz	1 Hz	...	499.5 Hz
Sxx index j	$N/2+1$	$N/2+2$	...	$N-1$	N
Frequency formula	f_{NQ}	$-(f_{NQ} - df)$	...	$-2\,df$	$-df$
Frequency value	500 Hz	−499.5 Hz	...	−1 Hz	−0.5 Hz

now let's simply use both quantities to consider how MATLAB relates the indices and frequencies of the vector `Sxx` (see table 3.1). For the first half of `Sxx`, the frequency axis increases in steps of the frequency resolution df (here, 0.5 Hz) until reaching the Nyquist frequency f_{NQ} (here, 500 Hz). This maximal frequency occurs just past the halfway point of the indices, at index $j = N/2 + 1$. Beyond this index, the frequency axis is negative, and the magnitude of the frequency becomes smaller and smaller until the value $-df$ is reached at index $j = N$.

We may now utilize a useful property of the Fourier transform. When a signal is real (i.e., the signal has zero imaginary component), the negative frequencies in the spectrum are redundant. So, the power we observe at frequency f_0 is identical to the power we observe at frequency $-f_0$. For this reason, we can safely ignore the negative frequencies; these frequencies provide no additional information. Because the EEG data are real, we conclude that the negative frequencies in the variable `Sxx` are redundant and can be ignored. As a specific example, the value of `Sxx` at index $j = 3$ is the same as the value of `Sxx` at index $j = N - 1$; these indices correspond to frequencies $2df$ and $-2df$, respectively (see table 3.1). We therefore need only plot the variable `Sxx` for the positive frequencies, more specifically, from index 1 to index $N/2 + 1$. This conclusion matches our visual inspection of `Sxx` in figure 3.10.

> Data collected in the world are almost always real (have zero imaginary component), so the negative frequencies in the spectrum are usually redundant and can be ignored.

Given the total duration of the recording (T) and the sampling frequency (f_0) for the data, we can define the frequency axis for the spectrum `Sxx`. Now, to compute and plot the spectrum in MATLAB, we again utilize the code first introduced early in section 3.2.4:

```
xf  = fft(x-mean(x));                %Compute Fourier transform of x.
Sxx = 2*dt^2/T*(xf.*conj(xf));       %Compute the spectrum.
Sxx = Sxx(1:N/2+1);                  %Ignore negative frequencies.
```

```
df = 1/max(T);                %Determine the frequency resolution.
fNQ = 1/ dt / 2;              %Determine the Nyquist frequency.
faxis = (0:df:fNQ);           %Construct frequency axis.
plot(faxis, Sxx)              %Plot spectrum vs frequency.
```

Notice that in the third line we select the first half of the vector `Sxx` up to index $N/2+1$, thereby ignoring the redundant negative frequencies.

In the next two sections, we focus on interpreting and adjusting the quantities df and f_{NQ}. Doing so is critical to develop further an intuition for the spectrum.

Nyquist Frequency, f_{NQ}. The formula for the Nyquist frequency is

$$f_{NQ} = \frac{f_0}{2}. \tag{3.13}$$

The Nyquist frequency is the highest frequency we can possibly hope to observe in the data. To illustrate this, let's consider a true EEG signal that consists of a very simple time series—a pure sinusoid that oscillates at some frequency f_s. Of course, we never observe the true signal. Instead, we observe a sampling of this signal, which depends on the sampling interval Δ. We consider three cases for different values of Δ. In the first case, we purchase a very expensive piece of equipment that can sample the true signal at a high rate, $f_0 \gg f_s$. In this case, we cover the true brain signal with many samples (figure 3.11, top), and given these samples, we can accurately reconstruct the underlying data.

Now, consider the case in which we purchase a cheaper piece of equipment that samples at a maximum rate equivalent to twice the frequency of the pure sinusoid: $f_0 = 2f_s$. In this case, we might collect sufficient samples to cover the underlying signal and approximate the oscillation frequency (figure 3.11, middle): the first sample resides on a peak of the sinusoid, the next sample on a trough, and so on. In this case, we collect two samples per cycle of the underlying true signal. Given only these sample points, we can connect the dots and still approximate the frequency of the true underlying sinusoid.

> **Q:** For the sampling rate $f_0 = 2f_s$, consider the case in which the first sample occurs on a zero crossing of the sinusoid. At what point does the next sample occur? and the next sample? If you connect the dots in this case, what do you find?

Finally, consider the case where our equipment records at a sampling rate less than the frequency of the pure sinusoid signal: $f_0 < 2f_s$. (figure 3.11, bottom). Assuming the first sample occurs at a peak of the sinusoid, the next sample occurs not at a trough (that would correspond to a sampling rate $f_0 = 2f_s$) but instead just after the trough. Connecting the samples with lines in this case produces something horrifying, an oscillation occurring at a different, lower frequency. Notice what has happened in this case. Sampling the sinusoid

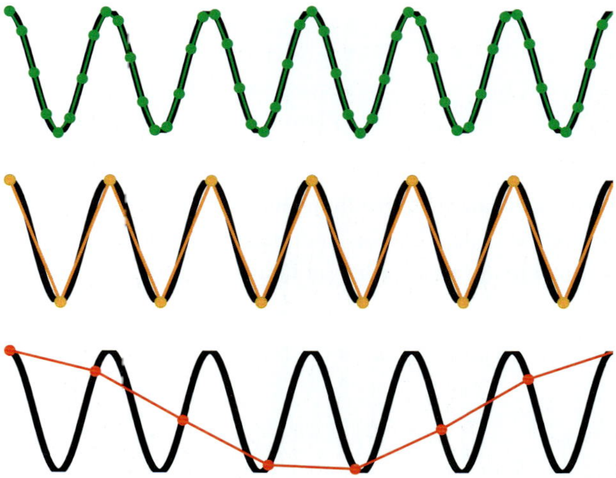

Figure 3.11
Sampling a sinusoid at different rates. When sampling rate is high enough (*top*), the sampled data provide a good approximation to the true data. When sampling rate is too low (*bottom*), the true high-frequency signal appears as a low-frequency oscillation.

at too low a frequency (i.e., at a frequency less than twice the signal's frequency, $f_0 < 2f_s$) causes this signal to manifest at a low-frequency upon sampling. This phenomenon—a high-frequency signal appearing as a low-frequency signal upon sampling—is known as *aliasing*. Once a signal has been aliased, it's impossible to distinguish from true signals oscillating at low frequencies.

To avoid aliasing, sample data at sufficiently high rates.

Typically, to prevent aliasing, recorded data are first analog-filtered before the digital sampling occurs. The analog filtering guarantees that activity at frequencies exceeding a threshold value (f_c, say) are dramatically reduced. The sampling rate can then be chosen to exceed this threshold value by at least a factor of 2 (i.e., $f_0 > 2f_c$). We note that in this case the EEG data were first analog-filtered at 200 Hz before digital sampling occurred at 1000 Hz. So, for our EEG data, aliasing is not a concern. For a more detailed investigation of aliasing, see appendix B at the end of this chapter.

Frequency Resolution, df**.** The frequency resolution is defined as

$$df = \frac{1}{T}, \tag{3.14}$$

where T is the total duration of the recording. For the EEG data in figure 3.1, $T = 2$ s, so the frequency resolution is $df = 1/(2\text{s}) = 0.5$ Hz.

Q: How do we improve the frequency resolution?

A: There's only one way to do it: increase T, that is, record more data. For example, if we demand a frequency resolution of 0.2 Hz, how much data must we record? We can rearrange (3.14) to solve for T,

$$T = \frac{1}{df} = \frac{1}{0.2\,\text{Hz}} = 5\,\text{s}. \tag{3.15}$$

So, record 5 s of data to obtain a frequency resolution of 0.2 Hz.

Q: We estimate the spectrum using the preceding MATLAB code. As we record more and more data, does the estimate of the spectrum improve?

A: Intuitively, you might answer yes. As we collect more and more data, we usually expect our estimate of a quantity (e.g., the mean or the standard deviation) to improve. However, that is not the case for the spectrum. As we collect more and more data, we acquire more and more points along the frequency axis (i.e., df becomes smaller). However, our estimate of the power at each frequency does not improve [8].

To gain some intuition for the frequency resolution formula (3.14), consider the case in which we collect T seconds of data. If the sampling interval is Δ, then we collect $N = T/\Delta$ data points; for example, for the EEG data of interest here, we collect $N = 2000$ data points. We know that the number of observations in the data equals the number of frequencies in the spectrum (where we now include negative frequencies); both the data vector `x` and the spectrum vector `Sxx` have length N. We also know that the maximum observable frequency in the spectrum, the Nyquist frequency, is fixed no matter how much data we collect. Recall that the Nyquist frequency depends only on the sampling interval: $f_{NQ} = 1/(2\Delta)$. Now, consider the case in which we increase T, or equivalently, increase N. As we collect more and more data, the maximum frequency remains fixed at the Nyquist frequency, while the length of the spectrum vector increases. We therefore need to fit more and more frequency values between 0 Hz and the Nyquist frequency as N increases (figure 3.12). This observation provides some intuition for the relation between the amount of data recorded (T or N) and the frequency resolution (df).

3.2.5 Reexamining the Spectrum—a Matter of Scale

Let's now return to the spectrum of the EEG data. As plotted in figure 3.6, the spectrum is dominated by a single peak at 60 Hz. Other, weaker rhythmic activity may occur in the data, but these features remain hidden from visual inspection because of the large 60 Hz peak; informally, we might state that the 60 Hz peak saturates the vertical scale in figure 3.6. One

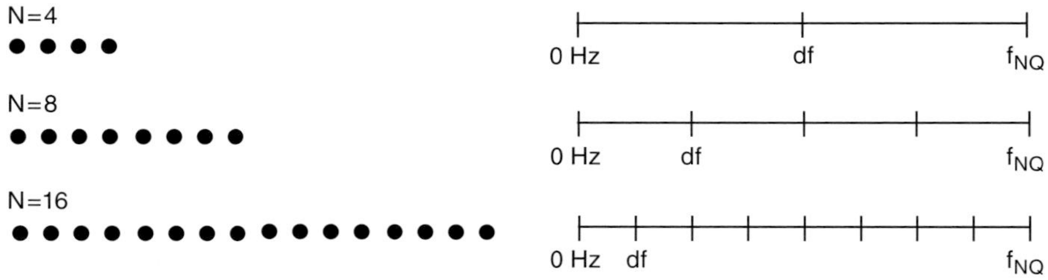

Figure 3.12

Cartoon representation of the relation between data duration and frequency resolution. Data (*left*) consist of different numbers of samples (N). As N increases, the number of values on the frequency axis increases (*right*), the maximal frequency (f_{NQ}) remains fixed, and the frequency resolution (df) decreases. Only non-negative frequencies are shown.

technique to emphasize lower-amplitude rhythms hidden by large-amplitude oscillations is to change the scale of spectrum to *decibels*. The decibel is a logarithmic scale and easily computed in MATLAB as follows:

```
plot(faxis, 10*log10(Sxx/max(Sxx)))  %Plot spectrum in decibels.
xlim([0 100])                        %Select frequency range.
ylim([-60 0])                        %Select decibel range.
xlabel('Frequency [Hz]')             %Label axes.
ylabel('Power [dB]')
```

To change to the decibel scale, we first divide the spectrum by the maximum value observed and then take the logarithm base 10 of this ratio and multiply the result by 10. The resulting spectrum is shown in figure 3.13a. The 60 Hz rhythm is still dominant and exhibits the most power.

Q: For this example, what is the value in decibels at 60 Hz?

A: Through our previous analysis, we know that the maximum value in the spectrum occurs at 60 Hz. By dividing the original spectrum by this maximum, we scale the spectrum at 60 Hz to a value of 1. The logarithm of 1 is 0, so we find a value of 0 at 60 Hz. Note that all other values are now smaller than 1 and therefore negative on the decibel scale.

Different conventions exist to define the decibel scale. Here we first divide by the maximum before computing the logarithm. Be sure to verify how the spectrum is scaled (if at all) to interpret the decibel axis.

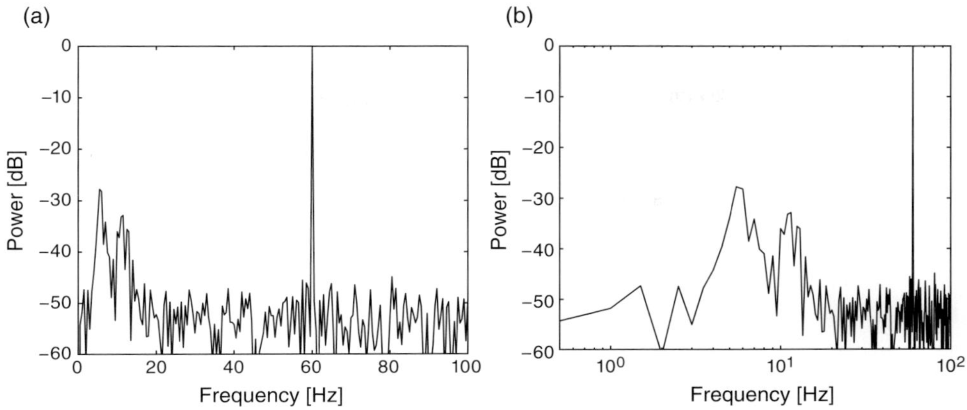

Figure 3.13

Power spectrum of EEG data on a decibel scale. Frequency axis is (*a*) linear or (*b*) logarithmic.

The decibel scale reveals new structure in the spectrum. In particular, two peaks have emerged at frequencies 5–15 Hz. These peaks are much weaker than the 60 Hz signal; both peaks are approximately 30 dB below the maximum at 60 Hz, or equivalently, three orders of magnitude weaker. Because these peaks are so small relative to the 60 Hz signal, neither was apparent in the original plot of the spectrum (figure 3.6). To further emphasize the low-frequency structure of the spectrum, we may also convert the frequency axis to a logarithmic scale:

```
semilogx(faxis, 10*log10(Sxx/max(Sxx))) %Log-log scale
xlim([df 100])                          %Select frequency range.
ylim([-60 0])                           %Select decibel range.
xlabel('Frequency [Hz]')                %Label axes.
ylabel('Power [dB]')
```

Notice the change in the first line to use the `semilogx` function in MATLAB. The resulting spectrum (in the log-log scale) is plotted in figure 3.13b. By using the logarithmic scale to stretch the low-frequency part of the horizontal axis, the two low-frequency peaks become more apparent. The changes in figure 3.13 compared to the original spectrum (figure 3.6) are purely cosmetic. However, these cosmetic changes have proved extremely useful. The two lower-frequency peaks were originally hidden from us, both in visual inspection of the raw data and in the original plot of the spectrum. In those cases, the large-amplitude 60 Hz activity masked the smaller-amplitude (three orders of magnitude smaller) rhythms.

3.2.6 The Spectrum Changing in Time—The Spectrogram

The results in figure 3.13 suggest that three rhythms appear in the EEG signal: 60 Hz, approximately 11 Hz, and approximately 6 Hz. Given only these results, we may reasonably conclude that these three rhythms appear simultaneously throughout the entire 2 s of EEG recording. That is an assumption we make in computing the spectrum of the entire 2 s interval. To further test this assumption in the EEG data, we compute a final quantity, the *spectrogram*. The idea of the spectrogram is to break up the time series into smaller intervals of data and then compute the spectrum in each interval. These intervals can be quite small and can even overlap. The result is the spectrum as a function of frequency and time.

> **Q:** Consider the 2 s of EEG data. If we break up these data into smaller intervals of duration 1 s, what is the resulting frequency resolution of each interval? What is the Nyquist frequency of each interval?

To compute and display the spectrogram in MATLAB, we use the (aptly named) function `spectrogram`:

```
Fs = 1/dt;                    %Define the sampling frequency.
interval = round(Fs);         %Specify the interval size.
overlap = round(Fs*0.95);     %Specify the overlap of intervals.
nfft = round(Fs);             %Specify the FFT length.

%Compute the spectrogram,
[S,F,T,P] = spectrogram(x-mean(x),interval,overlap,nfft,Fs);
imagesc(T,F,10*log10(P))      %... and plot it,
colorbar                      %... with a colorbar,
axis xy                       %... and origin in lower left,
ylim([0 70])                  %... set the frequency range,
xlabel('Time [s]');           %... and label axes.
ylabel('Frequency [Hz]')
```

The function `spectrogram` accepts five arguments. Briefly, these arguments specify the data (x), the interval size (specified in indices and here set to 1 s), the overlap between intervals (here set to 95%), the number of points to use in the Fourier transform (here set to the interval length), and the sampling frequency. More information about these options can be found in the MATLAB documentation. Notice that in the first four lines we use the `round` function to enforce integer values for three of these inputs. The resulting spectrogram is plotted in figure 3.14.

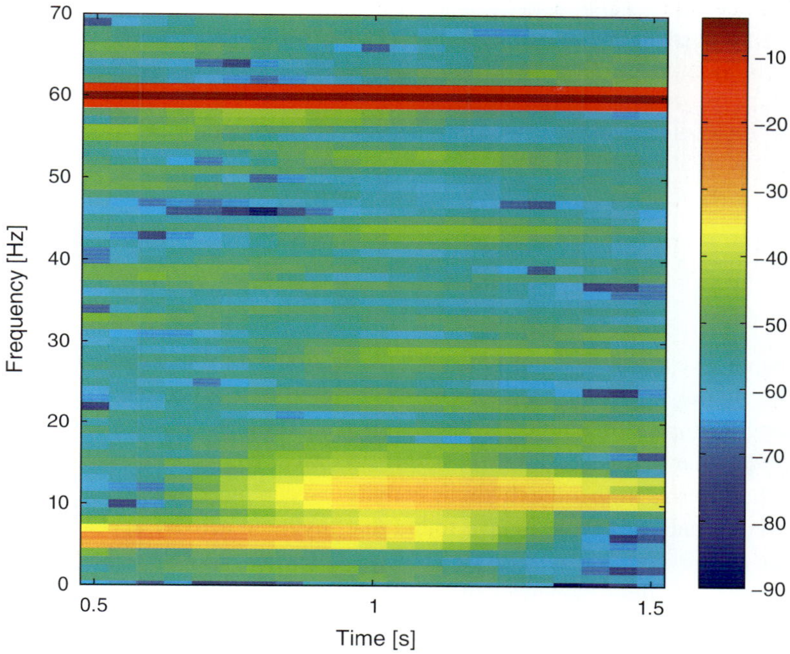

Figure 3.14

Spectrogram of EEG data. Power (in decibels, scale bar at right) is shown as a function of frequency and time.

Q: Consider the spectrogram in figure 3.14. What aspects of the spectrogram are consistent with our previous results? What aspects are new? Consider, in particular, the low-frequency rhythms and the conclusions deduced from this figure compared to figure 3.13.

A: The spectrogram displays the spectrum (in decibels) as a function of frequency (vertical axis) and time (horizontal axis). Values on the time axis indicate the center times of each 1 s window (e.g., 0.5 s corresponds to times [0, 1] s in the data). Intervals of high (low) values correspond to warm (cool) colors. Visual inspection immediately provides new insights into the observed EEG rhythms. First, we observe a band of high power at 60 Hz that persists for all time (red horizontal line in figure 3.14). This corresponds to the 60 Hz line noise present for the entire duration of the recording. Second, we observe intervals of increased power near 11 Hz and 6 Hz. Unlike the 60 Hz signal, the two low-frequency rhythms do not persist for the entire 2 s recording (as we may have incorrectly concluded from examination of the spectrum in figure 3.13). Instead, one weak rhythm (near 6 Hz) appears for the first half of the recording, while another weak rhythm (near 11 Hz) appears for the second half of the recording. Visualization via the spectrogram of how the rhythmic activity changes in time allows this important conclusion.

Summary

In this chapter, we analyzed 2 s of EEG data. We started with visual inspection of the EEG time series (figure 3.1). This is always the best place to start when analyzing new data and provides initial important intuition for the time series. Through the initial visual inspection, we concluded that rhythmic activity appeared and was dominated by a 60 Hz oscillation. Then, to characterize further the rhythmic activity, we computed two related quantities: the autocovariance and the spectrum. We found that rhythmic activity appeared in the autocovariance of the data. We then considered the spectrum. To do so, we first introduced the notion of the Fourier transform and discussed in detail how to compute the spectrum in MATLAB. We also defined two fundamental quantities—the frequency resolution and the Nyquist frequency—and explored how to manipulate these quantities. (We recommend you commit both quantities to memory. For every spectral analysis you encounter, ask: What is the frequency resolution? What is the Nyquist frequency?). We then considered how logarithmic scales can be used to emphasize features of the spectrum (figure 3.13), and finally how the spectrogram provides insight into spectral features that change in time (figure 3.14). We concluded that the EEG data are dominated by 60 Hz activity throughout the 2 s interval, and that weaker low-frequency activity emerges during two intervals: a 6 Hz rhythm from 0 s to 1 s, and an 11 Hz rhythm from 1 s to 2 s.

In this chapter, we only touched the surface of spectral analysis; many details and issues exist for further exploration. In chapter 4, we discuss the issues of windowing and zero padding. For those interested in exploring further, see [8, 9].

Problems

3.1. To improve the frequency resolution requires we record more data (i.e., increase T). Consider the case in which a collaborator demands a frequency resolution of 0.0001 Hz. How much data must be recorded? What are the disadvantages of recording the data required?

3.2. Explore the MATLAB function `periodogram`. How does this function compare to the computation of the spectrum described in this chapter?

3.3. A normalized measure of linear coupling in the time domain is the *autocorrelation*. Here is the formula for the autocorrelation evaluated at lag L,

$$\rho_{xx}[L] = \frac{r_{xx}[L]}{\sigma^2}.$$

Notice that the autocorrelation at lag L equals the autocovariance at lag L ($r_{xx}[L]$) divided by the variance of the signal (σ^2). Implement the autocorrelation in MATLAB, and apply it to the dataset considered in this chapter. How do the results of the autocovariance compare to the results of the autocorrelation?

3.4. Load the file `Ch3-EEG-2.mat`, available at

`http://github.com/Mark-Kramer/Case-Studies-Kramer-Eden`

into MATLAB. Then answer the following questions.

a. What is the sampling interval (Δ)? What is the total duration of the recording (T)? What is the frequency resolution (df)? What is the Nyquist frequency (f_{NQ})?

b. Plot the data and visually inspect them. Describe briefly (in a sentence or two) what rhythms, if any, you see in the data.

c. Plot the biased autocovariance versus lags. You will need to choose the maximum number of lags to investigate. What structure do you observe in the autocovariance, if any?

d. Plot the spectrum versus frequency. You may choose to plot the spectrum on a decibel scale, or not. Defend your choice.

e. Plot the spectrogram as a function of frequency and time. You will need to choose the interval size and the overlap between intervals. Do the rhythms in these data appear to change in time?

f. Interpret the autocovariance and spectrum, and describe the rhythms present in the signal. Compare your visual inspection of the data to the autocovariance and spectrum results. Do the analyses agree or disagree?

3.5. Load the file `Ch3-EEG-3.mat`, available at

`http://github.com/Mark-Kramer/Case-Studies-Kramer-Eden`

into MATLAB. Then answer the following questions.

a. What is the sampling interval (Δ)? What is the total duration of the recording (T)? What is the frequency resolution (df)? What is the Nyquist frequency (f_{NQ})?

b. Plot the data and visually inspect them. Describe briefly what rhythms, if any, you see in the data.

c. Plot the biased autocovariance versus lags. You will need to choose the maximum number of lags to investigate. What structure do you observe in the autocovariance, if any?

d. Plot the spectrum versus frequency. You may choose to plot the spectrum on a decibel scale, or not. Defend your choice.

e. Plot the spectrogram as a function of frequency and time. You will need to choose the interval size and the overlap between intervals. Do the rhythms in these data appear to change in time?

f. Interpret the autocovariance and spectrum, and describe the rhythms present in the signal. Compare your visual inspection of the data to the autocovariance and spectrum results. Do the analyses agree or disagree?

3.6. Load the file `Ch3-EEG-4.mat`, available at

`http://github.com/Mark-Kramer/Case-Studies-Kramer-Eden`

into MATLAB. Then answer the following questions.

a. What is the sampling interval (Δ)? What is the total duration of the recording (T)? What is the frequency resolution (df)? What is the Nyquist frequency (f_{NQ})?

b. Plot the data and visually inspect them. Describe briefly what rhythms, if any, you see in the data.

c. Plot the biased autocovariance versus lags. You will need to choose the maximum number of lags to investigate. What structure do you observe in the autocovariance, if any?

d. Plot the spectrum versus frequency. You may choose to plot the spectrum on a decibel scale, or not. Defend your choice.

e. Plot the spectrogram as a function of frequency and time. You will need to choose the interval size and the overlap between intervals. Do the rhythms in these data appear to change in time?

f. Interpret the autocovariance and spectrum, and describe the rhythms present in the signal. Compare your visual inspection of the data to the autocovariance and spectrum results. Do the analyses agree or disagree?

3.7. Load the file `Ch3-EEG-5.mat`, available at

`http://github.com/Mark-Kramer/Case-Studies-Kramer-Eden`

into MATLAB. Then answer the following questions.

a. What is the sampling interval (Δ)? What is the total duration of the recording (T)? What is the frequency resolution (df)? What is the Nyquist frequency (f_{NQ})?

b. Plot the data and visually inspect them. Describe briefly what rhythms, if any, you see in the data.

c. Plot the biased autocovariance versus lags. You will need to choose the maximum number of lags to investigate. What structure do you observe in the autocovariance, if any?

d. Plot the spectrum versus frequency. You may choose to plot the spectrum on a decibel scale, or not. Defend your choice.

e. Plot the spectrogram as a function of frequency and time. You will need to choose the interval size and the overlap between intervals. Do the rhythms in these data appear to change in time?

f. Interpret the autocovariance and spectrum, and describe the rhythms present in the signal. Compare your visual inspection of the data to the autocovariance and spectrum results. Do the analyses agree or disagree?

3.8. Consider the function $x(t) = \sin(2\pi t^2)$. Simulate this function in MATLAB using a sampling interval of $\Delta = 0.001$ ms, and $t = (0, 10)$ s. Analyze these data as you would an EEG time series collected in an experiment. Compute the spectrum, and compute the spectrogram. Explain the results you find in each case, and how these results compare to your expectations.

3.9. (*Advanced*) As stated in this chapter, to increase the frequency resolution of the spectrum, we must record more data (i.e., increase T). Consider a scenario in which the experiment that produced the data of duration T is now inaccessible. We therefore cannot perform another experiment of longer duration to improve the frequency resolution. So, we instead consider a work-around to improve the frequency resolution: let's simply append the data end to end. The result is a new dataset of duration $2T$. Note that the new dataset is redundant; the second half of the new dataset is a duplicate of the first half. Does this improve the frequency resolution in a meaningful way? Can we use this stratagem to append the observed data K times and thereby improve the frequency resolution by a factor of K?

Appendix A: The Spectrum and Autocovariance

To show that the spectrum is the Fourier transform of the autocovariance, we first make some reasonable assumptions that help simplify the analysis. Let's assume that x has zero mean (i.e., $\bar{x} = 0$), which we can always enforce by subtracting the mean from x as a first step in the analysis. In addition, let's assume we can evaluate x at any time index from negative infinity to positive infinity. Then, substituting (3.8) into (3.9), we get the sample power spectral density,

$$S_{xx,j} = \frac{2\Delta^2}{T}\Big(\sum_n x_n \exp(-2\pi i f_j t_n)\Big)\Big(\sum_m x_m^* \exp(2\pi i f_j t_m)\Big). \tag{3.16}$$

In the second term in parentheses, we set the dummy time index to another symbol, m. This second term is the complex conjugate of the Fourier transform of the data. The impact of the complex conjugate is that we replace i with $-i$, and x_m with x_m^*. Because the EEG data are real, $x_m^* = x_m$. Notice that we do not include the limits on the summations, which range from $-\infty$ to ∞. Replacing the expressions for f_j, t_n, and t_m with their definitions and simplifying, we rearrange the expression to find

$$S_{xx,j} = \frac{2\Delta^2}{T}\sum_n\sum_m x_n x_m \exp(-\frac{2\pi i}{N} j(n-m)). \tag{3.17}$$

Let's now define a new variable,

$$l = n - m. \tag{3.18}$$

Then

$$S_{xx,j} = \frac{2\Delta^2}{\Delta}\sum_l \Big(\frac{1}{N}\sum_m x_{m+l} x_m\Big) \exp(-\frac{2\pi i}{N} j\, l), \tag{3.19}$$

where we have replaced the n with l, used $T = N\Delta$, and reorganized the terms. Comparing the term in parentheses to the equation for the biased autocovariance (3.3), we see that the two are equivalent; remember that $\bar{x} = 0$. Therefore,

$$\begin{aligned} S_{xx,j} &= 2\Delta \sum_l r_{xx}[l] \exp(-\frac{2\pi i}{N} j\, l) \\ &= 2\Delta \sum_l r_{xx}[l] \exp(-\frac{2\pi i}{N\Delta} j\Delta l) \\ &= 2\Delta \sum_l r_{xx}[l] \exp(-2\pi i \frac{j}{T}\Delta l) \\ &= 2\Delta \sum_l r_{xx}[l] \exp(-2\pi i f_j t_l). \end{aligned}$$

Comparing this last expression to the equation for the Fourier transform (3.8), we find that the sample power spectral density ($S_{xx,j}$) is proportional to the Fourier transform of the sample autocovariance ($r_{xx}[l]$).

Appendix B: Aliasing

We illustrated in figure 3.11 the concept of aliasing. In this appendix, we further investigate the Fourier transform and the impact of sampling. Consider a signal $y(t)$. As illustrated in figure 3.2, we do not observe the signal directly. Instead, we only observe a discrete sampling in time of this signal. Let's make this sampling explicit, and write the observed signal $s(t)$ as

$$s(t) = y(t) \sum_{n \to -\infty}^{\infty} \delta(t - n\Delta), \tag{3.20}$$

where $\delta(t)$ is the Kronecker delta, and Δ is the sampling interval. This equation represents our discrete observations of $y(t)$ at integer multiples of the sampling interval Δ. Here, we assume an infinite number of observations (into the past and future).

To compute the Fourier transform of $s(t)$, we need to evaluate the Fourier transform of the product of the two functions in (3.20): the signal $y(t)$ and the summation of Kronecker delta functions. The Fourier transform of the product of two functions equals the convolution of the Fourier transform of each function (see chapter 4). We use this fact to express the Fourier transform of $s(t)$ as

$$FT\big[s(t)\big] = FT\big[y(t)\big] \star FT\Big[\sum_{n \to -\infty}^{\infty} \delta(t - n\Delta)\Big], \tag{3.21}$$

where $FT\big[a\big]$ denotes the Fourier transform of a, and $A \star B$ indicates the convolution of A and B. Let's define $Y(f) = FT\big[y(t)\big]$, and $S(f) = FT\big[s(t)\big]$. We can evaluate the Fourier transform of the repeated Kronecker delta functions as

$$\begin{aligned} FT\Big[\sum_{n \to -\infty}^{\infty} \delta(t - n\Delta)\Big] &= \frac{1}{\Delta} \sum_{n \to -\infty}^{\infty} \delta(f - \frac{n}{\Delta}) \\ &= \frac{1}{\Delta} \sum_{n \to -\infty}^{\infty} \delta(f - nf_0), \end{aligned}$$

where we've replaced $1/\Delta$ with the sampling frequency f_0. Then the Fourier transform of the observed signal $s(t)$ becomes

$$\begin{aligned} S(f) &= FT\big[s(t)\big], \\ &= Y(f) \star \frac{1}{\Delta} \sum_{n \to -\infty}^{\infty} \delta(f - nf_0) \end{aligned}$$

$$= \frac{1}{\Delta} \sum_{n \to -\infty}^{\infty} Y(f - nf_0)$$

$$\propto \sum_{n \to -\infty}^{\infty} Y(f - nf_0), \tag{3.22}$$

where, in the last line, we've dropped the leading constant term for convenience of presentation. This result relates the Fourier transform of the unobserved signal $y(t)$ to the Fourier transform of the observed (i.e., sampled) signal $s(t)$. Note that $S(f)$ is an infinite summation of terms. To make this example more concrete, let's consider the case that the (unobserved) signal is a cosine function at frequency f_*,

$$y(t) = \cos(2\pi f_* t),$$

with corresponding Fourier transform,

$$Y(f) \propto \delta(f - f_*) + \delta(f + f_*). \tag{3.23}$$

This result is perhaps intuitive; when the data consist of a pure sinusoid, we expect the power (in the ideal case) to be concentrated at the sinusoid's frequency. We usually ignore the negative frequency component because it is redundant, but here we've been careful to include both the positive frequency $f = f_*$ and the negative frequency $f = -f_*$ in the representation of the Fourier transform. We've also ignored the constant terms in (3.23) to focus on the quantities of interest. Substituting the expression for $Y(f)$ in (3.23) into the expression for $S(f)$ in (3.22) we find

$$S(f) \propto \sum_{n \to -\infty}^{\infty} \delta(f - f_* - nf_0) + \delta(f + f_* - nf_0). \tag{3.24}$$

This result relates the Fourier transform of the observed signal $S(f)$ at frequency f to the peaks we expect at the positive and negative frequencies of the sinusoid f_*. However, we observe not only these expected peaks (when $n = 0$) but also an infinite number of other peaks (for $n \neq 0$).

To explore the impact of these additional peaks, consider a scenario in which the sinusoid signal has frequency $f_* = 20$ Hz, and the sampling occurs with $\Delta = 0.01$ s or $f_0 = 1/\Delta = 100$ Hz. In this case,

$$S(f) \propto \sum_{n \to -\infty}^{\infty} \delta(f - 20 - n100) + \delta(f + 20 - n100),$$

and we find that $S(f)$ is nonzero when

$$f = \begin{cases} \mathbf{20}, -\mathbf{20}, & \text{if } n = 0, \\ 120, 80, & \text{if } n = 1, \\ -80, -120, & \text{if } n = -1, \\ \ldots . \end{cases}$$

At this sampling frequency, the Nyquist frequency is $f_0/2 = 50$ Hz. We find that only the $n = 0$ case provides frequencies f with magnitude less than the Nyquist frequency. All other frequencies appear outside of the interval bounded by the Nyquist frequencies $(-50, 50)$ Hz and therefore do not manifest in our analysis of $S(f)$. This is the scenario we desire: the Fourier transform of $s(t)$ is concentrated at the frequencies of the sinusoid $y(t)$.

Now, consider a second scenario in which the sinusoid signal has frequency $f_* = 80$ Hz, and the sampling again occurs at $f_0 = 100$ Hz. In this case,

$$S(f) \propto \sum_{n \to -\infty}^{\infty} \delta(f - 80 - n100) + \delta(f + 80 - n100), \tag{3.25}$$

and we find that $S(f)$ is nonzero when

$$f = \begin{cases} 80, -80, & \text{if } n = 0, \\ 180, \mathbf{20}, & \text{if } n = 1, \\ -\mathbf{20}, -180, & \text{if } n = -1, \\ \ldots . \end{cases} \tag{3.26}$$

In this case, at $n = 0$, both the positive and negative frequencies occur beyond the Nyquist frequency; we therefore do not observe these 80 Hz peaks in the Fourier transform of $s(t)$. However, we do find for $n = 1$ a nonzero value for $S(f)$ at $f = 20$ Hz. In this case, the negative frequency component has folded into the observable frequency range (here, from 0 Hz to 50 Hz). Similarly, for $n = -1$, we find that the positive frequency component has folded into the observable negative frequency range (here, from -50 Hz to 0 Hz). Therefore, when we compute the Fourier transform of the 80 Hz sinusoid, we instead find power concentrated at the much lower frequencies of ± 20 Hz. The impact of aliasing is to shift activity in the signal that occurs beyond the Nyquist frequency (i.e., high-frequency components of $y(t)$) into the observable frequency range (from negative Nyquist frequency to positive Nyquist frequency).

Appendix C: Numerical Scaling of the Spectrum

We defined in (3.9) the sample power spectral density, which we rewrite here for convenience,

$$S_{xx,j} = \frac{2\Delta^2}{T} X_j X_j^*.$$

In this appendix, we consider some motivation for the leading term $\frac{2\Delta^2}{T}$. We consider separately the three components of this leading term: Δ^2, $1/T$, and 2.

Term: Δ^2. Throughout this chapter, we've focused on the discrete-time Fourier transform, defined in (3.8). The discrete-time Fourier transform is an approximation to the continuous-time Fourier transform,

$$X(f) = \int_{-\infty}^{\infty} x(t) \exp(-2\pi i f t)\, dt. \tag{3.27}$$

In this equation, we evaluate the time integral of the continuous signal $x(t)$ multiplied by the continuous-time complex exponentials $\exp(-2\pi i f t)$. Let's consider a Riemann sum approximation to this integral. To do so, we replace the integral with a summation over discrete steps in time, and compute the area of the integrand between each step,

$$X(f) = \sum_{n=-\infty}^{\infty} \Big(x(\Delta n) \exp(-2\pi i f \Delta n) \Big) \Delta.$$

Notice that we replace the continuous-time variable t with the discrete-time Δn, where n includes all integers. We may think of $\Big(x(\Delta n) \exp(-2\pi i f\, \Delta n) \Big)$ as the height of each rectangle, and Δ as the width of each rectangle in the Riemann sum. If we truncate this summation to include a finite number of time points, we get

$$X(f) \approx \Big(\sum_{n=1}^{N} x_n \exp(-2\pi i f_j\, t_n) \Big) \Delta,$$

where x_n is the signal evaluated at discrete times, t_n is the discrete time, and f_j the discrete frequency. The impact of this truncation is to distort the spectrum (e.g., to produce side lobes; see chapter 4). We then conclude

$$X(f) \approx \Delta X_j,$$

which shows that the continuous-time Fourier transform is approximated by the discrete-time Fourier transform (3.8) multiplied by the sampling interval Δ. Then

$$\begin{aligned} X(f) X(f)^* &\approx \big(\Delta X_j\big)\big(\Delta X_j^*\big) \\ &\approx \Delta^2 X_j X_j^*. \end{aligned} \tag{3.28}$$

This result shows that the continuous-time Fourier transform multiplied by its complex conjugate is approximated by the corresponding product of the discrete-time Fourier transform multiplied by the constant Δ^2.

Term: $1/T$. To understand the scaling of the spectrum by $1/T$, we begin with (3.8), the definition of the discrete-time Fourier transform, with some additional notation,

$$X_{j,N} = \sum_{n=1}^{N} x_n \exp(-2\pi i f_j t_n),$$

where we've included the subscript N on $X_{j,N}$ to indicate the total number of time points (or equivalently the total time $T = N\Delta$) in the summation. We now examine how $X_{j,N}$ changes as we increase T. Consider the simple case in which the data x_n are a sinusoid. For the appropriate choice of f_j, we find an excellent match between the data x_n and the real or imaginary component of $\exp(-2\pi i f_j t_n)$ (figure 3.15a). Multiplying these two components point by point, and summing over time, results in $X_{j,N} = C$, where C is a number.

Now, imagine we increase T by a factor of 2. We assume that the data x are stationary, so that we observe the same sinusoid, now for a duration of $2T$. What is $X_{j,2N}$? Again, we find an excellent match between the data and a sinusoid of frequency f_j. To evaluate the product summed over time $2T$, we divide the sum into two intervals. We know that the first half of the summation (from $n = 1$ to $n = N$) produces a value of C; that's what we found for $X_{j,N}$. Now, the second half of the summation is identical to the first half (figure 3.15b).

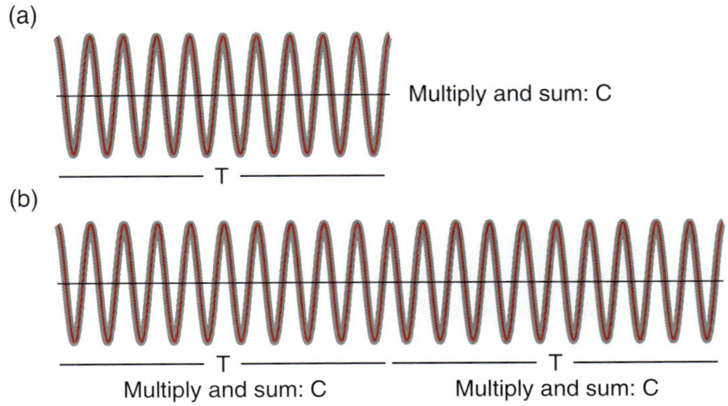

Figure 3.15

Impact of duration on the Fourier transform. (*a*) Artificial data *(gray curve)* are well matched by a sinusoid with appropriate frequency (*red curve*). Multiplying these two components element by element and summing over all time T yields a number, C. (*b*) Same data observed for a duration of $2T$. Element by element products of the data and matching sinusoid in first and second intervals of T are each equal to C, so total product over duration $2T$ is $2C$.

Therefore, we expect the second half of the summation (from $n = N + 1$ to $n = 2N$) to also equal C. We conclude that $X_{j,2N} = 2X_{j,N} = 2C$. In a similar way, $X_{j,3N} = 3C$, $X_{j,4N} = 4C$, and so on. The intuition is that because we observe the same signal for a longer time, the summed product increases.

To choose the scaling of the spectrum requires one additional fact:

$$\sigma^2 = \int_0^\infty S_{xx}(f)\, df, \tag{3.29}$$

which states that the (theoretical) variance σ^2 of a signal x equals the (theoretical) power spectral density $S_{xx}(f)$ of x integrated over all positive frequencies. This relation is consistent with the fact that the autocovariance is the inverse Fourier transform of the spectrum. To see this, consider the inverse Fourier transform of the spectrum evaluated at lag 0. In addition, this relation captures the idea that the spectrum decomposes the signal variance into different frequency bands; by summing this decomposition over all frequency bands, we calculate the total variance. To preserve this relation, we must choose an appropriate scaling of the spectrum. Let's replace $S_{xx}(f)$ in (3.29) with the discrete spectrum (3.9) and approximate the integral over frequencies using Riemann sums:

$$\begin{aligned}
\sigma^2 &= \sum_j \frac{2\Delta^2}{T} X_{j,N} X^*_{j,N}\, df \\
&= \sum_j \frac{2\Delta^2}{T} X_{j,N} X^*_{j,N} \left(\frac{1}{T}\right) \\
&= \frac{2\Delta^2}{T^2} \sum_j X_{j,N} X^*_{j,N}.
\end{aligned} \tag{3.30}$$

Notice that we've rewritten the frequency resolution df in terms of the total observation time T and that we've explicitly indicated the total number of time points (or equivalently, the total time $T = N\Delta$) in the Fourier transform $X_{j,N}$.

Now consider (3.30) and the impact of increasing the observation time. Doing so leaves the variance unaltered; if we observe a sinusoid for 1 s, and then the same sinusoid for K s, the variance is the same. Therefore, we require that the right–hand side of (3.30) remain unaltered. Let's replace $T \to KT$ and $N \to KN$ in (3.30),

$$\begin{aligned}
\frac{2\Delta^2}{(KT)^2} \sum_j X_{j,KN} X^*_{j,KN} &= \frac{2\Delta^2}{K^2T^2} \sum_j \left(K X_{j,N}\right)\left(K X^*_{j,N}\right) \\
&= \frac{2\Delta^2}{T^2} \sum_j X_{j,N} X^*_{j,N} \\
&= \sigma^2.
\end{aligned} \tag{3.31}$$

The factor of K^2 in the denominator (due to the T^2 term in the denominator) cancels the factor of K^2 in the numerator (due to the product of the Fourier transforms of x). Altering the duration of the recording does not alter the integrated spectrum; it remains equal to the variance. The scaling by $1/T$ preserves this relation.

Term: 2. The factor of 2 in the spectrum (3.9) is a conventional choice when the negative frequencies are ignored. We can think of this factor as accounting for the omitted negative frequencies. We note that this choice makes the variance equal to the spectrum integrated over positive frequencies in (3.29). If we were instead to remove this factor of 2 from the spectrum, we would need to integrate the spectrum of all positive and negative frequencies in (3.29) to equal the variance.

MATLAB Code. The ideas in this appendix are explored in the following MATLAB code. In this example, we generate artificial data as Gaussian noise. In your own exploration of this code, consider how the results change with the total duration of the recording (variable `T`) or nature of the signal (variable `d`).

```
dt = 0.001;            %Define the sampling interval.
T = 4;                 %Define the total duration of the recording.
t = (dt:dt:T);         %Define a time axis.
N = length(t);         %Define the number of samples.
d = randn(N,1);        %Create artificial data.
d = d - mean(d);       %Set mean to 0.

pow = 2*dt^2/T*(fft(d).*conj(fft(d)));  %Compute the power,
%... then sum spectrum over positive frequencies, multiply by df.
POW = sum(pow(1:N/2+1))*(1/T);
vr  = var(d);          %Compute variance of data.

%Print the results.
fprintf(['Summed spectrum ' num2str(POW) '\n'])
fprintf(['Variance       ' num2str(vr)   '\n'])
fprintf(['Difference     ' num2str(POW - vr) '\n'])
```

4 Analysis of Rhythmic Activity in an Invasive Electrocorticogram

Synopsis

Data Field data: 1 s of ECoG data sampled at 500 Hz.

Goal Characterize the observed rhythms in these data.

Tools Fourier transform, spectrum, tapers, multitaper method.

4.1 Introduction

4.1.1 Background

In chapter 3, we considered noninvasive recordings of brain electrical activity from the scalp surface. Although the scalp EEG provides fine temporal resolution of brain activity, the spatial resolution is poor because of the low conductivity of the skull [5]. An alternative, invasive approach to improve the spatial resolution of the scalp EEG is to record directly from the brain's surface [10]. This technique, known as electrocorticogram (ECoG), eliminates the distorting spatial blurring effect of the skull, at the cost of an invasive surgical procedure of implantation.

4.1.2 Case Study Data

A patient with epilepsy is scheduled to undergo resective surgery to remove the portion of her brain causing recurrent, unprovoked seizures. As part of her clinical workup, electrodes are implanted beneath the skull, directly on the brain's surface. We assume that our skilled neurosurgeon collaborator expertly implants the ECoG electrode, and that the ECoG data are collected with no artifacts. We receive from our clinical collaborator a 1 s segment of ECoG data recorded from a single electrode and sampled at 500 Hz.

4.1.3 Goal

Our collaborator (and we) would like to know what rhythms appear in these invasive brain voltage recordings. Our goal is to analyze the 1 s of ECoG data by characterizing the rhythmic attributes of the activity. To do so, we build upon the spectral analysis techniques developed in chapter 3 and focus on the impact of two important elements in computing the spectrum: windowing and zero padding.

4.1.4 Tools

In this chapter, we continue to develop understanding of the Fourier transform and spectrum. We apply the techniques introduced in chapter 3 to compute the spectrum. We also investigate the impact of windowing and zero padding on the spectrum, and explain how to apply and interpret the multitaper method.

4.2 Data Analysis

4.2.1 Visual Inspection

To access the data for this chapter, visit

```
http://github.com/Mark-Kramer/Case-Studies-Kramer-Eden
```

and download the file `Ch4-ECoG-1.mat`. We begin as always by looking at the data. In almost all cases, visual inspection is the first step in data analysis. We load the ECoG data into MATLAB and plot them by issuing the following commands:

```
load('Ch4-ECoG-1.mat')          %Load the ECoG data.
plot(t,ECoG)                    %Plot voltage vs time.
xlabel('Time [s]')              %Label time axis.
ylabel('Voltage [mV]')          %Label voltage axis.
```

Q: The ECoG data are plotted in figure 4.1. What do you see?

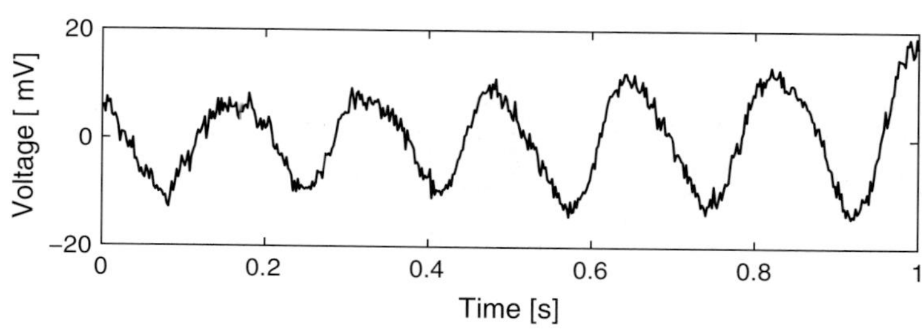

Figure 4.1
ECoG data provided by our collaborator.

You might notice a dominant rhythmic activity. We can approximate the frequency of this rhythm by counting the number of oscillations that occur in the 1 s interval. To do so, we may approximate the total number of large-amplitude cycles that we observe in the data. Through visual inspection of figure 4.1, we find that the first large-amplitude cycle occurs between ≈ 0 s and ≈ 0.175 s, the next between ≈ 0.175 s and ≈ 0.3 s, and so on. Counting this way, we approximate 6 full cycles per second, or a dominant 6 Hz rhythm.

4.2.2 Spectral Analysis: The Rectangular Taper and Zero Padding

Visual inspection, although essential to data analysis, is usually not enough. Visual inspection often guides intuition and reveals major features of the data, but it may also lead us astray; initial looks can be deceiving. To further explore the rhythmic activity of the ECoG data, we compute the spectrum.[1] We do so using the same approach implemented in chapter 3. The MATLAB code is nearly the same:

```
load('Ch4-ECoG-1.mat')              %Load the ECoG data.
x   = ECoG;                         %Relabel the data variable.
dt  = t(2)-t(1);                    %Define the sampling interval,
T   = t(end);                       %... and duration of data.

xf  = fft(x-mean(x));               %Compute Fourier transform of x,
Sxx = 2*dt^2/T*(xf.*conj(xf));      %... and the spectrum.
Sxx = Sxx(1:length(x)/2+1);         %Ignore negative frequencies.

df  = 1/T;                          %Define frequency resolution.
fNQ = 1/dt/2;                       %Define Nyquist frequency.
faxis = (0:df:fNQ);                 %Construct frequency axis.

plot(faxis, Sxx)                    %Plot spectrum vs frequency,
xlim([0 100])                       %... in select frequency range,
xlabel('Frequency [Hz]')            %... with axes labeled.
ylabel('Power [mV^2/Hz]')
```

> **Q:** For the ECoG data considered here, what is the frequency resolution *df*, and what is the Nyquist frequency (f_{NQ})? Compare your answers to the variables `df` and `fNQ` defined in this MATLAB code.

1. We could instead write the *sample* spectrum because we use the observed data to estimate the theoretical spectrum that we would see if we kept repeating this experiment. However, this distinction is not essential to the discussion here.

Q: Interpret the spectrum of the ECoG data plotted in figure 4.2a. What rhythms appear?

The visualization in figure 4.2a suggests a single dominant frequency near 6 Hz, consistent with the visual inspection of the ECoG trace (see figure 4.1). Other interesting structure may also appear, perhaps at frequencies near 10 Hz; note the tiny peak barely visible in the spectrum shown in figure 4.2a. These initial observations suggest we can more appropriately scale the spectrum to emphasize both the low-frequency bands and weaker signals. Let's utilize a logarithmic scale for both the power spectral density (decibels) and the frequency. In MATLAB,

```
semilogx(faxis, 10*log10(Sxx))  %Plot spectrum vs frequency,
xlim([0 100])                   %... in select frequency range,
xlabel('Frequency [Hz]')        %... with axes labeled.
ylabel('Power [dB]')
```

The first line of code is updated to plot the x-axis (frequency) on a logarithmic scale and to scale the spectrum to decibels. The resulting spectrum (figure 4.2b) confirms the initial observation of a strong rhythm at 6 Hz (the dominant peak in figure 4.2b) and also suggests interesting structure near 10 Hz. But what is really going on near 10 Hz? The spectra in figure 4.2 are hardly conclusive. Is the bump near 10 Hz an interesting feature or an artifact of the analysis? To address this question, we next consider the issue of windowing and computing the spectrum.

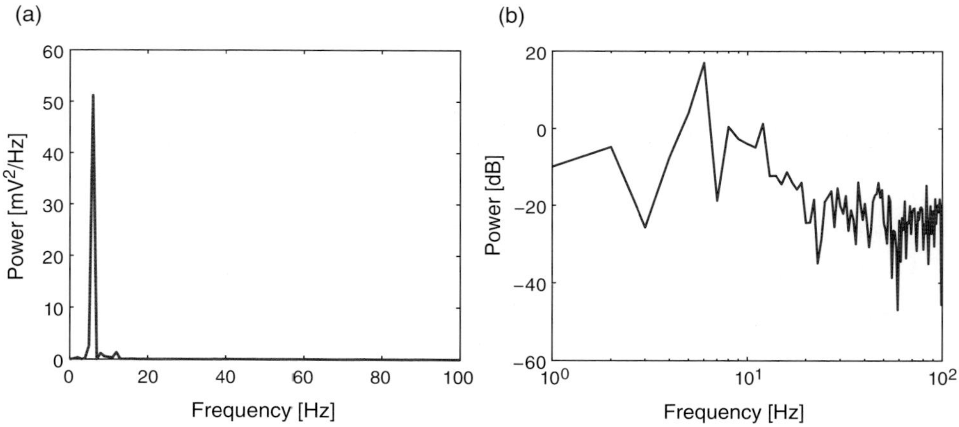

Figure 4.2
Power spectrum of ECoG data on (*a*) linear scale and (*b*) logarithmic scale.

Q: Are the terms *frequency resolution, Nyquist frequency, Fourier transform, decibel*, and *spectrum* familiar? Can you define or explain each term?

A: If not, we recommend reviewing the case study in chapter 3.

By Doing Nothing, We're Doing Something: The Rectangular Taper. ECoG time series continue for long durations. For example, an individual's brain voltage activity may persist for over 90 years, from birth until death. However, ECoG recordings are finite, limited by convenience, technology, or other factors. In the example here, we consider 1 s of ECoG data. Performing this finite observation (lasting 1 s) on a long duration (i.e., 90-year) time series can be understood as a *rectangular taper*. A rectangular taper multiplies the observed data by 1 and the unobserved data by 0 (figure 4.3). We can think of the value 1 as representing the time period when our recording device is operational; activating the ECoG recording device opens the rectangular taper (value 1), and deactivating the ECoG recording device closes the rectangular taper (value 0). The rectangular taper makes explicit our knowledge about the observed data (in this case, the 1 s interval of ECoG) and our ignorance about the unobserved data, which are assigned the value zero. Notice that the rectangular taper looks like a rectangle (figure 4.3).

So, by computing the spectrum of 1 s of observed ECoG data, we're actually computing the spectrum of the product of two functions: the many years of mostly unobserved ECoG data and the rectangular taper. We note that by "doing nothing" we have implicitly made the choice to use the rectangular taper. We have already plotted the resulting spectrum of the observed ECoG data (again, using the default rectangular taper) in figure 4.2.

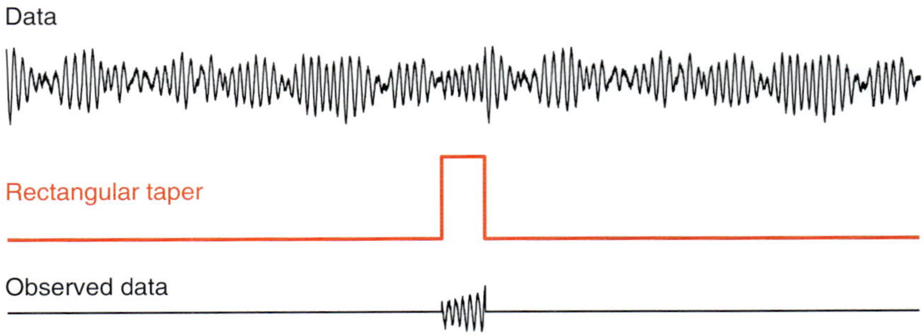

Figure 4.3

Example of rectangular taper application. Raw data continue for a long period of time (*top*). Most of these data are unobserved. Rectangular taper (*red*) specifies the interval of observation. Multiplying the raw data by the rectangular taper determines the observed ECoG data (*bottom*).

Exploring the Impact of the Rectangular Taper. Figure 4.3 illustrates how the rectangular taper impacts the observed data in the time domain, namely, the taper selects a region of observation. The rectangular taper also impacts the spectrum in the frequency domain. To see this, consider a perfect sinusoid at frequency 10 Hz that in theory continues forever. In this case, the energy concentrates at a single frequency—the frequency of the sinusoid (10 Hz)—and for the (theoretical) case of an infinite sinusoid, the power spectral density is infinite at that frequency (figure 4.4a). In mathematical language, we say that the spectrum of the infinite sinusoid is a *delta function*. However, we never observe an infinite sinusoid; to do so would require unlimited resources and unlimited time. Instead, let's assume we observe only 1 s of the sinusoid's activity; we imagine multiplying the infinite duration sinusoid by a rectangular taper and observing only a finite interval of time (figure 4.4b). The corresponding spectrum of the (now finite) sinusoid is shown in Figure 4.4b.

Q: Examine the spectrum of the finite sinusoid plotted in figure 4.4b. What do you see? Is the spectrum concentrated at one frequency (near 10 Hz), as we expected for an infinite sinusoid?

A: Visual inspection of the spectrum suggests an unexpected result: the spectrum is *not* concentrated at a single frequency. Instead, many peaks appear, centered at the expected frequency (10 Hz) but also spreading into neighboring frequency bands. The culprit responsible for this leakage outside of the 10 Hz peak is the rectangular taper applied to the true infinite-duration sinusoidal signal.

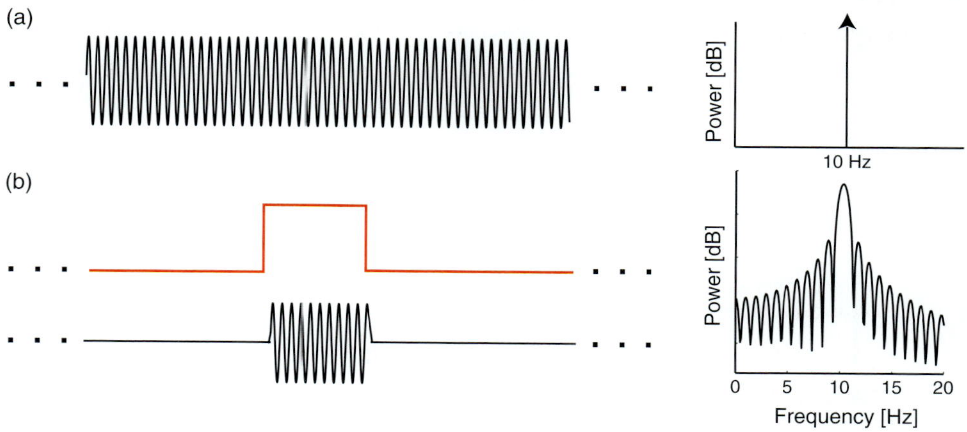

Figure 4.4

Example of rectangular taper application to a sinusoid. (*a*) Infinite sinusoid continues forever in time; energy concentrates at sinusoid's frequency of 10 Hz. (*b*) Multiplying the infinite duration sinusoid by a rectangular taper (*red*) yields a sinusoid of finite duration. Spectrum of the resulting signal (*black*) exhibits features at many frequencies, with peak at 10 Hz.

To understand further the impact of the rectangular taper, let's consider the spectrum of the rectangular taper itself.

Q: Examine the spectrum of the rectangular window plotted in figure 4.5b. What do you see? At what frequency is the spectrum concentrated?

To compute the spectrum of the rectangular taper, we consider the time series shown in figure 4.5a. In theory, the rectangular taper is infinite, and is preceded and followed by infinite intervals of zeros (i.e., there's an infinite period in which we do not observe the ECoG data). To represent the infinite extent of the rectangular taper, we add 10 s of zeros to the beginning and end of a 1 s interval of ones. Of course, 10 s of zeros is a poor representation of an infinite interval of time, but it's sufficient for our purposes here. We note that the rectangular taper consists of two sharp edges, when the observation interval opens and closes (i.e., transitions from 0 to 1, and then back from 1 to 0).

The spectrum of the rectangular taper is shown in figure 4.5b. Visual inspection suggests that most of the power spectral density is concentrated at a single frequency, 0 Hz. Notice that in this case we've plotted the spectrum at positive and negative frequencies. The negative frequencies are redundant (because the signal is real; see chapter 3), but the symmetry helps visualize the results. To understand why the spectrum concentrates at 0 Hz, consider the rectangular taper over the 1 s duration for which the data collection window is open (figure 4.5a). Within this window, the value of the taper is the constant 1. The only frequency present in this signal is 0 Hz (i.e., no oscillations occur). Therefore, the spectrum is concentrated at 0 Hz.

Although the spectrum is concentrated at 0 Hz, it exhibits regions of increased spectral density at nonzero frequencies, more specifically, the repeated peaks (or side lobes) in the

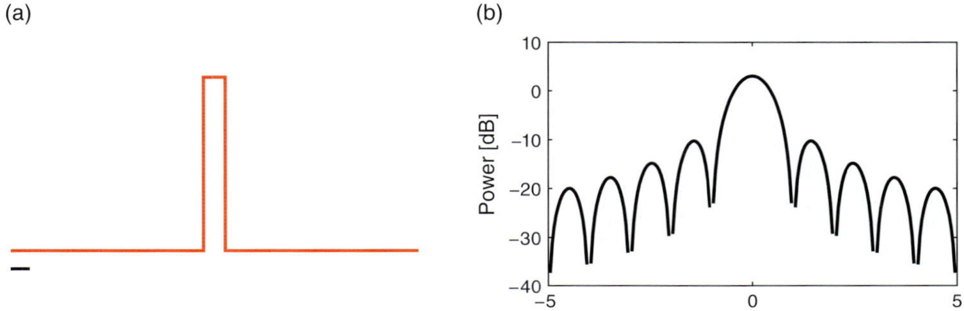

Figure 4.5

The rectangular taper possesses its own spectrum. (*a*) The rectangular taper; here, we set the taper equal to zero for 10 s, then equal to one for 1 s, then equal to zero for 10 s. The black line indicates a 1 s scale bar. (*b*) The spectrum of the rectangular taper; note in this case we plot both positive and negative frequencies.

spectrum near ±1.5 Hz, ±2.5 Hz, ±3.5 Hz, and so on. In fact, it's possible to work out exactly the functional form of the Fourier transform of the rectangular taper; it's the sinc function [8]. However, that's not particularly important for our purposes. The important result here is that the rectangular taper itself has a complicated spectrum, with features appearing across a range of frequencies.

Q: Why does the spectrum of the rectangular taper contain features at many frequencies?

A: The spectrum consists of features at many frequencies to represent the rapid increase and decrease of the rectangular taper (i.e., the rapid transition from 0 to 1, and then from 1 to 0). The intuition for this is that many sinusoids, aligned in a particular way, are required to represent a sharp transition in a time series. As an example, consider the square wave function in figure 4.6. This square wave consists of many repeated sharp transitions. Visual inspection suggests that the square wave is rhythmic, with a period of 2 Hz. Notice that the square wave begins at a sustained value of 1, then transitions to 0 and remains there for an interval of time, and then transitions back to a value of 1 in 0.5 s; the square wave therefore completes two cycles in 1 second.

So, the square wave is rhythmic, and we may try to represent this square wave with rhythmic sinusoids. In figure 4.6a, we plot a 2 Hz sinusoid. It's an okay match to the square wave but certainly not perfect; the 2 Hz sinusoid fails to capture the sharp transitions of the square wave. In figure 4.6b, we plot the sum of a 2 Hz sinusoid and a 6 Hz sinusoid. The combination of these two sinusoids better matches the square wave, although again the sharp transitions are not accurately captured. In figure 4.6c, we plot the sum of 2 Hz, 6 Hz, 10 Hz, 14 Hz, and 18 Hz sinusoid. These five sinusoids better match the square wave and begin to more accurately capture the sharp transitions. As more sinusoids are used to represent the square wave, the approximation of the sharp edges improves.

So, sharp transitions in data require many sinusoids to be accurately approximated. The rectangular taper consists of sharp transitions and therefore requires sinusoids at many frequencies for an accurate representation.

To understand why the spectrum of the rectangular taper is important, let's return to the example of an infinite sinusoid. The (theoretical) spectrum for this sinusoid concentrates all the spectral density at a single frequency (i.e., the frequency of the sinusoid), as illustrated in figure 4.4a. However, we never observe an infinite sinusoid; instead, we observe this sinusoid multiplied by the rectangular taper (figure 4.4b). This multiplication of the two functions impacts the spectrum we observe. To see why, we state an important property of the Fourier transform.

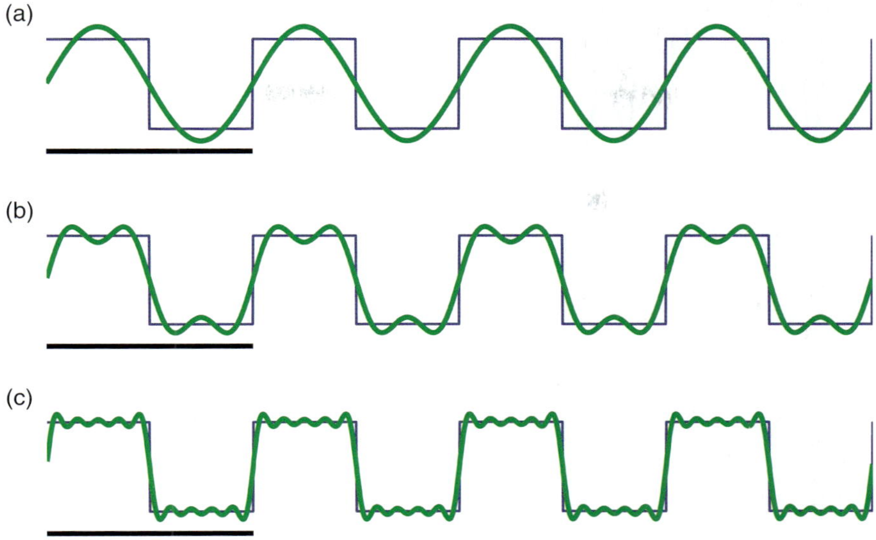

Figure 4.6

Approximation of a square wave with sinusoids. Square wave (*blue*) has a period of 2 Hz. As the number of sinusoids used to approximate the square wave increases from (*a*) one sinusoid, to (*b*) two sinusoids, to (*c*) five sinusoids, approximation improves. Scale bar indicates 0.5 s.

> Multiplication in the time domain is equivalent to *convolution* in the frequency domain.

Because we perform multiplication of two functions in the time domain (i.e., we multiply the rectangular taper and the infinite sinusoid element by element), we produce a convolution of the Fourier transforms of these functions in the frequency domain. Although the mathematical expression of convolution is somewhat complicated, we can understand the impact of convolution by examining a few plots. Let's start with the Fourier transform of the infinite sinusoid and the infinite rectangular taper (where we've appended zeros to the rectangular taper to make it infinite). We know (or can look up) the results of each Fourier transform. The Fourier transform of the infinite 10 Hz sinusoid, which we assume here is a cosine function, consists of two delta functions at ± 10 Hz (figure 4.7a). The Fourier transform of the rectangular taper is the sinc function (figure 4.7b). Now, let's imagine shifting in frequency the Fourier transform of the rectangular taper (i.e., shifting in frequency the sinc function). At each shift, we multiply element by element the two Fourier transforms (i.e., the unshifted Fourier transform of the sinusoid and the shifted Fourier transform of the rectangular taper) and sum the product. The result of this shifting, multiplying, and summing of one Fourier transform by the other is the convolution.

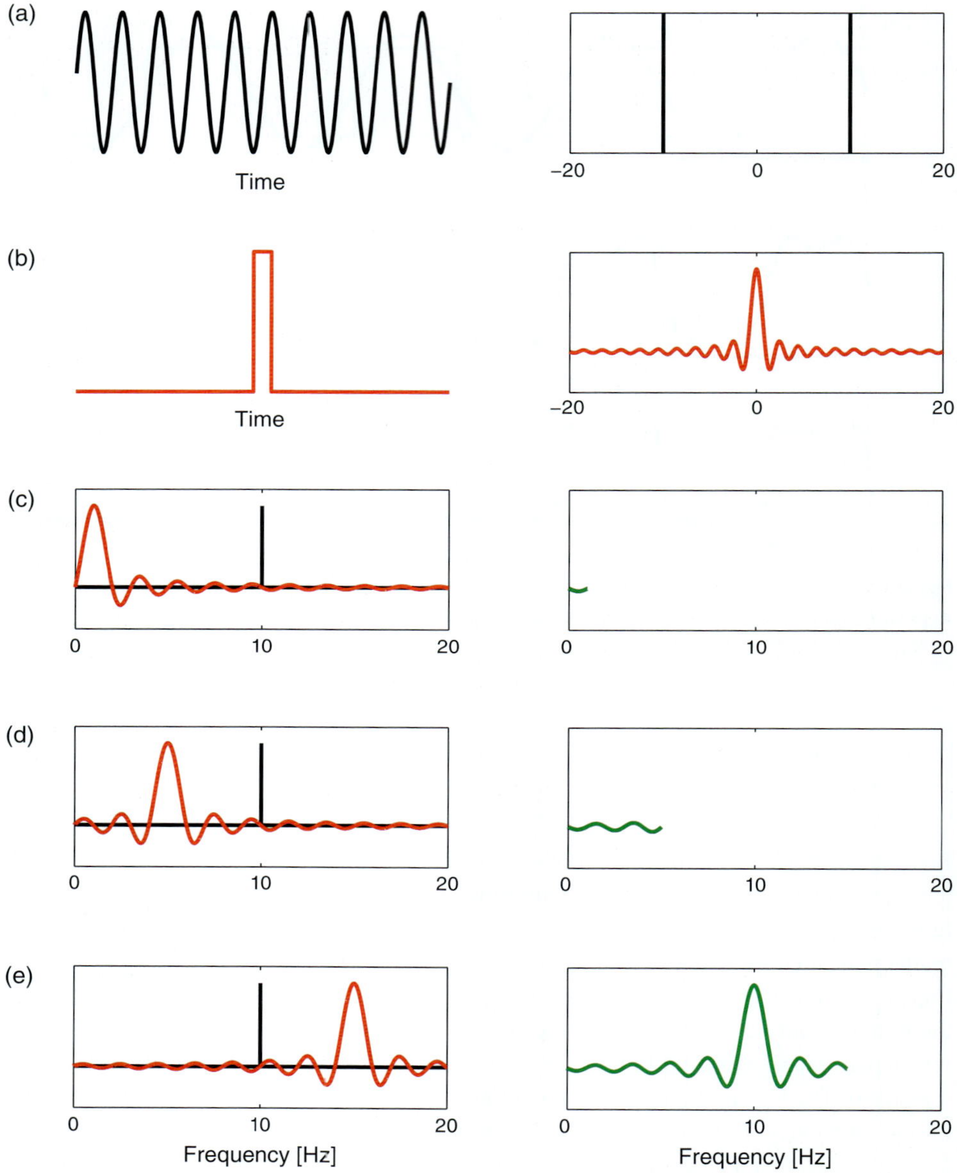

Figure 4.7

The impact of the rectangular taper on the Fourier transform of a sinusoid can be understood by considering the convolution between the two functions. (*a,b*) The sinusoid (*black*) and rectangular taper (*red*) time series (*left*) and Fourier transforms (*right*). (*c–e*) Illustration of the Fourier transform of each function (*left*) and their convolution evaluated at different shifts (*right in green*). Shifts up to 1 Hz (*c*), 5 Hz (*d*), and 15 Hz (*e*) are shown.

Q: What is the result of this convolution? How is the expected Fourier transform of the infinite sinusoid affected?

To illustrate this convolution procedure, we plot in figures 4.7c–e examples of shifting and multiplying the two Fourier transforms. Figure 4.7c shows the case in which the rectangular taper's Fourier transform is shifted by 1 Hz. Because the sharp peak of the sinusoid's Fourier transform does not overlap the large central peak of the rectangular taper's Fourier transform, the product of the two functions remains small.

As we continue to shift the rectangular taper's Fourier transform, the product of the two functions at first remains small (for shifts up to 5 Hz in figure 4.7d). However, once we shift the rectangular taper's Fourier transform up to and through the 10 Hz peak of the sinusoid, we begin to find larger deviations in the convolution. As the shifted rectangular taper's Fourier transform passes through the 10 Hz peak of the sinusoid, we multiply the very thin and very tall peak of the sinusoid (figure 4.7e) with the center portion of the shifted rectangular taper's Fourier transform. The result of this multiplication is to "smear" the sharp peak of the sinusoid in the frequency domain. Instead of a sharp peak at 10 Hz, the convolution produces a broad, wiggly peak in the frequency domain, with deviations at neighboring frequencies (i.e., in the side lobes around 10 Hz). In other words, because we observe the sinusoid over a finite interval of time (determined by the extent of the rectangular taper) the Fourier transform of the sinusoid becomes "smeared."

This description and the illustrations in Figure 4.7 provide some intuition for the relation between the multiplication of two signals in the time domain (here, a sinusoid and the rectangular taper) and their Fourier transform. For mathematical details and corresponding MATLAB code, see the appendix at the end of this chapter.

Zero Padding. An interesting issue to consider is how appending zeros impacts the spectrum of the rectangular taper. We know that increasing the signal length (T) improves the frequency resolution; recall the equation $df = 1/T$ from chapter 3. We therefore expect that adding more points to the signal (even noninformative points, such as zeros) will increase the number of points along the frequency axis. However, appending zeros to a time series is *not* equivalent to observing more data (and thereby increasing T). By appending zeros, we of course do not gain any additional information about the signal. Therefore appending zeros to a signal *cannot* improve the frequency resolution. Instead, the impact of appending zeros is to increase the number of points along the frequency axis in the spectrum. This can be useful in visualizing the spectrum; for example, by appending more and more zeros to the rectangular taper, we produce a less jagged spectrum (figure 4.8).

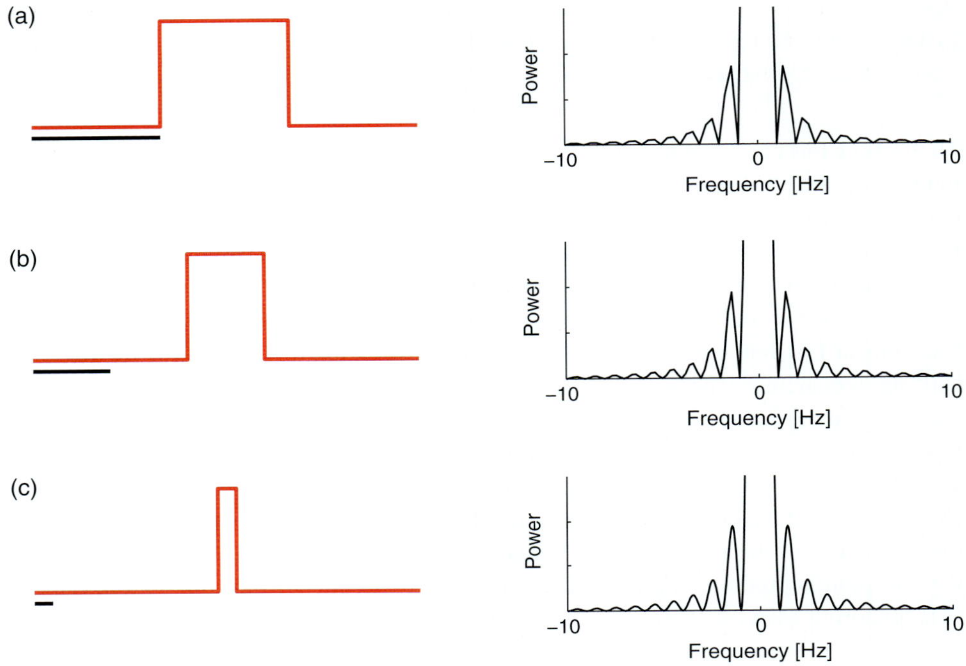

Figure 4.8

Impact of appending zeros on the spectrum of the rectangular taper. *Left*, Rectangular tapers vs. time; scale bars indicate 1 s. *Right*, Spectra of the rectangular tapers. (*a*) One second of zeros surrounding 1 s interval of ones; (*b*) two seconds of zeros surrounding 1 s interval of ones; (*c*) ten seconds of zeros surrounding 1 s interval of ones. As the number of zeros increases, the spectrum becomes smoother.

This procedure of appending zeros to a time series is called *zero padding*. It can be useful for visualizing a spectrum. But we must be careful to remember the following important fact.

> The frequency resolution of the spectrum is fixed by the amount of data recorded. The number of points along the frequency axis in the spectrum is adjustable and can be increased by zero padding.

With this understanding of the impact of the rectangular taper and zero padding, let's now return to the spectrum of the 1 s of sinusoidal activity (see figure 4.4b). We now expect that because we observe the infinite sinusoid for only a short duration (1 s), the spectral power at 10 Hz will leak into neighboring frequency bands. And as plotted in figure 4.4b, that's indeed what we find. In figure 4.4b the side lobe structure is clearly visible; in this example, we computed the spectrum with zero padding to evaluate the spectrum at many points along the frequency axis.

To explore further the impact of this zero padding, let's now consider an example in MATLAB. We define a 10 Hz sinusoid with duration 1 s, apply 10 s of zero padding and examine the impact on the spectrum:

```
Fs = 500;                          %Define sampling frequency.
dt = 1/Fs;                         %Define sampling interval.
t = (dt:dt:1);                     %Define time axis.
T = t(end);                        %Define total time.

d = sin(2.0*pi*t*10);              %Make a 10 Hz sinusoid,
d = [d, zeros(1,10*Fs)];           %... with 10 s of zero padding.

df = 1/(length(d)*dt);             %Define the frequency step size,
fNQ = Fs/2;                        %... and Nyquist frequency,
faxis = (0:df:fNQ);                %... to create frequency axis.

pow = 2*dt^2/T*(fft(d).*conj(fft(d)));%Compute spectrum.
pow = pow(1:length(d)/2+1);        %Ignore negative frequencies,
plot(faxis, 10*log10(pow))         %... and plot it,
xlim([0 20])                       %... in selected frequency range,
ylim([-60 10])                     %... and selected decibel range,
xlabel('Frequency [Hz]')           %... with axes labeled.
ylabel('Power [dB]')
```

In the first four lines of this code, we define the sampling frequency (`Fs`), sampling interval (`dt`), time axis (`t`), and duration of data (`T`) to simulate. We then define the 10 Hz sine function and include zero padding. The remaining lines evaluate the spectrum and display the results (as done for the preceding ECoG data). Notice that we plot the spectrum on a decibel scale, and focus on the frequency interval 0–20 Hz.

Q: We usually subtract the mean of a signal before computing the Fourier transform, but we did not do so here when defining the variable `pow`. Is that a problem?

A: In this case, the mean of `d` is zero, so subtracting the mean does not impact the results.

Q: How much zero padding do we add? How does this change the spacing on the frequency axis?

A: We append `10*Fs` zeros to the end of the simulated sinusoidal data. This corresponds to 10 s (or 5,000 points). The original spacing on the frequency axis for the data was $1/T = 1/1\text{ s} = 1$ Hz. After zero-padding, the spacing on the frequency axis becomes $1/(1 + 10\text{s}) = 1/(11\text{s}) \approx 0.091$ Hz.

The results for this case (using 10 s of zero padding) are shown in figure 4.9a. We find a dominant peak at 10 Hz, as expected, and large side lobes that extend throughout the 0–20 Hz frequency range. Choices of smaller-duration zero padding (5 s in figure 4.9c) and longer-duration zero padding (100 s in Figure 4.9d) produce similar results. With 5 s of zero-padding, the plotted spectrum appears less smooth; in this case, we evaluate the

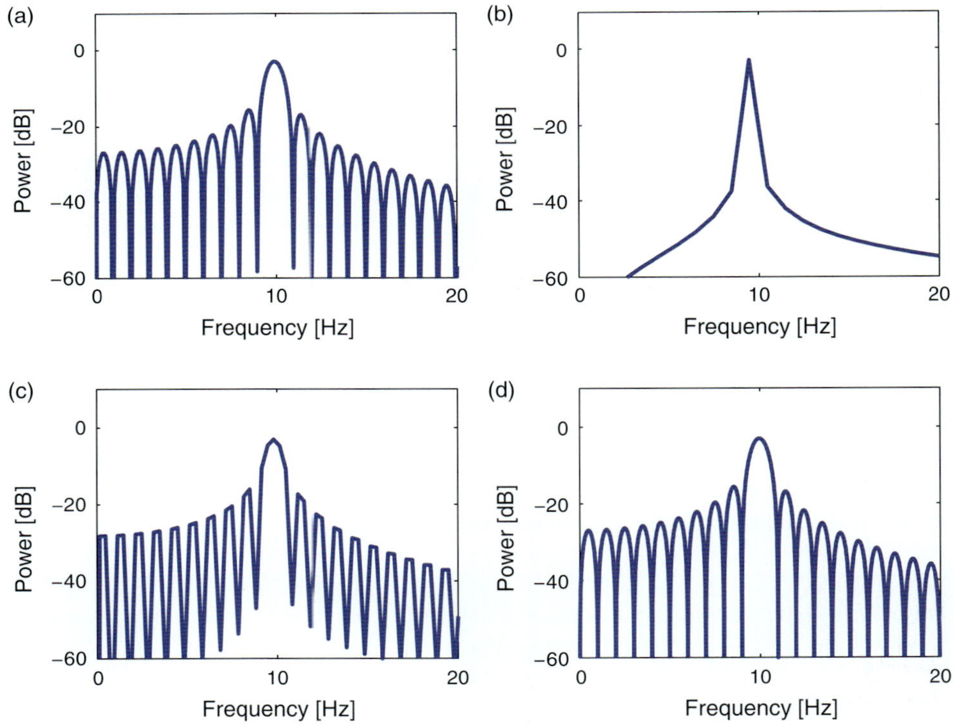

Figure 4.9
Impact of zero padding duration on the spectrum of a 1 s, 10 Hz sinusoid. Duration of zero paddings: (*a*) 10 s, (*b*) 0 s, (*c*) 5 s, (*d*) 100 s.

spectrum at fewer points on the frequency axis, and the side lobe peaks become more jagged. The choice of 100 s of zero padding does not produce a qualitative change compared to the 10 s of zero padding; in this case, evaluating the spectrum at 10 times as many points along the frequency axis is not particularly useful.

We also show in figure 4.9b the impact of no zero padding. To do so, we omit the sixth line in the preceding MATLAB code (i.e., `d = [d, zeros(1,10*Fs)];`). The spectrum is quite different from the original example. Without zero padding, the falloff from the 10 Hz peak appears rather smooth and gradual, and does not exhibit obvious side lobe structure, as expected after the discussion of the rectangular taper and example in figure 4.7. Leakage from the 10 Hz peak into neighboring frequency bands still occurs, but without zero padding we're sampling the frequency axis too coarsely to accurately make out the side lobes.

Zero Padding and Frequency Resolution: An Example. We stated in the previous section that zero padding does not improve the frequency resolution of the spectrum. As an example of this, consider a simple signal of duration 1 s that consists of two sinusoids: a 10 Hz sinusoid and a 10.5 Hz sinusoid.

> **Q:** Given this 1 s time series, can we distinguish the two rhythms in the spectrum?
>
> **A:** No. The frequency resolution is $df = 1/T = 1/(1\text{s}) = 1$ Hz. Because the two sinusoids are separated by less than 1 Hz, we cannot distinguish these two rhythms.

But, perhaps by zero padding the data, we can distinguish these two rhythms. After all, zero padding acts to increase T, right? We're appending lots of zeros to the time series, so the data become longer .... Let's create these synthetic data in MATLAB and investigate.

```
Fs = 500;                       %Define sampling frequency.
dt = 1/Fs;                      %Define sampling interval.
t = (dt:dt:1);                  %Define time axis.
T = t(end);                     %Define total time.

d1 = sin(2.0*pi*t*10);          %Make a 10 Hz sinusoid.
d2 = sin(2.0*pi*t*10.5);        %Make a 10.5 Hz sinusoid.
d  = d1+d2;                     %Make the summed signal,
d = [d, zeros(1,10*Fs)];        %... with 10 s of zero padding.

df = 1/(length(d)*dt);          %Define the frequency step size,
fNQ = Fs/2;                     %... and Nyquist frequency,
faxis = (0:df:fNQ);             %... to create frequency axis.
```

```
pow = 2*dt^2/T*(fft(d).*conj(fft(d)));%Compute spectrum.
pow = pow(1:length(d)/2+1);     %Ignore negative frequencies,
plot(faxis, 10*log10(pow))      %... and plot it,
xlim([0 20])                    %... in selected frequency range,
ylim([-40 10])                  %... and selected decibel range,
xlabel('Frequency [Hz]')        %... with axes labeled.
ylabel('Power [dB]')
```

In the first four lines, we define some useful parameters, including the sampling frequency and a time axis. We then create the two sinusoids (d1 and d2), sum these sinusoids to create a composite signal (d), and zero-pad this composite signal with 10 s of zeros. The rest of the code computes and plots the spectrum, which is shown in figure 4.10a.

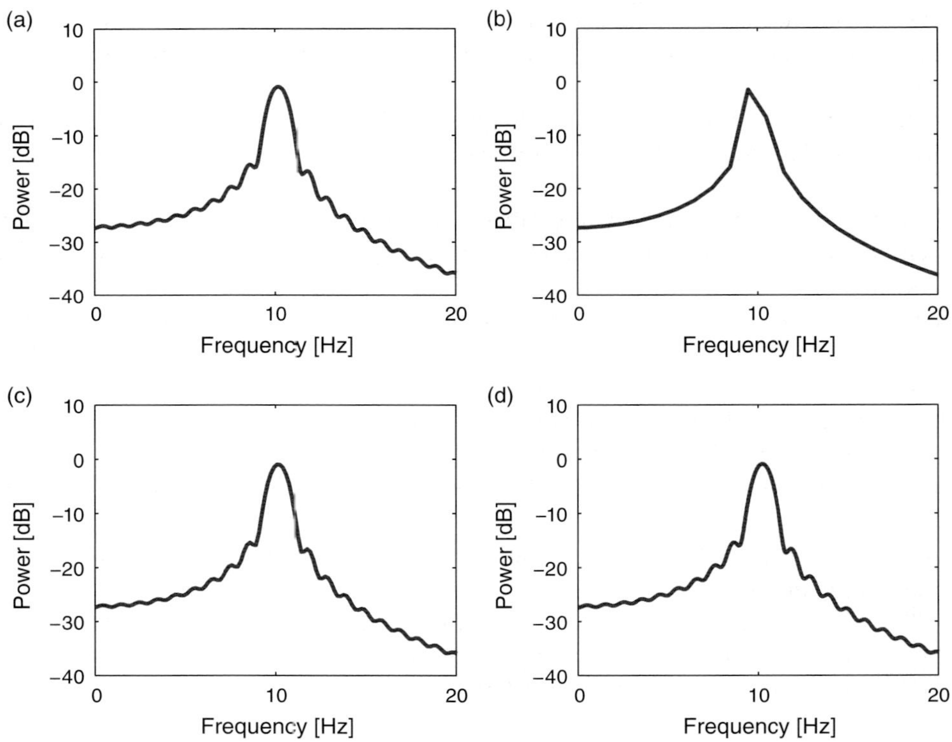

Figure 4.10

Impact of zero padding duration on the frequency resolution of the spectrum. The data consist of the sum of two sinusoids at frequencies 10 Hz and 10.5 Hz, and last for 1 s. Duration of zero padding is: (*a*) 10 s, (*b*) 0 s, (*c*) 5 s, (*d*) 100 s.

Q: Consider the spectrum in figure 4.10a. Can you identify the two rhythms present in the simulated time series? The other graphs show the spectrum with different durations of zero padding. Do any of these figures reveal the two separate rhythms?

A: No. No choice of zero padding resolves the two spectral peaks. In this case, the two rhythms are separated by less than the frequency resolution df. Zero padding the data (even with 100 s of zeros) does not resolve the two peaks. We only simulated 1 s of data, and therefore set $df = 1$ Hz; zero padding does not change this fact.

4.2.3 Beyond the Rectangular Taper—The Hanning Taper

We have considered so far a single type of taper (the default, a rectangular taper) and its impact on the spectrum. In particular, we noted the "smearing" of spectral peaks (i.e., side lobes) that impact neighboring frequency bands. At best these side lobes are distracting, and at worst they may contaminate our conclusions. For example, consider the spectrum for the 1 s of ECoG data plotted in figure 4.2. Is the small peak near 10 Hz representative of a true rhythm in these data, or is it a side lobe of the large peak near 6 Hz? Many different taper shapes have been developed with the goal of reducing the side lobes that contaminate the signals. Here we consider one of these tapers, the *Hanning taper*.

The problem with the rectangular taper is its sharp edges (i.e., the rapid transitions from 0 to 1, and from 1 back to 0). To represent these sharp edges in the frequency domain requires many sinusoids, oscillating at different frequencies (e.g., figure 4.6), which manifest as side lobes in the spectrum. The Hanning taper acts to smooth the sharp edges of the rectangular taper. To see this, we plot in figure 4.11a both the Hanning taper and the rectangular taper. Notice that the Hanning taper gradually increases from zero, reaches a maximum of 1 at the center of the interval, then gradually decreases to zero. The corresponding spectra of the Hanning taper and the rectangular taper are plotted in figure 4.11b. These spectra reveal two main differences between these tapers: (1) the central lobe of the Hanning taper is wider than the rectangular taper, and (2) the side lobes in the Hanning taper are reduced compared to the rectangular taper. These two features illustrate the trade-off between the two window choices. By accepting a wider central peak in the Hanning taper, we acquire side lobes with lower power.

The Hanning taper is applied to time series data in the same way as the rectangular taper. The data are multiplied element by element by the taper. Let's compute and apply the Hanning taper in MATLAB for the ECoG data:

```
load('Ch4-ECoG-1.mat')          %Load the ECoG data.
x  = ECoG;                      %Relabel the data variable.
N = length(x);                  %Define no. of points in data.
x = hann(N).*x;                 %Multiply data by Hanning taper.
```

The function `hann` returns the Hanning taper, which we multiply element by element with the ECoG data in variable `x`. The results of this multiplication are plotted in figure 4.11c. Notice that the slow increase in the Hanning taper reduces the amplitude of the ECoG activity near the taper edges and emphasizes the ECoG activity in the center of the interval.

Q: Under what conditions would reducing the activity near the taper edges be a bad idea? *Hint:* What if the signal features of interest occur at the very beginning or very end of the observed data?

Q: The spectra for the Hanning-tapered ECoG data and the rectangular tapered ECoG data are plotted in figure 4.11d. What conclusions do you now draw regarding the rhythms present in the ECoG data? Consider, in particular, the activity near 10–15 Hz.

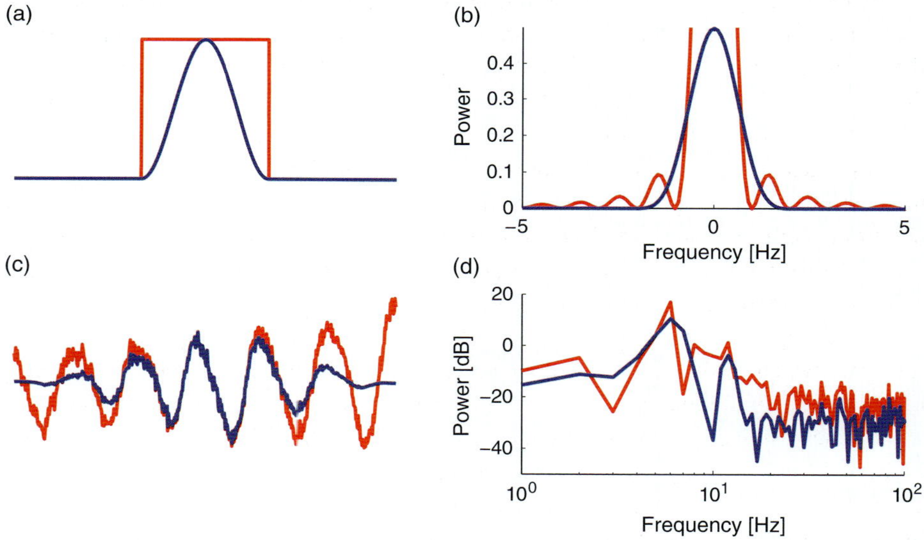

Figure 4.11

Hanning taper (*blue*) and the rectangular taper (*red*) have different trade-offs in the time and frequency domains. (*a*) The two tapers and (*b*) their corresponding spectra. (*c*) The tapers applied to the ECoG data, and (*d*) the resulting spectra.

A: The spectrum of the Hanning-tapered ECoG data reveals a peak at 10–15 Hz. This peak was hidden by the side lobes of the 6 Hz peak in the (default) rectangular-tapered ECoG data. The Hanning taper reduces the side lobes of the 6 Hz peak, and allows us to uncover the smaller 10-15 Hz peak that was originally obscured by these side lobes. This observation dramatically changes our interpretation of the ECoG data. We now propose that the ECoG activity consists of two rhythms: a rhythm near 6 Hz, and a second band of rhythms near 10–15 Hz. Without application of the Hanning taper, we might have missed the second rhythm by attributing it to side lobes of the 6 Hz peak.

4.2.4 Beyond the Hanning Taper—The Multitaper Method

The Hanning taper provides a nice alternative to the rectangular taper. If we're willing to allow slightly broader spectral peaks in the frequency domain, and lose some data near the taper edges in the time domain, then the Hanning taper helps reduce the impact of side lobes. We now consider a brief introduction to a more advanced approach to tapering, the multitaper method. The idea of the multitaper method is to apply multiple tapers to the data, each with a different shape. The spectrum is then computed for each taper, and the results averaged over the tapers. Each taper, which is given by portions of the discrete prolate spheroidal sequences, provides an independent estimate of the (theoretical) spectrum. Therefore, the variance of the average estimate over the tapers is $1/K$ times the variance of the estimated spectrum from a single taper, where K is the number of tapers used. The multitaper method has a number of additional advantages, such as minimizing the bias due to other spectral peaks outside of the frequency band being considered. We do not discuss these properties in detail, but more information can be found in [8, 9].

In the preceding examples, we estimated the spectrum from an observed time series. Often from these estimates we have difficulty identifying features that are significant (or not). For example, consider the small peak at 40–50 Hz in figure 4.11d. Is that peak significant or a random fluctuation we expect in an estimate from real-world ECoG data? To address this question, we might consider collecting more data and using these additional data to improve our estimate of power. However, as we mentioned in chapter 3, collecting more data does not automatically improve the spectral estimate. Instead, collecting more data (i.e., increasing the duration of data recorded, T) produces more spectral estimates at additional frequencies, yet the spectrum at each frequency remains just as variable. We would like a way to improve the spectral estimate—a way to reduce the variability of the estimate—so that we could more confidently identify interesting features. The multitaper method offers a procedure to do so. However, this comes at a cost. If we desire reduced variance in the spectral estimate, we must accept worse frequency resolution. Let's explore these issues in more detail.

To get a sense for the multitaper method, let's examine some of the tapers, as plotted in figure 4.12a. The first taper looks familiar—it's similar to the Hanning taper. Notice that the first taper starts at zero, increases throughout the interval, and then decreases back to zero. As we add more tapers, each contains more "wiggles." In particular, as the number of tapers increases, the edges of the data become better represented. Unlike the Hanning taper, which slowly approaches zero at the taper edges, some of these tapers increase near the window edges (see taper 5 in figure 4.12a). This can be useful, especially if we are interested in the spectral features near the beginning and end of the data. Just as with the Hanning taper, we multiply the time series data by each taper, which emphasizes different intervals of the data. We show examples of this multiplication for the ECoG data in figure 4.12b. The first taper emphasizes the middle of the ECoG data, and the fifth taper emphasizes the beginning and end of the ECoG data.

We only show the first five tapers in figure 4.12. There are an infinite number more. How do we choose the number of tapers to use in the multitaper method? To answer this, we utilize the equation

$$TW = X, \tag{4.1}$$

where T is the duration of the recording, $2W$ is the desired frequency resolution (or resolution bandwidth), and we're free to choose X, the aptly named *time-bandwidth product*. For concreteness, let's consider the 1 s of ECoG data.

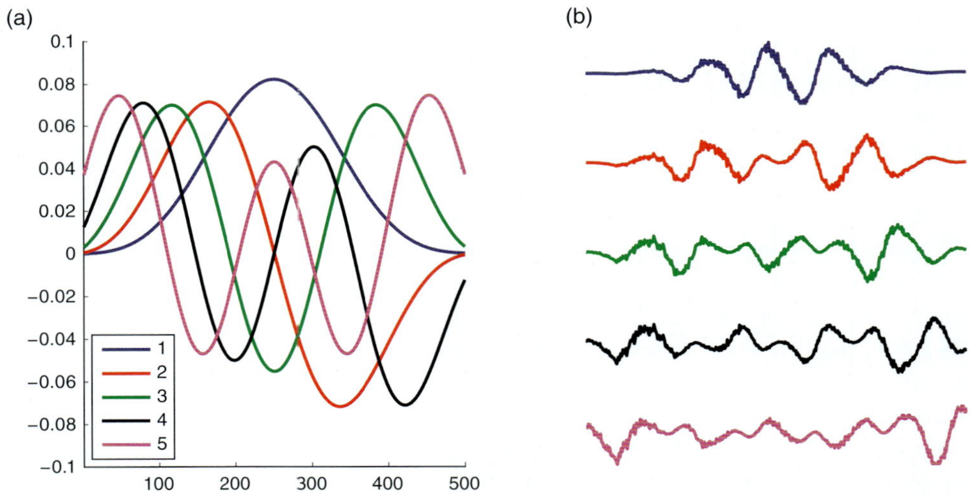

Figure 4.12

(*a*) Traces of the first five tapers, plotted as a function of indices, and (*b*) each taper applied to 1 s of ECoG data.

Q: Given 1 s of data, what is the frequency resolution?

A: Using the equation for the frequency resolution $df = 1/T$, we find $df = 1/(1\text{s}) =$ 1 Hz. Without applying the multitaper method, we begin with a frequency resolution of 1 Hz. We now concede some frequency resolution to apply the multitaper method and reduce the variability in the spectral estimate.

Remember that a frequency resolution of 1 Hz indicates that we can resolve features in the spectrum separated in frequency by 1 Hz or more. For example, at a frequency resolution of 1 Hz, we can distinguish 10 Hz activity from 9 Hz or 11 Hz activity. However, we are unable to distinguish 10 Hz activity from 10.5 Hz or 9.5 Hz; those frequencies lie within the frequency resolution.

Let's assume we do not require a frequency resolution of 1 Hz; instead, we are satisfied with a frequency resolution of 6 Hz. Using the multitaper method, we trade off a worse frequency resolution to improve the estimate of the spectrum. In this case, $T = 1$ s, and we are willing to accept a resolution bandwidth of $2W = 6$ Hz. So, from (4.1), we compute the time-bandwidth product and find $TW = 3$. Now, with this value, we select the number of tapers following this rule of thumb:

$$\text{No. of tapers} = 2TW - 1. \tag{4.2}$$

We choose the first $2TW - 1$ tapers because doing so allows us to preserve most of the information present in the original data. We could choose fewer tapers, but in most cases we follow this rule of thumb and pick as many tapers as we can. Choosing more tapers does not improve the multitaper estimate of the spectrum and may lead to spurious results; for more details, see [8]. So, for the ECoG data of interest here, we select the number of tapers to be $2 \times 3 - 1 = 5$. These five tapers are plotted in figure 4.12a. Applying these tapers to the ECoG data, we create the five time series shown in figure 4.12b. We then compute the spectrum of each tapered ECoG time series, and average the resulting spectra across the five tapers. Through this averaging procedure across tapers, we reduce the variability of the spectral estimate.

Multitaper method tradeoff: The multitaper method permits a trade-off between frequency resolution and variance of the spectrum. If we can tolerate worse frequency resolution, we can include more tapers and reduce the spectrum variance.

Q: Given 10 s of data, we demand a frequency resolution of 2 Hz or better. Using the multitaper method, what is the maximum number of tapers we could select and still maintain the desired frequency resolution?

The multitaper method is a sophisticated approach, and we have only touched on the surface in this brief discussion. There are many detailed references and important applications [8, 9]. Fortunately, there are also MATLAB software packages and functions to implement and apply the multitaper method. In subsequent chapters, we utilize the software package Chronux (chronux.org) [2]. But for now, let's use the MATLAB function pmtm to compute the multitaper spectrum of the ECoG data:

```
load('Ch4-ECoG-1.mat')         %Load the ECoG data.
x  = ECoG;                     %Relabel the data variable.
x  = x - mean(x);              %Set mean of x to zero.
dt = t(2)-t(1);                %Define the sampling interval.
Fs = 1/dt;                     %Define the sampling frequency.
TW = 3;                        %Choose time-bandwidth product of 3.

[Sxx,f]=pmtm(x,TW,length(x),Fs); %Compute MTM spectrum.

semilogx(f, 10*log10(Sxx))     %Plot spectrum vs. frequency,
xlim([0 100])                  %... in selected frequency range,
xlabel('Frequency [Hz]')       %... with axes labeled.
ylabel('Power [dB]')
```

We define the time-bandwidth product TW. By default, MATLAB sets the number of tapers to $2TW - 1$. We include as inputs to the function pmtm the data (x), the time-bandwidth product (TW), the amount of zero padding (here, set to none), and the sampling frequency (Fs). The function pmtm returns the multitaper estimate of the spectrum (Sxx) and the frequency axis (f). In figure 4.13 the spectrum of the ECoG data is computed in three ways, using the rectangular taper, the Hanning taper, and the multitaper method.

Q: Compare the spectra of the ECoG data in figure 4.13. How do the multitaper method results differ from the results computed with the other tapers? Do any new features appear in the spectrum using the multitaper method?

4.2.5 Confidence Intervals of the Spectrum

Another useful feature of the pmtm function is the ability to compute confidence intervals for the spectrum. The confidence interval for a multitaper spectral estimator with K tapers can be computed using a chi-square distribution with $2K$ degrees of freedom [8]. To return the confidence intervals, we recompute the spectrum with an additional returned output (Serr), as follows,

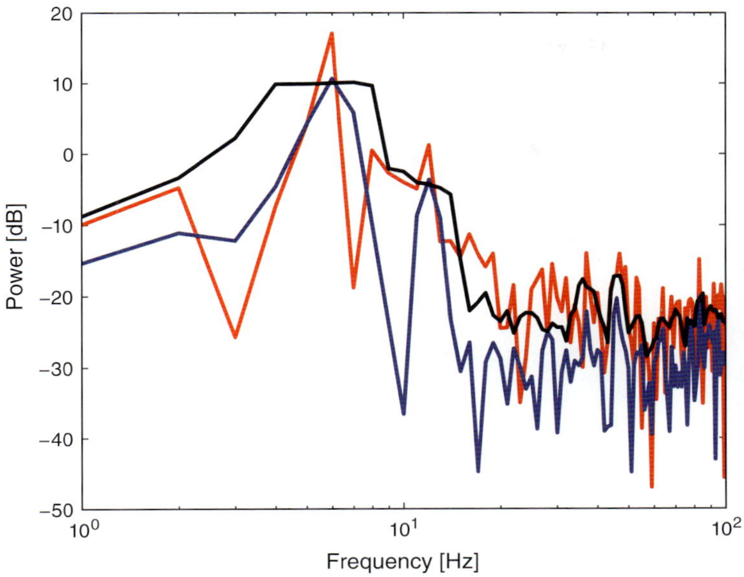

Figure 4.13

Spectrum of ECoG data computed using the rectangular taper (*red*), Hanning taper (*blue*), and the multitaper method (*black*).

```
%Compute MTM spectrum, and return CIs.
[Sxx,Serr,f]=pmtm(x,TW,length(x),Fs);

semilogx(f,10*log10(Sxx))       %Plot spectrum vs frequency,
hold on                         %... freeze the graphics window,
semilogx(f,10*log10(Serr))      %... plot the CI,
hold off                        %... release graphics window.
xlim([0 100])                   %Select frequency range,
xlabel('Frequency [Hz]')        %... and label the axes.
ylabel('Power [dB]')
```

The resulting spectra with confidence intervals are shown for different choices of time-bandwidth product in figure 4.14.

Q: Consider the multitaper spectra estimates of the ECoG data in figure 4.14. For each case determine the frequency resolution, and compare your answer to the plots in figure 4.14. How does changing the time-bandwidth product impact the plotted spectra?

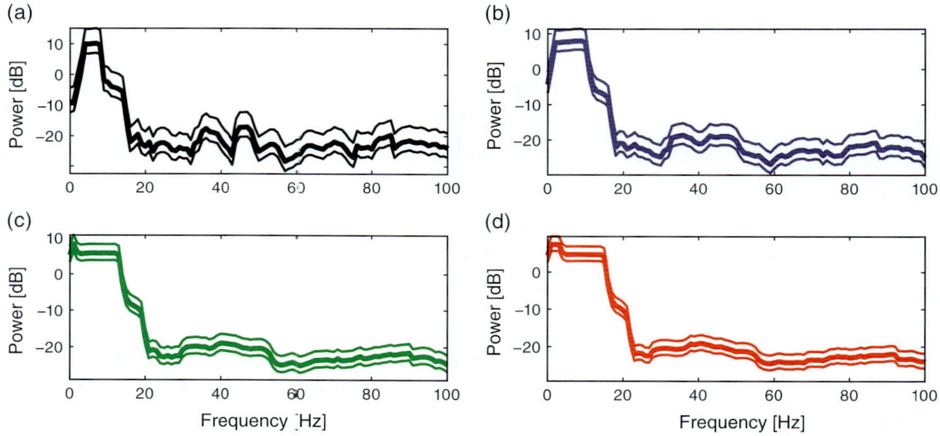

Figure 4.14

The multitaper spectral estimate of the ECoG data with different choices of the time-bandwidth product. The time-bandwidth products are (*a*) 3, (*b*) 5, (*c*) 8, (*d*) 10. The means are indicated with thick lines, and the 95% confidence intervals with thin lines.

Q: Use the spectral analysis results to draw some conclusions about the data. What rhythms do you think are present?

A: Our previous analysis using the Hanning taper suggested two peaks in the spectrum: a large peak near 6 Hz, and a second smaller peak near 10–15 Hz. Both peaks are still visible, at least partially, in the multitaper spectra estimates. We notice that as the time-bandwidth product increases, the frequency resolution becomes worse, and it becomes more difficult to resolve these two peaks; these neighboring peaks start to smear together. This is expected and the trade-off we accept in the multitaper method. The big payoff of the multitaper method occurs in the higher-frequency bands. As the time-bandwidth product increases, we observe a small elevation across a broad frequency range, at approximately 30–50 Hz. The spectral density in this interval is small and was hidden by the noise in the previous spectral estimates using the rectangular taper or the Hanning taper. However, by using many tapers in the multitaper method, we reduce this noise and reveal the broad elevation of power. Physiologically, the impact of this observation is enormous; the data exhibit a broad gamma band peak—one of the best studied and understood frequency bands in the brain—with implications for cognitive function and dysfunction [4].

Summary

In this chapter, we analyzed the rhythmic activity present in 1 s of ECoG data. We computed the spectrum and considered two issues: zero padding and tapering. We discussed that zero padding can increase the number of points along the frequency axis but cannot change the frequency resolution. We also explored the trade-off between three different tapers: the rectangular taper, the Hanning taper, and the multitaper method. The Hanning taper reduces the side lobes present in the rectangular taper, but fattens the spectral peaks. The multitaper method reduces the variance of the spectrum at the cost of worsened frequency resolution. Applying all three measures to the ECoG data allowed us to explore the rhythmic activity of these data in different ways. All three methods suggest rhythmic content at low frequencies, near 6–7 Hz, consistent with the visual inspection of the time series. Applying the Hanning taper, we uncovered activity in the 10–15 Hz range. Applying the multitaper method, we uncovered broadband activity at 30–50 Hz. This activity, hidden in the noisy spectrum, only became apparent upon increasing the number of tapers. However, this increase necessarily reduces the frequency resolution and can hide the low-frequency rhythms (compare figure 4.11d with figure 4.14d). In this case, we find it useful to examine the spectrum in a variety of ways.

Here we have only touched the surface of these concepts. Further discussions of zero padding, tapering, and the multitaper method may be found in [8, 9, 11].

Problems

4.1. Consider a 1 s sinusoid of 10 Hz activity (e.g., see figure 4.4). How does the spectrum differ when computed using the Hanning taper and rectangular taper? How does including 10 s of zero padding affect the results in each case?

4.2. Consider the file `Ch3-EEG-1.mat`, available at

```
http://github.com/Mark-Kramer/Case-Studies-Kramer-Eden
```

These data served as the case study data in chapter 3; see that chapter for a detailed analysis of these data. Load these data into MATLAB, and answer the following questions.

a. Plot the spectrum versus the frequency using a rectangular taper. Do so without zero padding the data and then after adding 10 s of zero padding to the data.

b. Plot the spectrum versus the frequency using a Hanning taper. Do so without zero padding the data and then after adding 10 s of zero padding to the data.

c. Plot the spectrum versus the frequency using the multitaper method. Do so using time-bandwidth products of 2 and of 10. In addition, for each choice of time-bandwidth product, compute the spectrum without zero padding the data and then after adding 10 s of zero padding to the data.

d. In a few sentences interpret the spectra and describe the rhythms present in the signal. Compare the three spectra results. Do the analyses agree or disagree? What choice of taper and zero padding do you prefer for these data?

4.3. Consider the file `Ch3-EEG-2.mat`, available at

`http://github.com/Mark-Kramer/Case-Studies-Kramer-Eden`

Load these data into MATLAB. If you have not done so, analyze these data as discussed in problem 3.4 of chapter 3. In addition, answer the following questions.

a. Plot the spectrum versus the frequency using a rectangular taper. Do so without zero padding the data and then after adding 10 s of zero padding to the data.

b. Plot the spectrum versus the frequency using a Hanning taper. Do so without zero-padding the data and then after adding 10 s of zero padding to the data.

c. Plot the spectrum versus the frequency using the multitaper method. Do so using time-bandwidth products of 2 and of 10. In addition, for each choice of time-bandwidth product, compute the spectrum without zero padding the data and then after adding 10 s of zero padding to the data.

d. Interpret the spectra and describe the rhythms present in the signal. Compare the three spectra results. Do the analyses agree or disagree? What choice of taper and zero padding do you prefer for these data?

4.4. Consider the file `Ch3-EEG-3.mat`, available at

`http://github.com/Mark-Kramer/Case-Studies-Kramer-Eden`

Load these data into MATLAB. If you have not done so, analyze these data as discussed in problem 3.5 of chapter 3. Then answer the following questions.

a. Plot the spectrum versus the frequency using a rectangular taper. Do so without zero padding the data and then after adding 10 s of zero padding to the data.

b. Plot the spectrum versus the frequency using a Hanning taper. Do so without zero padding the data, and then after adding 10 s of zero padding to the data.

c. Plot the spectrum versus the frequency using the multitaper method. Do so using time-bandwidth products of 2 and of 10. In addition, for each choice of time-bandwidth product, compute the spectrum without zero padding the data and then after adding 10 s of zero padding to the data.

d. Interpret the spectra and describe the rhythms present in the signal. Compare the three spectra results. Do the analyses agree or disagree? What choice of taper and zero padding do you prefer for these data?

4.5. Consider the file `Ch3-EEG-4.mat`, available at

`http://github.com/Mark-Kramer/Case-Studies-Kramer-Eden`

Load these data into MATLAB. If you have not done so, analyze these data as discussed in problem 3.6 of chapter 3. Also answer the following questions.

a. Plot the spectrum versus the frequency using a rectangular taper. Do so without zero padding the data and then after adding 10 s of zero padding to the data.

b. Plot the spectrum versus the frequency using a Hanning taper. Do so without zero padding the data and then after adding 10 s of zero padding to the data.

c. Plot the spectrum versus the frequency using the multitaper method. Do so using time-bandwidth products of 2 and of 10. In addition, for each choice of time-bandwidth product, compute the spectrum without zero padding the data and then after adding 10 s of zero padding to the data.

d. Interpret the spectra and describe the rhythms present in the signal. Compare the three spectra results. Do the analyses agree or disagree? What choice of taper and zero padding do you prefer for these data?

4.6. Consider the file `Ch3-EEG-5.mat`, available at

`http://github.com/Mark-Kramer/Case-Studies-Kramer-Eden`

Load these data into MATLAB. If you have not done so, analyze these data as discussed in problem 3.7 of chapter 3. Also answer the following questions.

a. Plot the spectrum versus the frequency using a rectangular taper. Do so without zero padding the data and then after adding 10 s of zero padding to the data.

b. Plot the spectrum versus the frequency using a Hanning taper. Do so without zero padding the data and then after adding 10 s of zero padding to the data.

c. Plot the spectrum versus the frequency using the multitaper method. Do so using time-bandwidth products of 2 and of 10. In addition, for each choice of time-bandwidth product, compute the spectrum without zero padding the data and then after adding 10 s of zero padding to the data.

d. Interpret the spectra and describe the rhythms present in the signal. Compare the three spectra results. Do the analyses agree or disagree? What choice of taper and zero padding do you prefer for these data?

4.7. Simulate a signal consisting of the sum of two sinusoids oscillating at 10.5 Hz and 10.8 Hz. Set $T = 1$ s and compute the spectrum. Can you resolve the two different frequencies? Zero-pad the signal by adding 19 s of zeros. Now can you resolve the two peaks? How much data would you need to simulate (i.e., how big should T be) to resolve the two frequencies? Show this in simulation.

Appendix: Multiplication and Convolution in Different Domains

We stated in this chapter the important fact that multiplication in the time domain is equivalent to convolution in the frequency domain. Mathematically, we may express this relation as,

$$FT[xw] = FT[x] \star FT[w], \tag{4.3}$$

where x and w are two time series, $FT[x]$ is the Fourier transform of x, and $X \star Y$ indicates the convolution of X and Y,

$$X \star Y[\beta] = \int_{-\infty}^{\infty} X[b] Y[\beta - b]\, db.$$

The convolution of two functions (with arguments b in this formula) is itself a function of the same argument (with symbol β in this formula). Equation (4.3) states that the Fourier transform of the element-by-element product of x and w equals the convolution of the Fourier transform of x and the Fourier transform of w. We consider here an equivalent, alternative statement: that convolution in the time domain is equivalent to multiplication in the frequency domain. Mathematically,

$$FT[x \star w] = FT[x]\, FT[w]. \tag{4.4}$$

This equation states that the Fourier transform of the convolution of x and w equals the product of the Fourier transform of x and the Fourier transform of w. To prove this relation, let's consider the Fourier transform of the convolution of x and w. We use the expression for the continuous-time Fourier transform (3.27) from chapter 3,

$$FT(x \star w[\tau]) = \int_{-\infty}^{\infty} \left(x \star w[\tau]\right) e^{-2\pi i f \tau}\, d\tau,$$

where the notation $[\tau]$ indicates that the convolution $(x \star w)$ is a function of time τ. Now, let's substitute the definition of convolution into this expression and simplify using an introduction of a second exponential expression,

$$\begin{aligned} FT(x \star w[\tau]) &= \int_{-\infty}^{\infty} \Big(\int_{-\infty}^{\infty} x[t] w[\tau - t] dt \Big) e^{-2\pi i f \tau} d\tau \\ &= \int_{-\infty}^{\infty} \int_{-\infty}^{\infty} x[t] w[\tau - t] \, dt \, e^{-2\pi i f(\tau - t)} e^{-2\pi i f t} d\tau \\ &= \int_{-\infty}^{\infty} \int_{-\infty}^{\infty} \big(x[t] e^{-2\pi i f t}\big)\big(w[\tau - t] e^{-2\pi i f(\tau - t)}\big) dt \, d\tau . \end{aligned}$$

Setting $T \equiv \tau - t$, we find

$$\begin{aligned} F(x \star w[\tau]) &= \int_{-\infty}^{\infty} \int_{-\infty}^{\infty} \big(x[t] e^{-2\pi i f t} dt\big)\big(w[T] e^{-2\pi i f(T)} dT\big) \\ &= \Big(\int_{-\infty}^{\infty} x[t] e^{-2\pi i f t} dt \Big) \Big(\int_{-\infty}^{\infty} w[T] e^{-2\pi i f(T)} dT \Big) \\ &= FT[x] \, FT[w] \end{aligned}$$

and therefore conclude that the Fourier transform of the convolution of x and w equals the element-by-element product of their Fourier transforms.

In MATLAB we may compute a simple example to illustrate this relation:

```
x = [3 4 5 6];              %Define a simple signal x,
w = [-1 0.1 -0.2 1];        %... and another simple signal w.
a=fft(conv(w,x));           %Take the FT of the convolution,
b=fft([w 0 0 0]).*fft([x 0 0 0]);%... and the product of the FTs.
```

In the first two lines, we define two simple signals; each consists of only four elements, which is enough to illustrate the relation. In the third line, we first compute the convolution of `w` and `x`, and then the Fourier transform. In the last line, we compute the Fourier transform of each variable, and then their element-by-element product. Notice that we zero-pad both variables before computing the Fourier transform in the last line. We do so to avoid computing circular correlations between the variables (i.e., wrapping around one variable when comparing it to another). Also, we make the lengths of variables `a` and `b` the same. Evaluating the statement, we find `a` equals `b`; to see this, print out both variables at the MATLAB command line.

5 Analysis of Coupled Rhythms in an Invasive Electrocorticogram

Synopsis

Data 1 s of ECoG data sampled at 500 Hz from two electrodes for 100 trials.

Goal Characterize the coupling of rhythmic activity between the two electrodes.

Tools Fourier transform, spectrum, amplitude, coherence, phase.

5.1 Introduction

5.1.1 Background

In chapters 3 and 4, we focused on field data recorded from a single electrode at the scalp (EEG) or cortical (ECoG) surface. However, typical brain voltage recordings consist of multiple electrodes. For example, the standard EEG recording consists of 21 electrodes spaced across the scalp surface, and sometimes many more [5]. The number of electrodes utilized in invasive ECoG recordings also range from a handful of contacts to over 100 implanted electrodes [10]. In this chapter, we continue our study of field data recorded from the cortical surface but now consider ECoG data recorded simultaneously from two electrodes during a task.

5.1.2 Case Study Data

We consider again the patient with epilepsy described in chapter 4. As part of her routine clinical workup before resective surgery, numerous electrodes were implanted directly on the cortical surface. The purpose of this invasive recording procedure was to monitor and localize her seizures for eventual surgical treatment. During this recording procedure, in which ECoG electrodes were implanted and recordings performed for one week, the patient volunteered to participate in an auditory task study administered by a collaborating researcher. The task required the patient to listen to individual phonemes through headphones and respond with a button click whenever she heard the phoneme “ba” (the other phonemes were different, e.g., “pa,” ”ma”). The tone presentation was repeated 100 times, and her ECoG recorded (sampling rate 500 Hz) from two cortical electrodes over the auditory brain area for 1 s.

5.1.3 Goal

Our goal is to understand the coupling between the voltage activity recorded from two brain areas during the auditory task. To do so, we compute the *cross-covariance* and *coherence* between the two electrodes. These coupling measures build upon our previous development of the autocovariance, Fourier transform, and spectrum.

5.1.4 Tools

In this chapter, we develop the cross-covariance and coherence measures. For the latter, we continue to explore and understand the Fourier transform and examine in detail the notion of phase. We also briefly discuss strategies to assess the coherence for a single trial of data.

5.2 Data Analysis

5.2.1 Visual Inspection

To access the data for this chapter, visit

`http://github.com/Mark-Kramer/Case-Studies-Kramer-Eden`

and download the file `Ch5-ECoG-1.mat`. We begin our analysis by visualizing the ECoG data; load the ECoG data into MATLAB and plot the data from the first electrode (variable `E1`) and second electrode (variable `E2`) versus time (variable `t`) for the first trial:

```
load('Ch5-ECoG-1.mat')  %Load the ECoG data.
%... and plot one trial from each electrode.
plot(t,E1(1,:), 'b', 'LineWidth', 2)
hold on
plot(t,E2(1,:), 'r', 'LineWidth', 2)
hold off
```

The results for this trial and three others are plotted in figure 5.1. Visual inspection immediately suggests a dominant rhythmic activity in each trial.

> **Q:** Approximate the dominant rhythmic activity in each electrode and trial by visual inspection of figure 5.1. A simple procedure is to count the number of peaks in each signal, then divide by the total length of the recording (in this case, 1 s). Does each electrode/trial exhibit approximately the same rhythms?

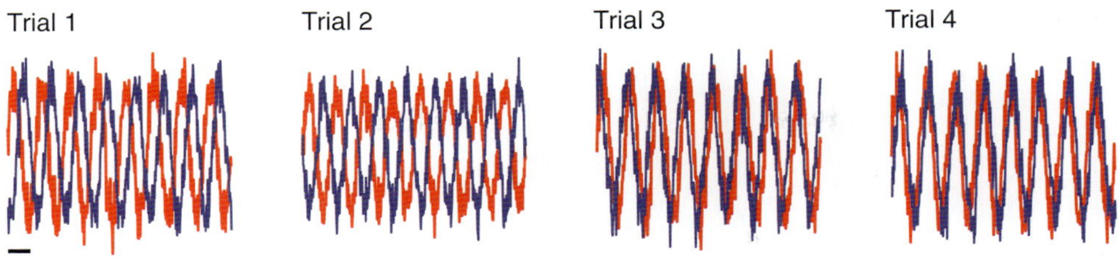

Figure 5.1
Traces of ECoG data recorded at two electrodes (*blue*, *red*) in four trials. Scale bar indicates 100 ms.

These techniques allow us to visualize the data one trial at a time. Doing so is often useful but can be time consuming, especially as the number of trials increases. Here we have 100 trials, and to visualize all of them in this way would require 100 plots. That's not so bad, but there's a better way. We can display the entire structure of the data across both time and trials as an *image*:

```
ntrials = size(E1,1);          %Define the number of trials,
imagesc(t,(1:ntrials),E1);     %... image all data,
xlabel('Time [s]')             %... and label the axes.
ylabel('Trial #')
```

The resulting image for the first electrode is shown in figure 5.2. Voltage is plotted as a function of time along the horizontal axis and trial number along the vertical axis. This allows us to visualize the voltage activity of the first electrode for all trials at once. We notice that each trial exhibits rhythmic structure, which manifests in figure 5.2 as repeating undulations of blue (low voltage), then red (high voltage) over time. We also observe variability in the alignment of these rhythms from trial to trial; from one trial to the next, the undulations appear not to align.

Q: Display an image of the activity for the second electrode and compare it to the image from the first electrode in figure 5.2. How do the two compare?

Visual inspection of the ECoG data allows us to draw some preliminary conclusions. First, the data appear to be rhythmic, with a particularly strong oscillation near 8 Hz. That's interesting but not the primary research objective. We would really like to understand whether the activity at the two electrodes is related. Many techniques exist to approach this problem [12], but let's begin with the most basic: visual inspection. By examining figure 5.1 we can attempt to deduce whether a consistent relation exists between the two ECoG signals

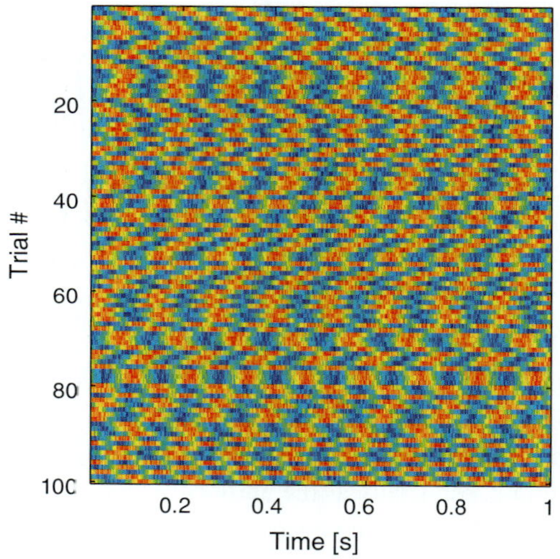

Figure 5.2
Image of ECoG data from first electrode.

across trials. We notice in the first two trials that the ECoG activity from the two electrodes appears nearly out of phase (i.e., when the blue curve is near a peak, the red curve is near a trough). However, for the next two trials, activity from the two electrodes nearly overlaps. From this initial visual inspection of four trials, it's difficult to conclude whether the ECoG activity at the two electrodes is interrelated; both electrodes display rhythmic activity across all trials, but the relation between these rhythms appears to change across trials: sometimes the activities overlap, and sometimes not.

> **Q:** Repeat this analysis by examining additional trials, and by inspecting the activity images for each electrode. What conclusions can you make about the relations between the ECoG activity at the two electrodes? Are they related? Are they not related?

Although visual inspection is a useful initial tool for analyzing data, assessing the relations between two electrodes across multiple trials is a difficult task. To go further, we employ a new data analysis tool that builds from our experience with the Fourier transform: the coherence.

5.2.2 Autocovariance and Cross-covariance

In chapter 3, we defined and applied the autocovariance to a single time series, equation (3.3), and found that this measure helped reveal dependent structure in the data. We could, of course, apply the autocovariance to each ECoG time series considered here. Let's do so, with a small update to the autocovariance formula that utilizes the trial structure of these data. We define the *trial-averaged autocovariance*[1] as

$$r_{xx}[L] = \frac{1}{K}\sum_{k=1}^{K}\frac{1}{N}\sum_{n=1}^{N-L}(x_{n+L,k} - \bar{x}_k)(x_{n,k} - \bar{x}_k)\,, \tag{5.1}$$

where $x_{n,k}$ indicates the data at time index n and trial k, and $\bar{x}_k$ is the mean value of x for trial k. Notice that we include a new term, $\frac{1}{K}\sum_{k=1}^{K}$, which instructs us to sum over all trials the autocovariance computed for each trial and then divide by the total number of trials K. To compute and display the trial-averaged autocovariance for the first electrode in MATLAB,

```
load('Ch5-ECoG-1.mat')   %Load the ECoG data.
dt = t(2)-t(1);          %Define the sampling interval.
K = size(E1,1);          %Define the no. of trials.
nlags = 100;             %Define the max no. of +/- lags.
ac = zeros(1,2*nlags+1);%Declare empty vector for autocov.

for k=1:K                %For each trial,
 x = E1(k,:)-mean(E1(k,:));            %...subtract the mean,
 [ac0,lags]=xcorr(x,100,'biased');  %... compute autocovar,
 ac = ac + ac0/K;        %...and add to total, scaled by 1/K.
end
plot(lags*dt,ac)         %Plot autocovar vs lags in time.
xlabel('Lag [s]')        %Label the axes.
ylabel('Autocovariance');
```

Q: In using the function `xcorr`, we set the third input to `'biased'`. Why do we compute the biased autocovariance? *Hint:* See chapter 3 for a detailed discussion.

Q: Consider the results for the trial-averaged autocovariance plotted in figure 5.3. What do these results suggest about the rhythmic structure in these data?

A: The trial-averaged autocovariance in figure 5.3 helps characterize the rhythmic activity at this electrode. Visual inspection of this figure reveals three large positive peaks. The largest peak occurs at a lag of 0 s, as expected; the signal matches itself

1. We could instead write the trial-averaged *sample* autocovariance because this equation uses the observed data to estimate the theoretical covariance that we would see if we kept repeating this experiment. However, this distinction is not essential to the discussion here.

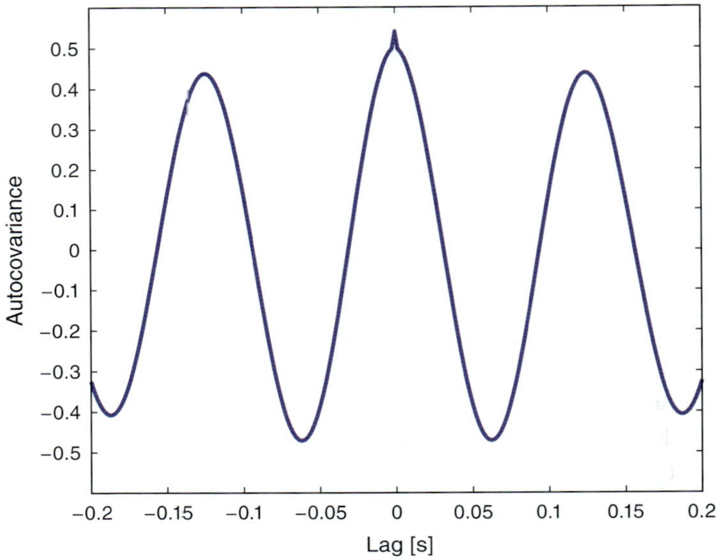

Figure 5.3
Trial-averaged autocovariance of ECoG data recorded at one electrode.

at zero lag. The two other peaks occur at lags of approximately ±0.125 s. These peaks reveal that the data, and a version of the data shifted by +0.125 s or −0.125 s, are a good match. Notice that a shift of ±0.125 s is consistent with periodic activity of approximate frequency $1/(0.125\,\text{s}) = 8$ Hz. For example, imagine a sinusoid of frequency 8 Hz; if we shift the sinusoid by its period (0.125 s) and compare it to the original (unshifted) sinusoid, the match will be excellent. Our data are more complicated than a simple sinusoid, but our visual inspection of the voltage traces (figure 5.1) did reveal a dominant 8 Hz rhythm consistent with these autocovariance results.

Q: Repeat the analysis to compute the trial-averaged autocovariance for the second electrode. What do you find? How do the trial-averaged autocovariances for the two electrodes compare?

The trial-averaged autocovariance results for each electrode are interesting, but our primary scientific question for these data is whether dependent structure exists *between* the ECoG activity recorded from the two electrodes. In other words, are the time series recorded from the two electrodes coupled? Many tools exist to characterize coupling between time series, and in this chapter we focus on two such tools. The first is the cross-covariance, $r_{xy}[L]$, an extension of the autocovariance to include two time series, defined as,

$$r_{xy}[L] = \frac{1}{N} \sum_{n=1}^{N-L} (x_{n+L} - \bar{x})(y_n - \bar{y}) , \tag{5.2}$$

where x and y are two time series with time index n. Notice what we've done; compared to the autocovariance, defined in (3.3), the cross-covariance formula simply replaces the x's in the second term in parentheses with y's.

The intuition for understanding the cross-covariance is similar to that for the autocovariance (see chapter 3). To calculate the cross-covariance, we multiply y with x shifted in time by lag L (figure 5.4). The cross-covariance is large at lag L if the two shifted time series x and y match. If we're interested in determining the coupling between x and y, finding these matches could be particularly useful. To illustrate an application of the cross-covariance, let's compute it between the two electrodes during the first trial of the ECoG data:

```
load('Ch5-ECoG-1.mat')                    %Load the ECoG data.
dt = t(2)-t(1);                           %Define sampling interval.
x = E1(1,:) - mean(E1(1,:));              %Define one time series,
y = E2(1,:) - mean(E2(1,:));              %... and another.
[xc,lags]=xcorr(x,y,100,'biased');        %Compute trial 1 cross cov.
```

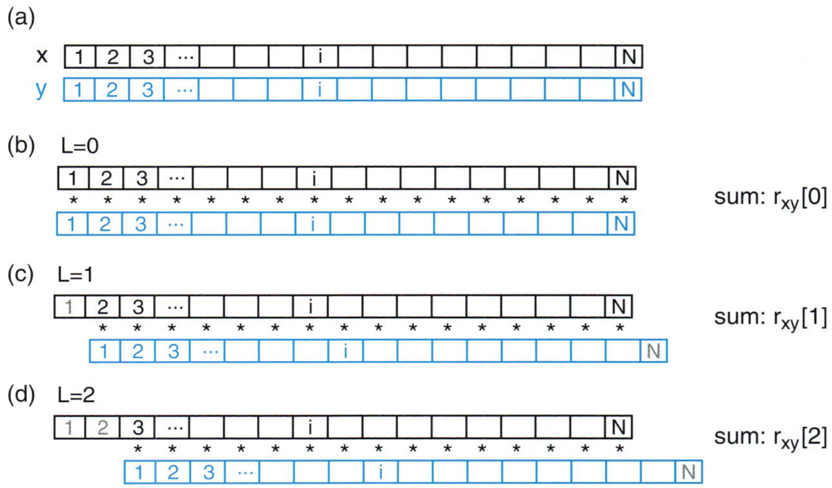

Figure 5.4

Cartoon representation of cross-covariance between two time series x and y. Data x and y are visualized as one-dimensional vectors, x in black and y in blue. The cross-covariance at (*b*) lag 0, (*c*) lag 1, and (*d*) lag 2 requires different alignments between the two vectors. To compute the cross-covariance at each lag, we multiply the overlapping elements of the two vectors, and sum the product. Non-overlapping elements are not included in the computation.

```
plot(lags*dt,xc)                    %Plot cov vs lags in time.
xlabel('Lag [s]')                   %... with axes labeled.
ylabel('Cross-covariance');
```

We subtract the mean from each electrode (the third and fourth lines) before computing the cross-covariance (in the fifth line) using the MATLAB function `xcorr`. In this case, we supply the `xcorr` function with four inputs, beginning with the two time series, `x` and `y`, and including the maximum number of lags to consider (`100`), and the keyword specifying calculation of the biased cross-covariance.

> **Q:** Examine the cross-covariance between the ECoG data from the two electrodes in the first trial (figure 5.5a). What do you observe? At what lags are the largest and smallest values of the cross-covariance? How do these results compare to the trial-averaged autocovariance? How do these results compare to the voltage traces from each electrode in the first trial (figure 5.5b)?

Like the trial-averaged autocovariance for a single electrode (figure 5.3), the cross-covariance between the two ECoG electrodes in the first trial reveals periodic variations (figure 5.5a). To understand the structure of this cross-covariance, let's return to the voltage traces from the two electrodes in this trial (figure 5.5b). The largest peak in the cross-covariance occurs near a lag of 0.04 s. Now, imagine shifting the blue time series

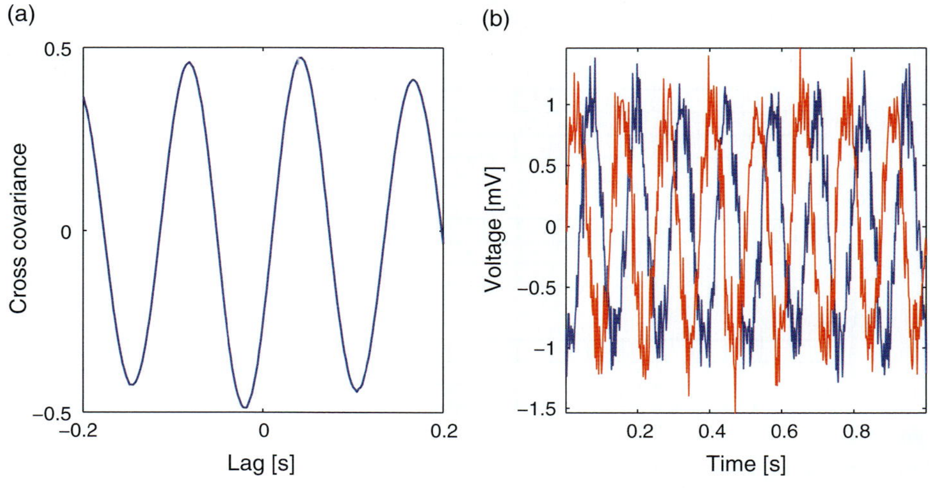

Figure 5.5

Cross-covariance between the two ECoG electrodes for the first trial. (*a*) Rhythmic structure is apparent. (*b*) Voltage data (with mean subtracted) for electrode 1 (*blue*) and electrode 2 (*red*).

(corresponding to electrode 1) in figure 5.5b by 0.04 s to the left. Doing so, we find that the red and blue traces approximately match; at this lag, when one time series is positive, so is the other, and when one time series is negative, so is the other. Because of this strong match, the cross-covariance is large; the sum in (5.2) at this lag involves many positive terms, so $r_{xy}[L]$ is a positive number. The largest *trough* in the cross-covariance occurs near a lag of -0.02 s. To understand this feature, imagine shifting the blue time series in figure 5.5b by 0.02 s to the right. After this shift, the red and blue time series match, but in a different way; when one voltage trace is positive, the other is negative, and vice versa. Therefore, the sum in (5.2) at this lag involves many negative terms, so $r_{xy}[L]$ is a negative number.

Q: Continue this exercise of comparing the cross-covariance with the voltage traces in figure 5.5. At what lags is the cross-covariance near zero? Can you explain these points in terms of shifted versions of the ECoG traces? Can you explain the repeated appearance of peaks (and troughs) at positive and negative lags in terms of shifted versions of the ECoG traces?

Let's also define the *trial-averaged cross-covariance*. The formula is similar to the trial-averaged autocovariance in (5.1):

$$r_{xy}[L] = \frac{1}{K}\sum_{k=1}^{K}\frac{1}{N}\sum_{n=1}^{N-L}(x_{n+L,k}-\bar{x}_k)(y_{n,k}-\bar{y}_k)\,. \tag{5.3}$$

Notice that compared to the trial-averaged autocovariance in (5.1), we have replaced the x's in the last term with y's to compute the trial-averaged cross-covariance in (5.3). To implement the trial-averaged cross-covariance in MATLAB,

```
load('Ch5-ECoG-1.mat')          %Load the ECoG data.
K = size(E1,1);                 %Define the number of trials.
dt = t(2)-t(1);                 %Define the sampling interval.
maxlags = 100;                  %Define variable with max lags.

XC = zeros(K,2*maxlags+1);      %Create variable to store cross cov.
for k=1:K                       %For each trial ...
    x=E1(k,:)-mean(E1(k,:));%...get data from one electrode,
    y=E2(k,:)-mean(E2(k,:));%...and the other electrode,
    [xc0]=xcorr(x,y,maxlags,'biased');%...compute cross cov,
    XC(k,:) = xc0;              %...and store result.
end
XC = mean(XC,1);                %Average cross cov over trials.
```

```
%Plot trial-averaged cross cov vs lags in units of time,
plot((-maxlags:maxlags)*dt,XC)
xlabel('Lag [s]')              %... with axes labeled.
ylabel('Trial-Averaged Cross-Covariance');
```

The implementation of the trial-averaged cross-covariance is similar to the implementation of the single-trial cross-covariance. The main difference is the inclusion here of the `for` statement, which we use to compute and store the cross-covariance of each trial. We then average these results across trials using the `mean` command. The trial-averaged cross-covariance for the ECoG data is shown in figure 5.6a.

Q: Compare the trial-averaged cross-covariance in figure 5.6a to the example single-trial cross-covariances in figure 5.6b. What differences and similarities do you notice between the two cross-covariances?

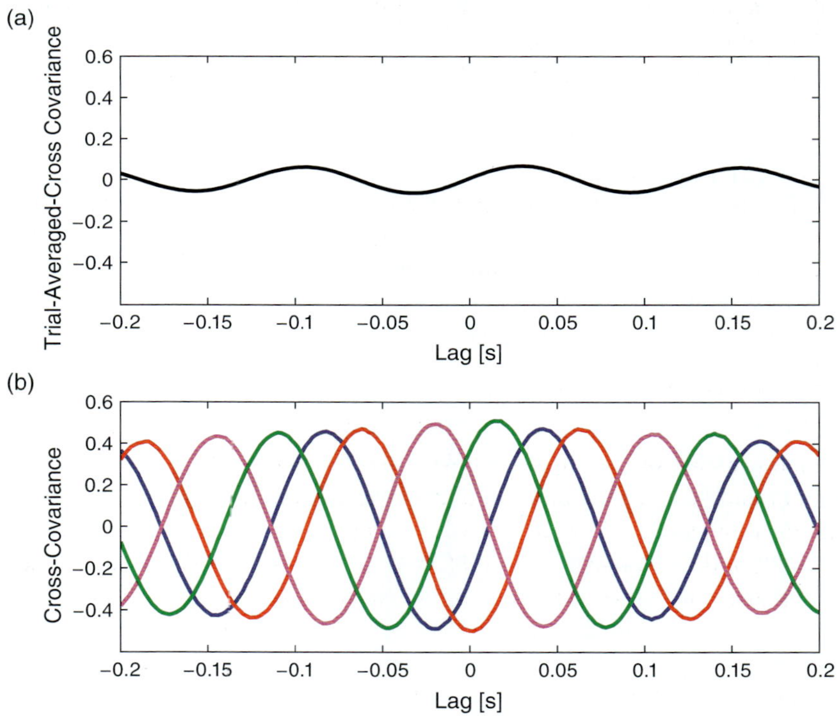

Figure 5.6

(*a*) Trial-averaged cross-covariance, and (*b*) four example single-trial cross-covariances between the two electrodes.

A: Perhaps the most striking difference between the two cross-covariances is their magnitude; the single-trial cross-covariances are much larger—approximately an order of magnitude—than the trial-averaged cross-covariance. To understand why this difference occurs, consider the impact of averaging the four example single-trial cross-covariances in figure 5.6b. At each lag, we find both positive and negative cross-covariance values. We therefore expect that, upon averaging these values across trials, we will obtain a value near zero at each lag. In fact, that's just what we find in the trial-averaged cross-covariance. Because the single-trial cross-covariance functions lack alignment across trials, the averaging procedure acts to cancel out the individual (large) fluctuations of each single-trial cross-covariance.

We may therefore conclude the following. At the single-trial level we find strong cross-covariance that is periodic with period near 0.125 s (examples in figure 5.6b). However, we find much weaker trial-averaged cross-covariance; the cross-covariance structure that exists at the single-trial level does not persist when averaged across trials.

Why are the prominent cross-covariance features in the single-trial analysis lost in the trial-averaged cross-covariance? We discuss this issue in more detail in the chapter summary.

5.2.3 Trial-Averaged Spectrum

One goal of this chapter is to characterize the relations (if any) between the data recorded at the two ECoG electrodes. To do so, let's review a vital tool in this characterization, the Fourier transform. We defined in chapter 3 the Fourier transform of a signal x; we repeat that definition here:

$$X_j = \sum_{n=1}^{N} x_n \exp(-2\pi i f_j t_n)\,. \tag{5.4}$$

Recall that x_n are the data evaluated at time index n. For the ECoG data of interest here, we have 1 s of data sampled at 500 Hz; therefore n ranges from 1 to $N = 500$, and $t_n = \Delta\, n$ denotes the discrete time steps, where Δ is the sampling interval. The discrete frequencies are $f_j = j/T$, where $j = \{-N/2+1, -N/2+2, \ldots, N/2-1, N/2\}$. Replacing the expressions for f_j and t_n with their definitions and simplifying, we can rewrite (5.4) as

$$X_j = \sum_{n=1}^{N} x_n \exp(\frac{-2\pi i}{N} j n)\,. \tag{5.5}$$

In general, X_j can be a complex quantity (i.e., the Fourier transform of x_n can have both real and imaginary parts). We can therefore think of X_j as residing in the two-dimensional *complex plane* (figure 5.7). As you may remember from a geometry or calculus class, we can represent a point in the plane using another coordinate system: polar coordinates. In polar coordinates, we imagine connecting each point to the origin. The resulting line has a length, called the radius or *amplitude*, and forms an angle with the real axis, called the *phase*. Like the real and complex parts, the amplitude and phase uniquely specify each point in the complex plane.[2] These two coordinate systems are shown for an example point in the complex plane in figure 5.7.

Using polar coordinates, we can then express the complex quantity X_j as

$$X_j = A_j \exp(i\phi_j)\,, \tag{5.6}$$

where A_j is the amplitude and ϕ_j is the phase at frequency index j. Notice that both the amplitude and phase are functions of frequency. Recall that to compute the spectrum, we multiply the Fourier transform of the data by its complex conjugate and scale the result (see chapter 3). The spectrum of x_n then becomes

$$S_{xx,j} = \frac{2\Delta^2}{T} X_j X_j^* \tag{5.7}$$

$$= \frac{2\Delta^2}{T} \big(A_j \exp(i\phi_j)\big)\big(A_j \exp(-i\phi_j)\big)\,, \tag{5.8}$$

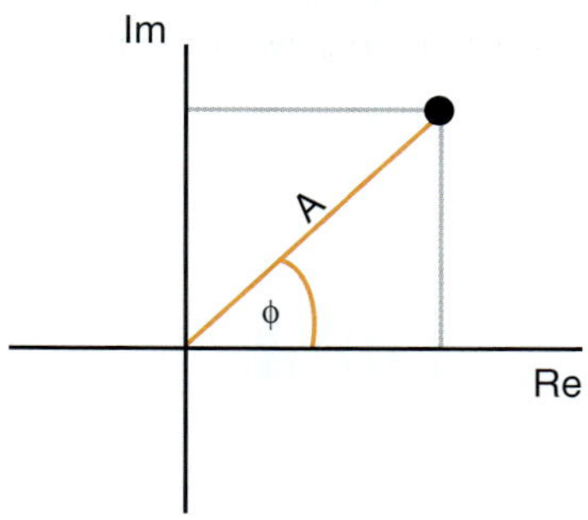

Figure 5.7
Points in the complex plane can be specified in two coordinate systems: Cartesian coordinates (*gray*) or polar coordinates (*orange*). The complex plane contains the real part (horizontal axis) and imaginary part (vertical axis) of every point.

2. This statement is mostly correct. Can you think of the exception? *Hint:* Think small.

where, to compute the complex conjugate in the second term, we replace i with $-i$. The last expression simplifies rather nicely:

$$\begin{aligned} S_{xx,j} &= \frac{2\Delta^2}{T} A_j^2 \exp(i\phi_j - i\phi_j) \\ &= \frac{2\Delta^2}{T} A_j^2 \exp(0) \\ &= \frac{2\Delta^2}{T} A_j^2 . \end{aligned} \tag{5.9}$$

This expression provides a new, and perhaps more direct, interpretation of the spectrum as proportional to the squared amplitude of the point X_j in the complex plane. We can extend this simplified expression in one additional way to make explicit the trial structure of the ECoG data analyzed here. Because we possess multiple trials, and we assume that each trial represents an instantiation of the same underlying process, we average the spectra across trials to compute the *trial-averaged spectrum*,

$$< S_{xx,j} >= \frac{2\Delta^2}{T} \frac{1}{K} \sum_{k=1}^{K} A_{j,k}^2 , \tag{5.10}$$

where k indicates the trial number, K the total number of trials, and $A_{j,k}$ the amplitude of the signal at frequency index j and trial index k. Notice how we implement the trial averaging: we simply average the squared amplitude at frequency index j across the K trials. We use the angular brackets ($< >$) to denote that the spectrum ($S_{xx,j}$) has been averaged across trials. We can compute the trial-averaged spectrum in MATLAB:

```
load('Ch5-ECoG-1.mat')          %Load the ECoG data.
K = size(E1,1);                 %Define the number of trials.
N = size(E1,2);                 %Define the number of time indices.
dt  = t(2)-t(1);                %Define the sampling interval.
T   = t(end);                   %Define the duration of data.

Sxx = zeros(K,N);           %Create variable to store each spectrum.
for k=1:K                       %For each trial,
    x = E1(k,:);                %... get the data,
    xf  = fft(x-mean(x));       %... compute Fourier transform,
    Sxx(k,:) = 2*dt^2/T *(xf.*conj(xf));%... compute spectrum.
end
Sxx = Sxx(:,1:N/2+1);           %Ignore negative frequencies.
Sxx = mean(Sxx,1);              %Average spectra over trials.
```

```
df = 1/max(T);                  %Define frequency resolution,
fNQ = 1/dt/2;                   %... and Nyquist frequency.
faxis = (0:df:fNQ);             %... to construct frequency axis.

plot(faxis, 10*log10(Sxx))     %Plot spectrum in decibels vs
xlim([0 100]);                  %... frequency, in select frequency
ylim([-50 0])                   %... range, in select decibel
xlabel('Frequency [Hz]')        %... range, with axes labeled.
ylabel('Power [ mV^2/Hz]')
```

Q: Are the terms *frequency resolution, Nyquist frequency*, and *decibel* familiar to you? Can you define each in words and equations?

A: If not, we recommend reviewing the case study in chapter 3.

The resulting trial-averaged spectrum is shown in figure 5.8. Compared to the example spectrum from a single trial, the variability is greatly reduced. By reducing the variability in this way, interesting structure in the data may become more apparent.

Q: Upon examining the trial-averaged spectrum from one electrode, what additional conclusions can you now make about the data beyond visual inspection of the voltage traces? Repeat this computation of the trial-averaged spectrum for the second electrode. What do you find? *Hint:* The 8 Hz peak is obvious and consistent with our visual inspection of the data. Do you notice any other (smaller) peaks?

5.2.4 Introduction to the Coherence

Coherence is a measure of association between two time series. Briefly, two signals are coherent at some frequency if there exists a constant phase relation between them at this frequency. To compute the coherence, we use the simplified expression for the spectrum (5.9) and an additional term, the *cross-spectrum*. Consider two signals $x_{n,k}$ and $y_{n,k}$, with time index n and trial index k. These signals have corresponding Fourier transforms $X_{j,k}$ and $Y_{j,k}$. Then the trial-averaged cross-spectrum between these two signals is

$$< S_{xy,j} >= \frac{2\Delta^2}{T}\frac{1}{K}\sum_{k=1}^{K} X_{j,k} Y^*_{j,k} , \tag{5.11}$$

where compared to (5.7) we replace X^*_j with Y^*_j and include the average over the trial index k. Let's modify and clean up this expression by using polar coordinates. We first define

$$Y_{j,k} = B_{j,k} \exp(i\,\theta_{j,k}) , \tag{5.12}$$

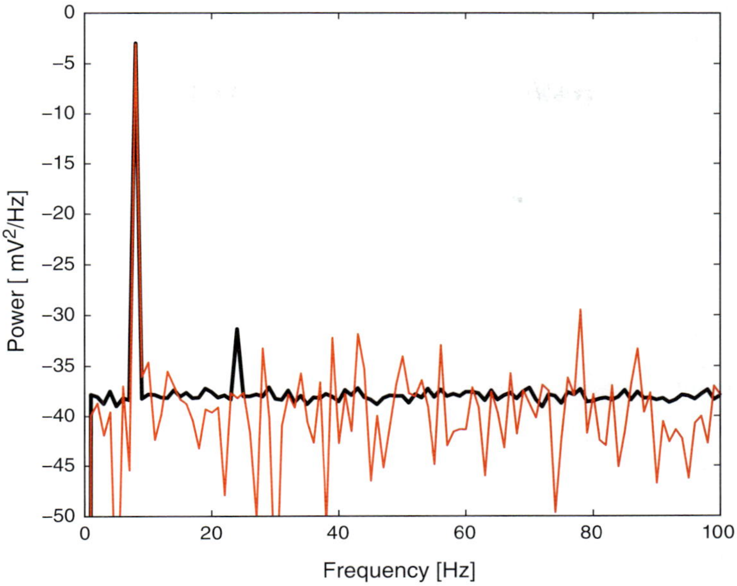

Figure 5.8

Trial-averaged spectrum reduces the variation of the power. Compare the trial-averaged spectrum (*black*) to an example spectrum from an individual trial (*red*).

where $B_{j,k}$ is the amplitude and $\theta_{j,k}$ is the phase at frequency index j and trial index k for the signal $y_{n,k}$. A similar expression exists for $X_{j,k}$, with amplitude $A_{j,k}$ and phase $\phi_{j,k}$. Then replacing $X_{j,k}$ and $Y^*_{j,k}$ in (5.11) with their polar coordinate expressions, we find

$$< S_{xy,j} >= \frac{2\Delta^2}{T}\frac{1}{K}\sum_{k=1}^{K} A_{j,k}B_{j,k}\exp\left(i\Phi_{j,k}\right), \tag{5.13}$$

where we have defined the *phase difference* between the two signals as $\Phi_{j,k} = \phi_{j,k} - \theta_{j,k}$. Equation (5.13) is the trial-averaged cross-spectrum of the two signals $x_{n,k}$ and $y_{n,k}$. We note that the trial-averaged cross-spectrum ($< S_{xy,j} >$) can be complex (i.e., may have nonzero real and imaginary parts).

At last we can define the coherence,

$$\kappa_{xy,j} = \frac{|< S_{xy,j} >|}{\sqrt{< S_{xx,j} >}\sqrt{< S_{yy,j} >}}, \tag{5.14}$$

where $|< S_{xy,j} >|$ indicates the magnitude of the trial-averaged cross-spectrum. In words, the coherence is the magnitude of the trial-averaged cross-spectrum between the two signals at frequency index j divided by the magnitude of the trial-averaged spectrum of each signal at frequency index j.

To further our understanding of the mathematical expression of the coherence in (5.14), let's replace the trial-averaged spectra in the numerator and denominator with their corresponding expressions in polar coordinates:

$$\kappa_{xy,j} = \frac{\left| \sum_{k=1}^{K} A_{j,k} B_{j,k} \exp\left(i\Phi_{j,k}\right) \right|}{\sqrt{\sum_{k=1}^{K} A_{j,k}^2} \sqrt{\sum_{m=1}^{K} B_{j,m}^2}}. \tag{5.15}$$

The expression in (5.15) is complicated. So, to gain some intuition for the behavior of $\kappa_{xy,j}$, let's make the simplifying assumption that at each frequency the amplitude is identical for both signals and all trials, that is, $A_{j,k} = B_{j,k} = C_j$. In using only the expression C_j for the amplitude, we've eliminated the trial index k and only preserved the frequency index j. With this simplifying assumption, the expression for the coherence (5.15) becomes,

$$\begin{aligned} \kappa_{xy,j} &= \frac{\left| \sum_{k=1}^{K} C_j C_j \exp\left(i\Phi_{j,k}\right) \right|}{\sqrt{\sum_{k=1}^{K} C_j^2} \sqrt{\sum_{m=1}^{K} C_j^2}} \\ &= \frac{C_j^2}{C_j^2} \frac{\left| \sum_{k=1}^{K} \exp\left(i\Phi_{j,k}\right) \right|}{\sqrt{\sum_{k=1}^{K} 1} \sqrt{\sum_{m=1}^{K} 1}} \\ &= \frac{1}{K} \left| \sum_{k=1}^{K} \exp\left(i\Phi_{j,k}\right) \right|. \end{aligned} \tag{5.16}$$

Under the simplifying assumption that the amplitude is identical at each frequency for both signals and all trials, the coherence simplifies to (5.16). In this special case, the expression for the coherence only involves the phase difference between the two signals averaged across trials; the amplitudes in the numerator and denominator have canceled out.

Now, let's interpret the simplified expression in (5.16). To do so, we consider two scenarios.

Scenario 1: Phases Align across Trials. We assume that at a specific frequency index j, the two signals possess a *constant* phase difference across trials. Under this assumption, the phase difference in the first trial ($\Phi_{j,1}$) equals the phase difference in the second trial ($\Phi_{j,2}$), which equals the phase difference in the third trial ($\Phi_{j,3}$), and so on for all trials. To denote this equivalence in the phase difference across trials, let's replace the symbol for the phase difference $\Phi_{j,k}$ with $\Phi_{j,0}$; here, we have replaced the subscript k with the subscript 0 to remind ourselves that the phase difference does not depend upon the choice of trial. Now consider the expression

$$\exp\left(i\Phi_{j,0}\right).$$

This term defines a point in the complex plane with amplitude 1, which we can visualize as a vector leaving the origin at angle $\Phi_{j,0}$ to the real axis (figure 5.9a). The summation of these terms across trials then becomes

$$\sum_{k=1}^{K}\exp\left(i\Phi_{j,0}\right).$$

This expression defines a sum of vectors in the complex plane, each of radius 1. Because the phase difference is the same for each trial, these vectors point in the same direction for each trial (figure 5.9a). Then by summing up these vectors end to end across trials, we produce a long vector in the complex plane that terminates far from the origin (figure 5.9a).

Q: How long is the summed vector in this case?

A: We add K vectors (one for each trial) each of length 1, and each pointing in the same direction ($\Phi_{j,0}$). So the total length of the vector (i.e., the total distance from the origin to the termination point of the summed vector) is K.

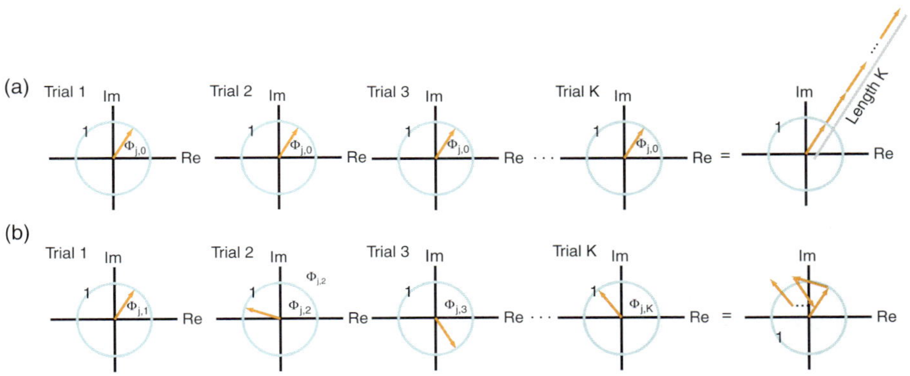

Figure 5.9

Cartoon illustration of the complex plane for two coherence scenarios. (*a*) For each trial, phase difference is the same, and summed vector (last column) terminates far from the origin. (*b*) For each trial, phase difference is random, and summed vector terminates near origin. Blue circles indicate radius 1.

The coherence (5.16) is this vector length, divided by K, so we conclude in this case that

$$\kappa_{xy,j} = 1,$$

which indicates strong coherence between the two signals. The strong coherence in this case results from the constant phase relation between the two signals across trials at frequency index j.

Q: Does the conclusion $\kappa_{xy,j} = 1$ depend upon the value of the phase difference $\Phi_{j,0}$? For example, does this result require that the phase difference between the two signals in each trial ($\Phi_{j,0}$) equal 0, or $\pi/4$, or π?

Scenario 2: Phases Are Random across Trials. As a second scenario, consider another specific frequency j in which the two signals have a *random* phase difference in each trial. In this case, the phase difference can assume any value between 0 and 2π for each trial. To visualize this, let's imagine the phase differences in the complex plane (figure 5.9b); in this scenario, the vectors point in different (random) directions from trial to trial.

Q: Consider the sum of these vectors end to end in the complex plane. What is the approximate length of this summed vector across trials?

A: We expect the length of this vector to be small. Because the angles lack organization from trial to trial, the vectors are equally likely to point in any direction. Therefore, when we sum these vectors across trials, the length fails to accumulate in any particular direction (figure 5.9b).

Under the simplifying assumption that the amplitude is identical at this frequency for both signals and all trials, the coherence (5.16) is this summed vector length, divided by K. Our visual inspection of figure 5.9b suggests that this summed vector length will be small. Therefore, for this scenario we conclude that

$$\kappa_{xy,j} \approx 0,$$

which indicates weak coherence between the two signals. The weak coherence in this case results from the random phase relation over trials between the two signals.

Summary of Coherence. These two examples illustrate in simplified scenarios the behavior of the coherence. To summarize, the coherence (5.14) is a measure of the relation between x and y at the same frequency. The coherence ranges between 0 and 1:

$$0 \leq \kappa_{xy,j} \leq 1,$$

in which 0 indicates no coherence between signals x and y at frequency index j, and 1 indicates strong coherence between signals x and y at frequency index j.

> The coherence is a measure of the phase consistency between two signals at frequency index j across trials.

We note that because computing the coherence requires the Fourier transform, the notions of frequency resolution and Nyquist frequency are identical to those described for the spectrum. In other words, the frequency resolution of the coherence is $1/T$, and the Nyquist frequency is half of the sampling frequency; see chapter 3 for details.

> **Q:** What are the units of the coherence? *Hint:* Consider (5.14) and the units of the terms in the numerator and denominator. You should find that the coherence is unitless.

Cross-Covariance and Cross-Spectrum. Although we defined the cross-spectrum in (5.11) and used it to define the coherence in (5.14), the cross-spectrum may appear somewhat unmotivated. Fortunately, there is additional insight to be gained. We showed in appendix A of chapter 3 that the spectrum is the Fourier transform of the autocovariance. Conceptually, the spectrum and autocovariance provide a frequency domain and time domain measure of a signal's rhythms, respectively. In the same way, the cross-spectrum and cross-covariance are partners.

> The cross-spectrum is the Fourier transform of the cross-covariance.

The cross-spectrum and cross-covariance form a Fourier transform pair. The cross-spectrum is a frequency domain measure of coupling, while the cross-covariance is a time domain measure of coupling. To move back and forth between these two measures, we use the Fourier transform. In practice, we rarely examine the cross-spectrum directly; it's a complex quantity and so requires two dimensions (i.e., the complex plane) to visualize. However, the cross-spectrum is fundamental to the coherence, so in that sense it's an important actor in the analysis.

Computing the Coherence. With that introduction, we are now equipped to compute the coherence. We expect the coherence to reveal the frequencies at which the two ECoG signals exhibit a constant phase relation across trials.

Q: Before we compute the coherence, hypothesize whether you expect to observe coherence between the two ECoG signals. If so, at what frequencies? Your hypothesis should be based on the previous visual analysis and spectral analysis of these data (see, for example, figures 5.1 and 5.8).

Q: To plot the coherence versus frequency, we must identify the frequency resolution and Nyquist frequency appropriate for the analysis of the ECoG data. What are they?

There are a variety of alternatives to compute the coherence. To start, let's compute the coherence by hand. The reason for doing so is that we can implement the preceding mathematical expressions and in that way gain more understanding of their features. Here's the MATLAB code:

```
load('Ch5-ECoG-1.mat')   %Load the ECoG data.
K = size(E1,1);          %Define the number of trials.
N = size(E1,2);          %Define the number of indices per trial.
dt = t(2)-t(1);          %Define the sampling interval.
T  = t(end);             %Define the duration of data.

Sxx = zeros(K,N);        %Create variables to save the spectra,
Syy = zeros(K,N);
Sxy = zeros(K,N);
for k=1:K                %... and compute spectra for each trial.
    x=E1(k,:)-mean(E1(k,:));
    y=E2(k,:)-mean(E2(k,:));
    Sxx(k,:) = 2*dt^2/T * (fft(x) .* conj(fft(x)));
    Syy(k,:) = 2*dt^2/T * (fft(y) .* conj(fft(y)));
    Sxy(k,:) = 2*dt^2/T * (fft(x) .* conj(fft(y)));
end

Sxx = Sxx(:,1:N/2+1);    %Ignore negative frequencies.
Syy = Syy(:,1:N/2+1);
Sxy = Sxy(:,1:N/2+1);

Sxx = mean(Sxx,1);       %Average the spectra across trials.
Syy = mean(Syy,1);
Sxy = mean(Sxy,1);       %... and compute the coherence.
cohr = abs(Sxy) ./ (sqrt(Sxx) .* sqrt(Syy));
```

```
df = 1/max(T);               %Determine the frequency resolution.
fNQ = 1/dt/2;                %Determine the Nyquist frequency,
faxis = (0:df:fNQ);          %... and construct frequency axis.

plot(faxis, cohr);           %Plot coherence vs frequency,
xlim([0 50])                 %... in chosen frequency range,
ylim([0 1])
xlabel('Frequency [Hz]')%... with axes labeled.
ylabel('Coherence')
```

Q: That's quite a bit of code. Look through it line by line, and confirm that each step makes sense. Can you identify the calculation of the cross-spectrum? of the trial averaging?

Q: Consider the coherence between the two ECoG electrodes in figure 5.10. At what frequencies do strong coherences appear? How do these frequencies compare to the trial-averaged spectra, shown for one electrode in figure 5.8?

A: The coherence measures the phase consistency at a chosen frequency between two signals across trials. For the ECoG data, both electrodes possess trial-averaged spectra with similar features: a large peak near 8 Hz and a smaller peak near 24 Hz (see the trial-averaged spectrum for one electrode in figure 5.8). However, the coherence between the two ECoG signals reveals a peak only at 24 Hz (figure 5.10). We conclude that the two ECoG signals both exhibit a dominant oscillation at 8 Hz, yet this rhythm is not coherent across trials; only the smaller-amplitude rhythm at 24 Hz is coherent between the two electrodes.

5.2.5 Visualizing the Phase Difference across Trials

The coherence results suggest for the two ECoG recordings a constant phase relation across trials at 24 Hz and a random phase relation across trials at 8 Hz. To further explore these relations, let's visualize the distribution of phase differences at the two frequencies, as implemented in the following MATLAB code:

```
load('Ch5-ECoG-1.mat')  %Load the ECoG data.
K = size(E1,1);         %Define the number of trials.
N = size(E1,2);         %Define the number of indices per trial.
dt = t(2)-t(1);         %Define the sampling interval.
```

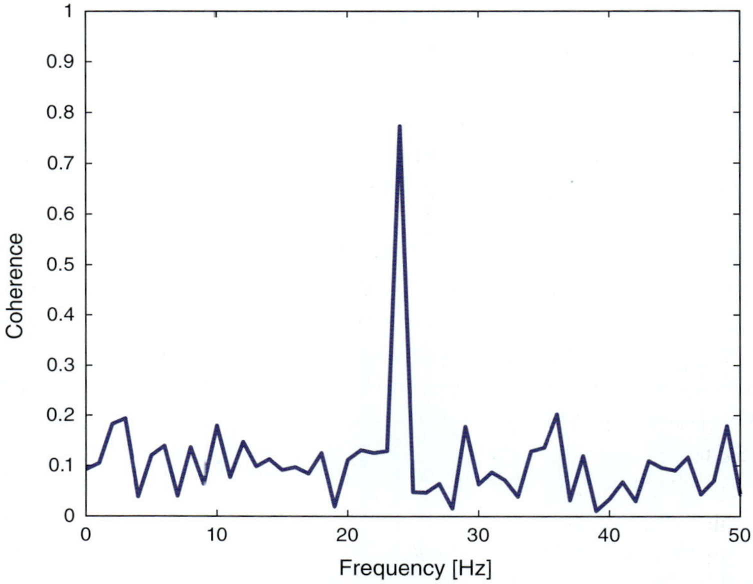

Figure 5.10
Coherence between the two ECoG signals.

```
T   = t(end);               %Define the duration of data.
df  = 1/max(T);             %Determine the frequency resolution.
fNQ = 1/dt/2;               %Determine the Nyquist frequency,
faxis = (0:df:fNQ);         %... and construct frequency axis.

j8 = find(faxis == 8);  %Determine index j for frequency 8 Hz.
j24= find(faxis == 24); %Determine index j for frequency 24 Hz.

phi8=zeros(K,1);            %Variables to hold phase differences.
phi24=zeros(K,1);

for k=1:K                   %For each trial, compute cross spectrum,
    Sxy = fft(E1(k,:)).*conj(fft(E2(k,:)));
    phi8(k)  = angle(Sxy(j8));  %... and the phases.
    phi24(k) = angle(Sxy(j24));
end

subplot(1,2,1)              %Display the phase differences.
rose(phi8);    title('\Phi at 8 Hz')
```

```
subplot(1,2,2)
rose(phi24);  title('\Phi at 24 Hz')
```

Again, we're encountering quite a bit of MATLAB code. Fortunately, large chunks of this code are familiar. We begin by defining useful quantities, like the number of trials (`K`), the number of indices per trial (`N`), and the frequency axis (`faxis`). Then, within the frequency axis variable (`faxis`), we use the `find` function to identify the indices corresponding to a frequency of 8 Hz and a frequency of 24 Hz. For each trial, we then compute the cross-spectrum (`Sxy`). The cross-spectrum is a complex quantity at each frequency, and we identify the angle in the complex plane corresponding to the frequencies 8 Hz and 24 Hz using the MATLAB function `angle`. We store these results in two vectors, `phi8` and `phi24`.

The function `rose` displays a histogram of the phase differences (figure 5.11). By default, the phase axis is divided into 20 bins of equal size from 0 to 2π radians, or equivalently, 0 to 360 degrees. At 8 Hz, we observe that phase differences appear in all angular intervals; notice that the number of phase differences located in each angular interval remains small, typically less than 10. At 24 Hz, the angular differences concentrate near 0 degrees; all of the angles lie between -60 and 60 degrees. This visualization is consistent with the strong coherence at 24 Hz, indicative of a consistent phase difference across trials between the two electrodes.

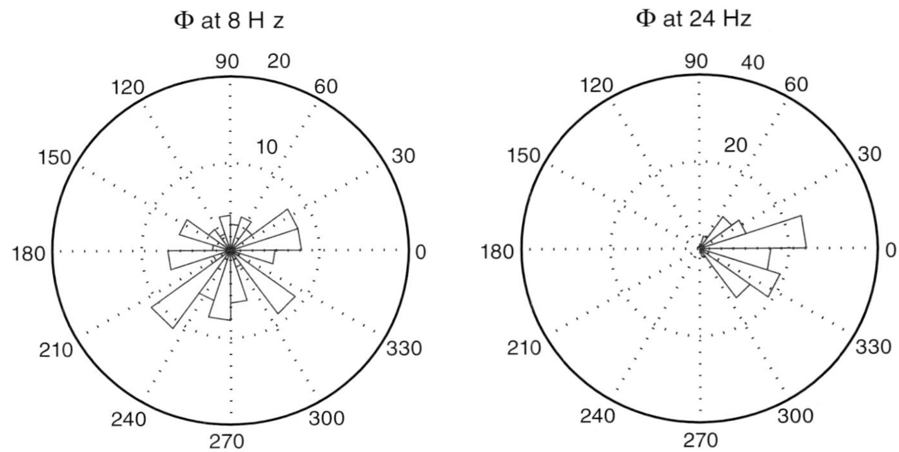

Figure 5.11
Distribution of phase differences between the two ECoG signals depends on the frequency. Angular histograms of phase differences at 8 Hz (*left*) and 24 Hz (*right*). Number of counts in each phase bin is indicated by labeled circles.

Q: Compute and display the distribution of phase differences at other frequencies. What do you find? Are these results consistent with the coherence plotted in figure 5.10?

5.2.6 Single-Trial Coherence

We have emphasized that coherence is a measure of phase consistency between two signals at some frequency *across trials*. This type of analysis is appropriate in many instances in which data are collected in a trial structure. However, we might also be interested in computing the coherence between two signals recorded in a single observation or trial.

Q: Is it possible? Can we compute the coherence between two signals for a single trial?

To address this question, consider the equation for the coherence written in polar coordinates (5.15). Remember that in writing this equation, we have made no assumptions about the data; instead, all we have done is express the complex quantities in polar coordinates. Now consider (5.15) for the case in which we possess only one trial, so that $K = 1$. Then

$$\kappa_{xy,j} = \frac{\left| A_{j,1} B_{j,1} \exp\left(i\Phi_{j,k}\right) \right|}{\sqrt{A_{j,1}^2}\sqrt{B_{j,1}^2}} = \left| \exp\left(i\Phi_{j,k}\right) \right| = 1\,. \tag{5.17}$$

So, we find here perfect coherence ($\kappa_{xy,j} = 1$) for any choice of signals x and y and for any frequency (index j). For example, we could choose x to be the price of a publicly traded stock (e.g., GE) and y to be an ECoG recording, both sampled at 500 Hz for 1 s. Even in this case, we will find perfect coherence between the two signals.

Q: Can we use an ECoG signal to predict the stock price of GE? If so, then we're rich! How can any two arbitrary signals be perfectly coherent at all frequencies?

The answer is that the coherence measure requires a trial structure. Recall that the coherence measures the phase consistency between two signals *across trials*. If only one trial is observed, then the two signals are trivially coherent; the two signals have some phase difference between 0 and 2π and because we have no other trials with which to compare this difference, the two signals are "coherent."

But what if we only collect one trial of data? We can still attempt to compute the coherence in (at least) two ways. First, we could divide the single trial of data into smaller intervals and then treat each interval as a trial. This approach can be effective if we believe the phase relation persists in time, and if we possess a long enough recording. Note that by dividing the data into smaller intervals, we impact the frequency resolution.

Q: Imagine we collect 10 s of ECoG data (sampling frequency 500 Hz) from two electrodes and would like to compute the coherence. To do so, we divide the data into ten nonoverlapping 1 s intervals, and treat each interval as a trial to compute the coherence. What is the frequency resolution of the coherence? If instead we divide the data into 100 nonoverlapping frequency intervals, what is the frequency resolution? In both cases, what is the Nyquist frequency?

A second approach to compute the coherence from a single trial of data is to use the multitaper method. In this case, each taper acts like a trial. Therefore, to acquire more trials for an accurate estimate of the coherence, we include more tapers. But remember that to increase the number of tapers we worsen the frequency resolution (see chapter 4). Computing the coherence using a multitaper method is made relatively easy by software packages like Chronux [2]. Because the ECoG data of interest here consist of multiple trials, we do not focus on measures of single-trial coherence. An example of using the multitaper method to compute the coherence is described in the problems section at the end of this chapter.

Relation between Statistical Modeling and Coherence. Before concluding the discussion of coherence, let's briefly consider a complementary statistical modeling approach. In developing this statistical model, our goal is to capture the (linear) relation between two signals x and y observed simultaneously for multiple trials. We begin by proposing a statistical model that predicts one signal (y) as a linear function of the other (x):

$$\begin{aligned} y_n &= \sum_{m=-\infty}^{\infty} \beta_m x_{n-m} + \epsilon_n \\ &= (\beta \star x)\big[n\big] + \epsilon_n , \end{aligned}$$

where we express the predicted signal (y_n) as a function of x_n, coefficients β_m, and a Gaussian noise term ϵ_n, and where n is a discrete-time index. In the first equation, the summation limits indicate that the predicted signal at time index n may depend on x at any past or future time. The second equality expresses the summed product of β and x as their convolution. Taking the Fourier transform of both sides of this equation, and remembering that convolution in the time domain is equivalent to multiplication in the frequency domain, we find

$$Y_j = \gamma_j X_j + \Upsilon_j ,$$

where Y_j is the Fourier transform of y_n, γ_j is the Fourier transform of β_n, X_j is the Fourier transform of x_n, Υ_j is the Fourier transform of ϵ_n, and j indicates a discrete frequency index. Multiplying both sides of this equation by the complex conjugate of the Fourier transform of x,

$$Y_j X_j^* = \gamma_j X_j X_j^* + \Upsilon_j X_j^* ,$$

and averaging this result across the trials of data, we find

$$< Y_j X_j^* > = \gamma_j < X_j X_j^* > + < \Upsilon_j X_j^* > ,$$

where we use the notation $<>$ to indicate the trial average. Assuming that the noise term and signal x are unrelated, their trial average is zero (i.e., $< \Upsilon_j X_j^* >= 0$). Solving for γ_j, we find

$$\gamma_j = \frac{< Y_j X_j^* >}{< X_j X_j^* >} = \frac{< S_{xy,j} >}{< S_{xx,j} >} . \tag{5.18}$$

Then, comparing (5.18) to the equation for coherence (5.14), we find

$$\kappa_{xy,j} = \mid \gamma_j \mid \frac{\sqrt{< S_{xx,j} >}}{\sqrt{< S_{yy,j} >}} . \tag{5.19}$$

We conclude that the coherence ($\kappa_{xy,j}$) is a scaled version of the frequency domain representation of the statistical model coefficients (γ_j) for predicting y from x. We note that γ_j is a complex quantity that allows us to model both the magnitude and phase of the relation between x and y. The phase difference computed from the model and the coherence is the same as well.

Summary

In this chapter, we analyzed ECoG data recorded from two electrodes during an auditory task. The task involved the repeated presentation of auditory stimuli, resulting in 100 trials of 1 s duration recorded simultaneously from the two electrodes. We began the analysis with visual inspection of individual trials and of all trials at once. Then, to assess the relations between the two recordings, we computed the cross-covariance. We discussed how the cross-covariance is an extension of the autocovariance, and found that the single-trial cross-covariance between the ECoG signals exhibited periodic structure, consistent with rhythmic coupling of period 0.125 s. However, the trial-averaged cross-covariance provided less evidence for consistent rhythmic coupling across trials. We then computed the trial-averaged spectrum and found a large peak near 8 Hz and a much smaller peak near 24 Hz.

To further assess the relation between the two electrodes, we computed the coherence. The coherence is strong (approaches 1) at a chosen frequency f_0 when there exists a constant phase relation at frequency f_0 between two electrodes over trials. We found a strong coherence between the two ECoG electrodes only at 24 Hz. We concluded that although both ECoG signals possessed dominant rhythms at 8 Hz, these rhythms were not coherent between the two electrodes. The strong coherence appeared only at the small-amplitude 24 Hz rhythm. Finally, we implemented a technique to visualize the distribution of phase differences between the two electrodes across trials, and provided some suggestions for how to compute the coherence for a single trial of data.

Caution! Large amplitude does not imply large coherence.

In this example, only the coherence revealed the low-amplitude coupling at 24 Hz between the two ECoG electrodes. This coupling was not obvious in the single-trial or trial-averaged cross-covariance. In fact, the single-trial cross-covariance was deceiving; we found strong single-trial cross-covariance with period 0.125 s, or 8 Hz (figure 5.6b), yet no coherence at 8 Hz.

To understand this discrepancy, consider two unrelated signals, each dominated by the same rhythm. By *unrelated* we mean that the signals do not communicate in any way. Yet both are rhythmic and happen to oscillate at the same frequency. If we compute the cross-covariance between these two unrelated signals, we will find periodic lags at which two signals nearly overlap and the cross-covariance is large. The period of these cross-covariance peaks corresponds to the period of the common rhythm shared by the two signals. Here the periodic, large cross-covariance values occur because the two signals happen to both exhibit a similar rhythm, not because one signal influences the other.

This example illustrates a point of caution in the interpretation of cross-covariance results. Unrelated signals that happen to share a similar dominant rhythm will exhibit large periodic structure in the cross-covariance. One approach to defend against such cross-covariance results is to compute the trial-averaged cross-covariance. If two signals are unrelated—to one another and to the trial structure—then we do not expect similar cross-covariance functions across trials. Therefore, although each single-trial cross-covariance may have large values at some lags, their average across trials will be small. This is just what we found for the ECoG data examined here (figure 5.6). We note that the unrelated 8 Hz signals, which dominate the ECoG activity at each electrode, mask the much smaller amplitude 24 Hz activity that is coupled between the two electrodes. The coupling at 24 Hz is not apparent in the trial-averaged cross-covariance (figure 5.6). The coherence, which normalizes by the power at each frequency, uncovers this relation.

As is true for the Fourier transform and spectrum, there exists a vast literature on computing and interpreting the coherence. Some references for further reading include [8, 9, 11].

Problems

5.1. Consider two signals x and y, where x is a cosine function and y is a sine function. Both signals are of duration 2 s and of frequency 10 Hz. Simulate both signals (each with a sampling interval of 0.001 s) and compute their cross-covariance. What do you find, and how do you interpret the results? Imagine that the signal x was collected from the scalp EEG of a human subject two years ago, while signal y was collected from a voltage recording made in rat hippocampus yesterday. Would you expect these two signals—collected from very diverse preparations—to be related? How does this knowledge impact your interpretation of the cross-covariance results? Consider your answer in terms of the cautions issued in the chapter summary.

5.2. Generate synthetic data consisting of Gaussian noise. More specifically, generate 100 trials of 1 s data sampled at 500 Hz. Do this twice to generate two synthetic datasets, and the compute the following:

a. The trial-averaged spectrum of each synthetic dataset.

b. The trial-averaged cross-covariance between the two synthetic datasets.

c. The coherence between the two synthetic datasets.

Describe your results for each analysis. What cross-covariance and coherence results do you expect to find between these noisy, unrelated sets of data? Do your results match your expectations?

5.3. Generate synthetic data consisting of a sinusoid oscillating at frequency f plus additive Gaussian noise. More specifically, generate 100 trials of 1 s data sampled at 500 Hz. For each trial, set the initial phase of the sinusoid to a random value between 0 and 2π. Repeat this procedure to create a second dataset, but in this case fix the initial phase of the sinusoid to π. Then compute the coherence between these two synthetic datasets. What do you expect to find (i.e., do these two signals possess a constant phase relation across trials at any frequency)? Do your coherence results match your expectations?

5.4. Generate synthetic data consisting of Gaussian noise. More specifically, generate 1 s of data sampled at 500 Hz. Do this twice, and then compute the coherence between these two synthetic signals. Notice, in this case, the data consist of *single trials*. What do you expect to find (i.e., are these noisy signals coherent at any frequency)? Do your coherence results match your expectations?

5.5. In chapter 4, we discussed the notion of tapering in the context of computing the spectrum. Tapers are also applicable in computing the coherence. To analyze the ECoG data in this chapter, we applied the (default) rectangular taper. To (briefly) investigate the application of an alternative tapering procedure, let's consider the Hanning taper and use it to compute the coherence. The Hanning taper is discussed in detail in chapter 4. To apply the Hanning taper, update the MATLAB code provided in section "Computing the coherence." You will need to apply the Hanning taper to the data from each trial before computing the Fourier transforms and the spectra. Compare the results of your coherence analysis using the Hanning taper to the coherence analysis using the (default) rectangular taper shown in figure 5.10. How does the coherence change?

5.6. Load the file `Ch5-ECoG-2.mat`, available at

`http://github.com/Mark-Kramer/Case-Studies-Kramer-Eden`

into MATLAB. You will find three variables in your workspace. The variables `x` and `y` correspond to two simultaneous recordings of ECoG activity from two electrodes. Both of these variables are organized so that the rows correspond to trials and the columns to time. You should find 100 trials, with 1,000 time points per trial. The variable `t` corresponds to the time axis for these data, in units of seconds. Use these data to answer the following questions.

a. Visualize the data from each electrode. What rhythms do you observe?
b. Plot the trial-averaged spectrum versus frequency for each electrode. Are the dominant rhythms in the spectrum consistent with your visual inspection of the data?
c. Plot the trial-averaged cross-covariance between the two datasets. What features do you observe?
d. Plot the coherence between the two datasets. At what rhythms, if any, is the coherence large?
e. Summarize the results of your data analysis. What are the important features of these data that you would communicate to a colleague?

5.7. Load the file `Ch5-ECoG-3.mat`, available at

`http://github.com/Mark-Kramer/Case-Studies-Kramer-Eden`

into MATLAB. You will find three variables in your workspace. The variables `x` and `y` correspond to two simultaneous recordings of ECoG activity from two electrodes. Both of these variables are organized so that the rows correspond to trials and the columns to time. You should find 100 trials, with 1,000 time points per trial. The variable `t` corresponds to the time axis for these data, in units of seconds. Use these data to answer the following questions.

a. Visualize the data from each electrode. What rhythms do you observe?
b. Plot the trial-averaged spectrum versus frequency for each electrode. Are the dominant rhythms in the spectrum consistent with your visual inspection of the data?

c. Plot the trial-averaged cross-covariance between the two datasets. What features do you observe?

d. Plot the coherence between the two datasets. At what rhythms, if any, is the coherence large?

e. Summarize the results of your data analysis. What are the important features of these data you would communicate to a colleague?

5.8. (*Advanced*) As an illustration of using the multitaper method to compute the coherence, we consider the following synthetic data: two time series each consisting of 10 s of Gaussian noise, generated with a sampling interval of 0.001 s. We generate these data in MATLAB as follows:

```
T = 10;                   %Define total duration of data,
dt = 0.001;               %... sampling interval,
N = T/dt;                 %... and no. of pts in data.
x = randn(N,1);           %Generate Gaussian noise data,
y = randn(N,1);           %... and additional Gaussian noise data.
```

Using the approach described in this chapter to compute the coherence between these two signals, we find a value of 1 at all frequencies; see section "Single-trial Coherence." Of course, that's not the answer we want. The two simulated time series are Gaussian noise, and we expect no coherence between them at any frequency. Let's instead use the multitaper method to compute the coherence between these two synthetic signals. First download and install the software package Chronux; see section 1.24 of chapter 1. We use the function `coherencyc` from this software package. When using the multitaper method, we need to choose the time-bandwidth product. Let's assume we demand a resolution bandwidth of 4 Hz. Then the time-bandwidth product is $(10\,\text{s}) \times (2\,\text{Hz}) = 20$, and we choose $2 * 20 - 1 = 39$ tapers. Conceptually, we can think of each taper as acting like a trial; each of these trials selects a different chunk of the synthetic data. The coherence then evaluates the phase consistency of these data across the trials. In MATLAB,

```
%Set the parameters of the MTM.
TW = 20;                  %Choose time-bandwidth product of 20.
ntapers = 2*TW-1;         %...which sets the no. of tapers.
params.Fs = 1/dt;         %Define sampling frequency,
params.tapers = [TW,ntapers];%... time-band product,
                             %... no. of tapers.
params.pad = -1;          %Specify no zero padding,
                          %... and compute the coherence.
[C,phi,S12,S1,S2,f]=coherencyc(x, y, params);
```

```
plot(f,C)                  %Plot the coherence vs frequency,
ylim([0 1])                %... set the vertical axis,
xlabel('Frequency [Hz]')%... and label the axes.
ylabel('Coherence')
```

We set the parameters used by the multitaper method with the variable `params`. This variable is a *structure array*, with data containers called fields; see MATLAB Help for more details about structure arrays. We use the Chronux function `coherencyc` to compute the coherence using the multitaper method. This function returns a variety of outputs, including the cross-spectrum (`S12`) and the spectra (`S1` and `S2`). We focus here only on the coherence, and plot the results in figure 5.12. We find that using the multitaper method, the coherence is not 1 for all frequencies.[3] Instead, the coherence remain small and fluctuates between 0.0 and approximately 0.4. That's closer to the answer we expect for two unrelated noisy time series. This example illustrates that the multitaper method can serve to evaluate the coherence for a single trial of data and does not simply produce a trivial result. To compute this coherence, we accept

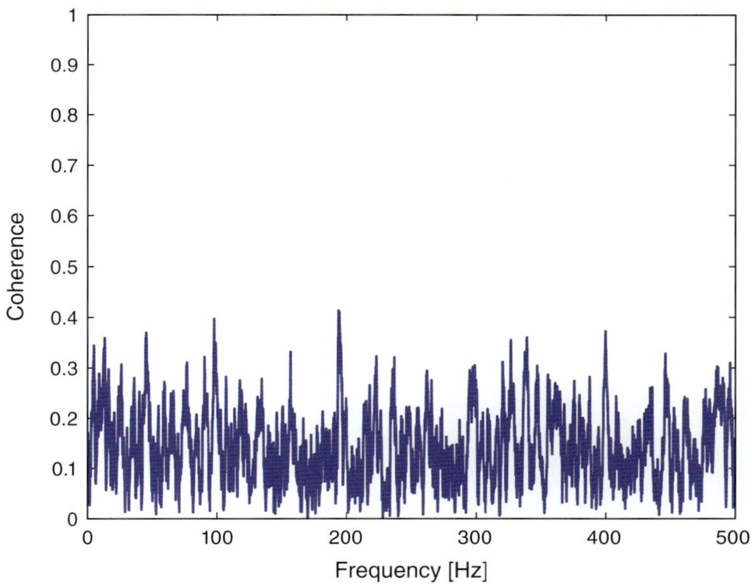

Figure 5.12

Coherence between two Gaussian noise time series.

3. The plot you generate will differ slightly in detail from the plot in figure 5.12 because the synthetic noise data will differ.

worse frequency resolution to acquire the tapers over which to detect constant phase relations between the two signals.

To further examine this assessment of single-trial coherence, consider the following synthetic signals: to the variables `x` and `y` defined in preceding code, add a sine function of amplitude 1 and frequency 10 Hz. Recompute the single-trial coherence using the multitaper method for these new data and examine the results. Does the computed coherence match your expectations for these new data?

6 Application of Filtering to Scalp Electroencephalogram Data

Synopsis

Data Ten 1 s trials of EEG data sampled at 1000 Hz.

Goal Filter these data to identify an evoked response.

Tools Fourier transform, convolution, magnitude response, frequency response, phase response.

6.1 Introduction

6.1.1 Background

In previous case studies, we analyzed brain rhythms and discussed techniques to characterize these rhythms. Observed brain rhythms are often corrupted by noise. Sometimes this noise is obvious (e.g., electrical noise). In general, however, identifying the components of a brain signal that constitute signal versus noise is a difficult problem.

In this chapter, we develop techniques to isolate, emphasize, or remove rhythmic activity in neural field data. To do so we introduce a broad area of study and research: filtering. In general, filtering is a very common procedure in the analysis of neural data. Although it is typically considered a preprocessing step, how filtering is performed may make or break subsequent analysis. This is a vast area, and we focus here on some of the important concepts and tools.

6.1.2 Case Study Data

Our colleague recorded the electroencephalogram (EEG) from a human subject during a task. After performing the experiment, he analyzed the data with the hope of finding an evoked response over visual cortex. However, his initial analysis suggested no evoked response. He therefore asked us to assist in his data analysis. He provided us with the EEG data recorded on the scalp surface above the left occipital lobe of one subject. He would like to understand the rhythmic features that appear in these data during the recording, and in particular whether an evoked response can be detected. He provided us with ten trials of

EEG data, each of duration 1 s, recorded during the subject's response to a visual stimulus (a small flash of light).

6.1.3 Goal

Our goal is to better understand whether an evoked response appears in the data. We first make a visual inspection of the data, using techniques we developed to study evoked responses (see chapter 2). However, our main goal is to understand the fundamental procedures of filtering neural field data, and we examine filtering methods applied to these example EEG data. We start by developing an intuitive approach to filtering and then implement and apply more sophisticated methods. We explain procedures to visualize filter properties and the resulting impact on the input signal.

6.1.4 Tools

In this chapter, we rely on the Fourier transform (and associated measures) to develop a basic understanding of filters. If you are not confident using the Fourier transform, we strongly recommend reviewing chapters 3 and 4. This case study reinforces concepts in those chapters and provides another opportunity to compute and to examine spectra and to examine the relations between the time and frequency domain representations of a time series. Upon completing this chapter, readers should be familiar with basic filtering principles and methods to visualize the impact of filters, and equipped for further study and development of filtering procedures.

6.2 Data Analysis

6.2.1 Visual Inspection

To access the data for this chapter, visit

`http://github.com/Mark-Kramer/Case-Studies-Kramer-Eden`

and download the file `Ch6-EEG-1.mat`. As always, let's begin by looking at the data. To do so, we load the EEG data into MATLAB and plot it:

```
load('Ch6-EEG-1.mat')     %Load the EEG data.
plot(t,EEG(1,:))          %... and plot it for trial 1,
xlabel('Time [s]');       %... with axes labeled.
ylabel('Voltage [mV]')
```

We note that the variable `EEG` is a matrix, with each row corresponding to a single trial. Inspection of the variable `EEG` reveals there are 10 trials (i.e., 10 rows) each consisting of 1,000 indices. We also note that the visual stimulus is delivered just after the start of the trial (at time 0.001 s, corresponding to index 1 of variable `t`), and the response is recorded for the subsequent 1 s (the time 1 s corresponds to index 1000 of variable `t`).

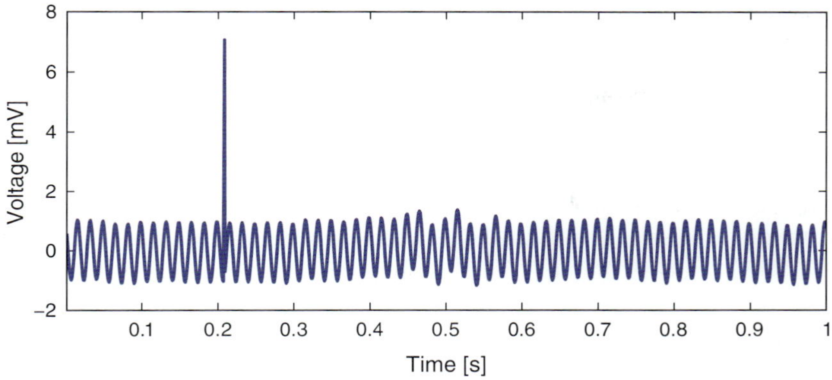

Figure 6.1
Example trace of EEG data.

Q: What are the sampling interval and sampling frequency of the EEG data?

A: Inspection of the variable `t`, loaded into MATLAB, reveals that the sampling interval is 0.001 s, or 1 ms, and the sampling frequency is therefore 1/(0.001 s), or 1000 Hz.

Visual inspection of the 1 s interval of the first trial suggests at least two distinct features (figure 6.1). First, a rapid oscillation appears to occur, with amplitude near 1 mV. Second, a large deviation in voltage occurs (near 0.2 s). Additional slower oscillations may occur, although it's difficult to tell from visual inspection alone.

Q: Consider the fast rhythmic activity that dominates the EEG data plotted in figure 6.1. What is the frequency of this dominant rhythm? Do you observe an evoked response in this single trial? If so, where in time, and what features characterize the evoked response?

A: Careful counting of the number of peaks (or troughs) in the signal reveals that the fast rhythmic activity has a frequency of approximately 60 Hz. Visual inspection of the single-trial data does *not* suggest an evoked response (at least to the authors).

Q: Examine other individual trials of the EEG data. Do you find features similar to those in figure 6.1?

6.2.2 Spectral Analysis

Initial visual inspection of the single-trial data suggests that a 60 Hz rhythm dominates each individual trial. To further characterize this observation, let's compute the spectrum of a single trial of EEG data.[1] We do so here for the first trial, and apply a Hanning taper before computing the spectrum (see chapter 4):

```
load('Ch6-EEG-1.mat')                  %Load the EEG data.
x = EEG(1,:);                          %... and analyze first trial.
x = x - mean(x);                       %Subtract the mean from data.
dt = t(2)-t(1);                        %Define the sampling interval.
T = t(end);                            %Define the duration of trial.
N = length(x);                         %Define no. of points in trial.

xh = hann(N).*transpose(x);            %Multiply data by Hanning window,
xf = fft(xh);                          %... compute Fourier transform.
Sxx = 2*dt^2/T*(xf.*conj(xf));         %... compute spectrum,
Sxx = Sxx(1:length(x)/2+1);            %... ignore negative frequencies.

df = 1/max(T);                     %Determine frequency resolution.
fNQ = 1/dt/2;                      %Determine Nyquist frequency.
faxis = (0:df:fNQ);                %Construct frequency axis.

semilogx(faxis,10*log10(Sxx)) %Plot decibels vs frequency,
xlim([0 100])                      %... in limited frequency range,
xlabel('Frequency [Hz]')           %... with axes labeled.
ylabel('Power [dB]')
```

Notice that in the third line of code, we remove the mean from the single trial of data. This is not always necessary but often useful; we are usually most interested in the rhythmic behavior of the EEG activity, not the changes in the mean signal. Also notice the use of `transpose` when multiplying the data from trial 1 (variable `x`) by the Hanning taper (MATLAB function `hann(N)`). The transpose is required to align the dimensions of the two variables before performing the element-by-element multiplication.

Q: What is the resulting frequency resolution?

1. We could instead write the *sample* spectrum because this equation uses the observed data to estimate the theoretical spectrum that we would see if we kept repeating this experiment, but this distinction is not essential to the discussion here.

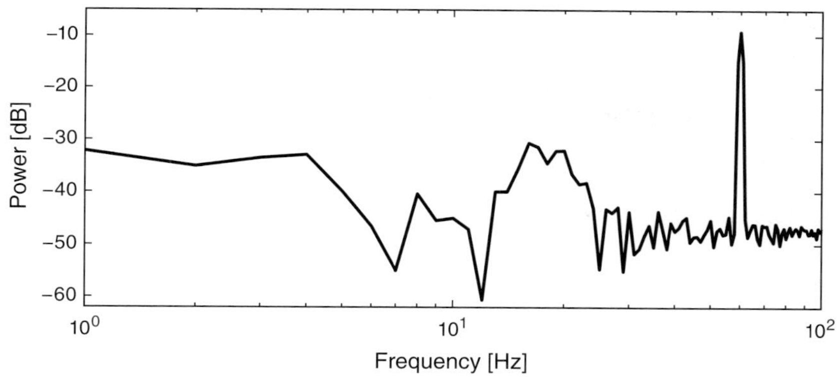

Figure 6.2
Spectrum of EEG data from first trial.

A: The total duration of a trial is 1 s. Therefore the frequency resolution is $1/(1\text{ s}) =$ 1 Hz. If this answer makes no sense, we recommend reviewing the case study in chapter 3.

The resulting spectrum, shown in figure 6.2, reveals two important features. First, the power spectral density appears increased at lower frequencies compared to higher frequencies, that is, the spectrum tends to decrease with increasing frequency. This distribution of power is common in neural field data (e.g., [13]) and in other biological systems [14]. Second, a large peak occurs at 60 Hz. This peak is consistent with the dominant rhythm apparent through visual inspection of the data in figure 6.1. A 60 Hz peak is common in EEG data recorded in North America, where the alternating current in an electrical socket has frequency 60 Hz. We might also perhaps observe a second small peak between approximately 15 and 25 Hz. However, it's not immediately apparent whether this peak represents a rhythm or a random fluctuation in the (noisy) spectrum. In any case, these initial spectral results are somewhat reassuring. The data exhibit characteristics we expect in typical scalp EEG data. Namely, the spectrum tends to decrease with frequency, and a large, sharp 60 Hz peak occurs, consistent with electrical noise.

Q: Examine the spectrum of other individual trials. Do you find features similar to those in figure 6.2?

6.2.3 Evoked Response and Average Spectrum

Initial inspection of the features observable in the individual trials of EEG data has not been encouraging; the data appear to be dominated by 60 Hz line noise, with an occasional large-amplitude deviation. Through this initial inspection, it's not clear whether an evoked response occurs in these data. We may therefore conclude that if an evoked response does occur in the data, the effect is weak and not apparent in a single trial. To further search for a weak evoked effect, let's average the EEG responses across trials. In doing so, we hope that events unrelated to the stimulus will be reduced while responses evoked by the stimulus will be enhanced (see chapter 2). More specifically, let's compute the mean and standard deviation of the mean EEG response at each time across trials:

```
K = size(EEG,1);          %Define variable to record no. of trials.
mn = mean(EEG,1);         %Compute mean EEG across trials (ERP).
sd = std(EEG,1);          %Compute std of EEG data across trials.
sdmn = sd / sqrt(K);      %Compute the std of the mean.

plot(t, mn)               %Plot the ERP,
hold on                   %... and the confidence intervals,
plot(t, mn+2*sdmn);
plot(t, mn-2*sdmn);
hold off
xlabel('Time [s]')        %... and label the axes.
ylabel('Voltage [mV]')
```

The resulting event-related potential (ERP) and 95% confidence intervals are plotted in figure 6.3. Visual inspection suggests three important features. First, the 60 Hz rhythm, which dominated the individual trial data, is less prominent here. Second, large and brief increases in voltage appear in the ERP throughout the 1 s interval (e.g., at times near 0.2 s and at times near 0.8 s). Third, an interesting event appears near time 0.5 s, where we observe a brief interval of rhythmic fluctuations. The mean voltage appears to increase and decrease approximately twice in 125 ms, corresponding to an approximate 16 Hz rhythm. These mean results suggest an evoked response. However, we also observe that the 95% confidence intervals of the ERP include zero; we therefore do not find evidence for a significant ERP.

In addition to the ERP, we also compute the trial-averaged spectrum (see chapter 5). Compared to the single-trial spectrum (e.g., figure 6.2), the variability is greatly reduced in the trial-averaged spectrum (figure 6.3). By reducing the variability in this way, the rhythmic peak between 15 Hz and 25 Hz becomes more apparent. We also observe both the 60 Hz electrical noise peak and the trend of decreasing power spectral density with increasing frequency, as in the single-trial spectrum.

Q: Given the initial analysis, what conclusions do you make regarding the EEG data?

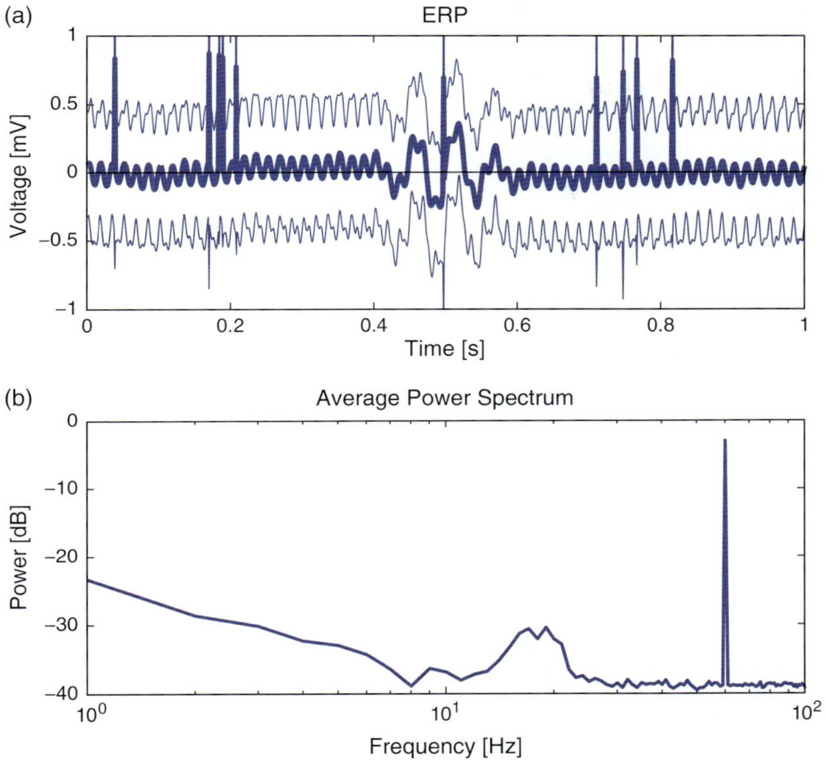

Figure 6.3

ERP and trial-averaged spectrum of EEG data. (*a*) ERP (*thick curve*) plotted with 95% confidence intervals. (*b*) Trial-averaged spectrum versus frequency.

> **A:** The analysis of the single-trial and trial-averaged results suggests that 60 Hz electrical noise dominates the EEG signal. We observe this electrical noise directly in the voltage traces and as a prominent sharp peak in the spectrum. We have some suggestive observations that an interesting evoked response may occur in the data but no conclusive (i.e., significant) evidence. Perhaps if we can reduce the 60 Hz electrical noise, we can uncover a weak evoked response that is currently hidden by the 60 Hz signal. To reduce the 60 Hz signal, we next try to filter these data.

6.2.4 Naive Filtering

In this section, we introduce some fundamental concepts related to filtering. An easy approach to filtering is simply to use built-in MATLAB functions to implement standard filtering methods. However, this approach can be dangerous. If we treat the filter as a black box, then we may not understand what a filter actually does and when it might fail to perform as we hope. To gain intuition for filtering, we therefore begin with a naive approach. This approach uses understanding of the Fourier transform and spectrum, and allows us to investigate how a simple filter performs. We then attempt to improve this naive approach, again using notions familiar from the study of spectra in previous chapters (see chapters 3 and 4). Through this approach we do not implement the most sophisticated or useful filters; in fact, we do not recommend using this approach in practice. But we develop a deeper understanding of what a filter actually does. Later, we use built-in MATLAB functions to implement standard filtering approaches. Ideally, these built-in functions will seem more interpretable with the knowledge gained through the initial naive filtering investigations.

A Naive Rectangular Filter. Initial analysis of the data suggests that the dominant rhythmic activity is 60 Hz electrical noise. This activity is reassuring (we expect it) but also a nuisance; because the line noise is so dominant, other interesting features of lower amplitude may be masked. To search further for a weak evoked response in the data, we must reduce the dominant 60 Hz rhythm.

We build our own filter to achieve this goal. The idea is simple: eliminate the 60 Hz rhythm from the EEG data. Recall that the Fourier transform converts a time domain representation of a signal to a frequency domain representation. Schematically, in symbols,

$$x_n \xrightarrow{FT} X_j, \tag{6.1}$$

where x_n is the data at each time index n, X_j is the data at each frequency index j, and FT denotes the Fourier transform. We also note that the *inverse* Fourier transform (iFT) converts the frequency domain representation of the data back to the time domain:

$$x_n \xleftarrow{iFT} X_j. \tag{6.2}$$

With these concepts, we may define three steps for a naive rectangular filter:

1. Move the observed EEG data to the frequency domain by computing the Fourier transform.
2. Set the frequency domain components of the EEG signal at 60 Hz to zero.
3. Move the altered data back to the time domain by computing the inverse Fourier transform.

We call this initial approach a *naive rectangular filter*. We call it "naive" because we're using naive intuition to construct the filter; we propose to eliminate the 60 Hz activity in the frequency domain in the simplest way and explore the consequences. We call it "rectangular" because we isolate abrupt intervals in the frequency domain to eliminate. Let's attempt this procedure and examine the impact on the EEG data.

Step 1. Our first step is to compute the Fourier transform of the EEG data. We focus specifically on the first trial; the same analysis can be performed on any individual trial. For completeness, we recompute some quantities from the previous sections:

```
load('Ch6-EEG-1.mat')           %Load the EEG data.
x  = EEG(1,:);                  %Relabel the data from trial 1,
x = x - mean(x);                %... subtract mean from the data,
xf = fft(x);                    %... and compute the FT.
```

Q: What are the dimensions of `xf`?

A: Note that `xf` is a vector with the same dimensions as the original EEG data from trial 1 (labeled `x`). Remember that computing the Fourier transform of a signal does not alter its dimensions. However, also note that `xf` is a complex vector, consisting of both real and imaginary parts.

Step 2. We first define the frequency axis that corresponds to `xf`. We do so in the standard way implemented in chapters 3–5:

```
dt  = t(2)-t(1);            %Define the sampling interval.
N   = length(x);            %Define no. of points in single trial.
df  = 1/(N*dt);             %Determine the frequency resolution.
fNQ = 1/dt/2;               %Determine the Nyquist frequency.
faxis = fftshift(-fNQ:df:fNQ-df);   %Construct frequency axis.
```

Notice that the frequency axis (`faxis`) consists of both positive and negative frequencies. When examining the spectrum, we typically ignore the redundant negative frequencies. However, when developing a filter, we must be careful to include all frequencies. The frequency domain representation of the EEG data requires both positive and negative frequencies, and we must adjust both to filter the signal. Also notice that we've applied the `fftshift` command to orient the frequencies appropriately for the vector `xf` (see chapter 3).

Q: Consider the real and imaginary components of `xf` as a function of frequency. For each component, is the negative frequency axis a redundant representation of the positive frequency axis? *Hint:* Consider a plot of the imaginary component of `xf` versus frequency:

```
plot(faxis, imag(xf))
```

What do you observe?

With the frequency axis defined, let's now identify the indices corresponding to the 60 Hz rhythm we'd like to remove from the signal. Of the many approaches to perform this search, here's one:

```
%Find interval near 60 Hz.
indices = find((abs(abs(single(faxis))-60)) <= 1);
```

In words, we find the indices where the difference between the absolute value of the entries in `faxis` and 60 is small (i.e., less than, or equal to, 1 Hz). We use the MATLAB function `single`, which reduces the precision of `faxis` and helps ensure we find all values within 1 Hz of the target frequency and do not exclude any values because of small numerical errors. Note that the absolute value function is called twice, once to convert the frequencies to positive values, and a second time to convert the differences from 60 to positive values.[2]

Q: Confirm that the values of `faxis` at the determined indices are less than, or equal to, 1 Hz from ± 60 Hz.

A: Executing the command: `faxis(indices)`, we find two intervals of values: $(-61, -60, -59)$ Hz, and $(59, 60, 61)$ Hz, consistent with our expectations.

With the indices surrounding the line noise frequency located, we're ready to set the frequency domain components of the EEG signal at 60 Hz to zero. Let's first define the filter in the frequency domain. This filter will have a value of 1 at all frequencies except near ± 60 Hz, where we set the filter to 0. We then apply this filter to the frequency domain representation of the EEG data. By doing so, we set the (complex) values of `xf` (i.e., the frequency domain representation of the EEG data) to zero at the indices surrounding the line noise frequency. At all other indices, we leave `xf` unaltered (i.e., we multiply by 1). In MATLAB,

```
rectangular_filter = ones(1,N); %Define filter in freq domain,
rectangular_filter(indices)=0;  %...set filter @ line noise to 0,
xf_filtered = xf.*rectangular_filter;   %...apply filter to data.
```

Before continuing, let's visualize the filter and anticipate the impact on the frequency domain representation of the EEG data. We plot the variable `rectangular_filter`, the real part of `xf`, and their element-by-element product versus frequency in figure 6.4.

2. Aesthetically, this procedure to find the indices is unappealing. It is the authors' intuition that there is often an aesthetically appealing alternative to aesthetically unappealing code.

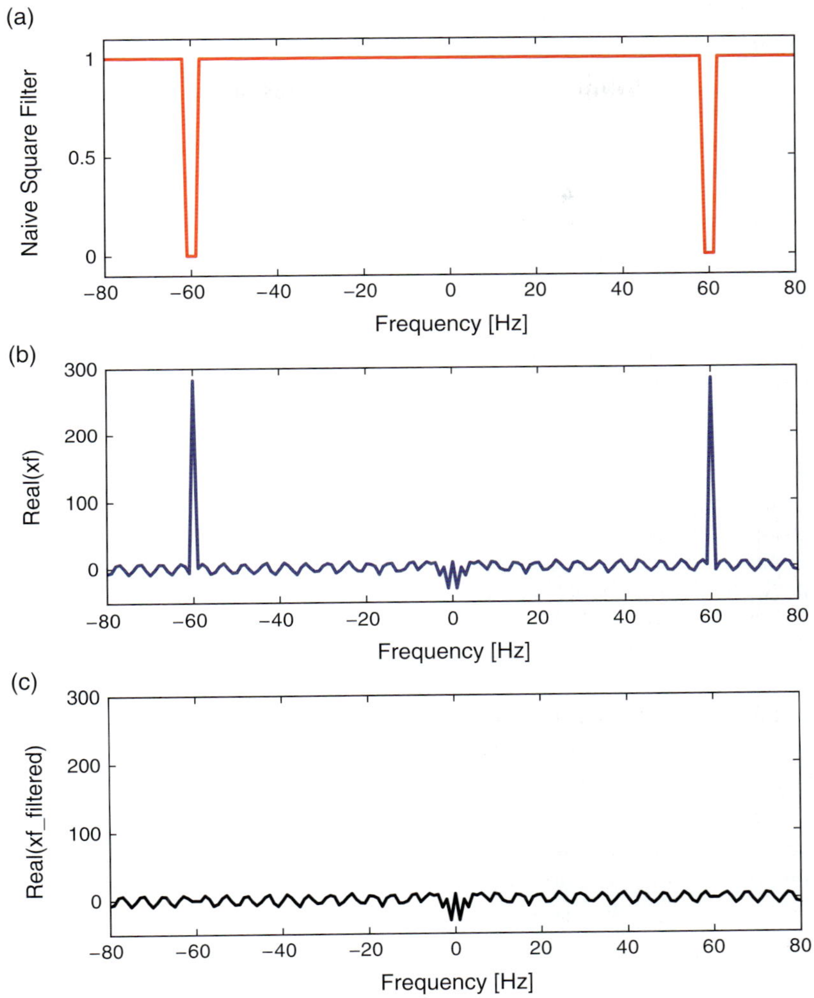

Figure 6.4

Naive rectangular filter applied to frequency domain representation of EEG data from first trial. *(a)* Value of the filter (*red*) as a function of frequency. (*b*) Real part of frequency domain representation (*blue*): imaginary part is not shown. At each frequency, the filter multiplies real and imaginary part of frequency domain representation of EEG data. (*c*) Result is modified frequency domain representation of EEG data.

Because of the way the frequencies are arranged, displaying these plots requires an additional step:

```
[~, isorted] = sort(faxis, 'ascend');
plot(faxis(isorted), rectangular_filter(isorted))
```

In these two lines of code we first define the indices (`isorted`) that sort the frequency axis in ascending order and then use these indices to plot the frequency axis and the rectangular filter.

We find that the filter maintains a value of 1 at all frequencies except for small intervals near ±60 Hz, where the filter has a value of 0 (figure 6.4a). Considering the frequency domain representation of the EEG data, we observe that the real part of `xf` exhibits large peaks at ±60 Hz (figure 6.4b); these peaks correspond to the dominant 60 Hz rhythm apparent in the time domain EEG data. To apply the filter, we multiply the filter (figure 6.4a) by the frequency domain representation of the data (figure 6.4b) at each frequency. The result is shown in figure 6.4c. The peaks at ±60 Hz are eliminated because the value of the filter is set to 0 near these frequencies. All other frequency components in the EEG data are preserved, unaltered by the filter. The frequency domain representation of the EEG data also consists of an imaginary component, not shown in figure 6.4. However, the filter is applied to both the real and imaginary parts of `xf` through the element-by-element multiplication that defines `xf_filtered`.

Step 3. Having eliminated the line noise in the frequency domain, we're now ready to perform the third step of the naive rectangular filter. We apply the inverse Fourier transform, and transform the manipulated frequency domain signal back to the time domain:

```
xnew = ifft(xf_filtered);    %Compute iFT of freq domain data.
```

Q: If the procedure behaves as expected, the resulting time domain signal `xnew` should be real and contain no imaginary components, consistent with the original EEG data. Is this so?

A: To verify this, consider the MATLAB code `max(imag(xnew))`. You should find a value equal to (or within numerical precision of) zero.

To understand the behavior of the new filtered signal in the time domain, let's plot it. An initial visual inspection (figure 6.5a) suggests that the filter has the desired effect; the 60 Hz line noise is dramatically reduced. The same is true in the frequency domain (figure 6.5b). The spectrum of the original signal matches the spectrum of the filtered signal at all frequencies except near 60 Hz, where the power spectral density of the filtered signal is dramatically reduced.

These initial observations suggest that the naive filter is performing quite well and achieving its intended purpose. Yet something is amiss. Consider an expanded view of the filtered signal near the large, brief increase in voltage at $t \approx 0.2$ s (figure 6.5c).

Q: Do any features stand out in figure 6.5c? Consider the voltage fluctuations near the large, brief increase in voltage in the filtered EEG data. What do you observe?

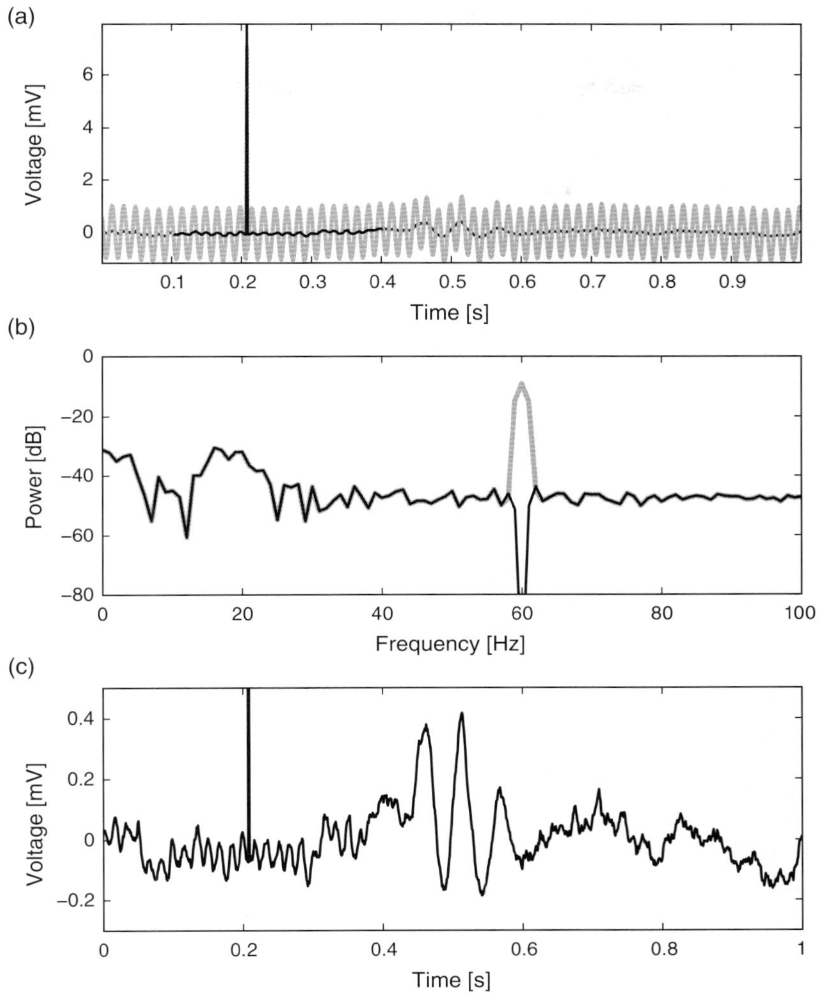

Figure 6.5

Naive rectangular filter applied to EEG data. Filtered signal from first trial (*black*) and original signal (*blue*) in (*a*) time domain and (*b*) frequency domain. (*c*) Expanded view of filtered signal, showing low-amplitude voltage fluctuations near the brief large discharge at t $\approx$ 0.2 s.

A: Careful visual inspection suggests that near the abrupt voltage increase, small-amplitude oscillations emerge in the filtered signal. These oscillations appear to persist for an extended time interval around the abrupt amplitude deviation (at least 100 ms before and after the deviation) and have a period near 60 Hz.

We might interpret the small-amplitude, approximately 60 Hz transient rhythms surrounding each large-amplitude deviation in the EEG data as a biological phenomenon. Perhaps the brain generates these coupled dynamics (i.e., the spike and surrounding rhythmic activity) to achieve a particular function. If so, this would be an important scientific result. However, let's maintain some skepticism regarding the initial filter we've developed. Perhaps our filtering procedure is producing these small-amplitude rhythms around each spike; if so, these rhythms are an artifact of our analysis, not a biologically generated phenomenon. In what follows, we continue to explore the impact of the initial filter and suggest ways to further test the naive rectangular filter's performance.

Impulse Response. So far we've characterized the naive rectangular filter by its impact in the frequency domain. To further characterize this filter, we examine its behavior in the time domain by computing the *impulse response*. As the name suggests, the impulse response indicates the filter's *response* to a simple input signal consisting of a single brief *impulse*. The impulse, although simple in the time domain, possesses spectral content across a wide frequency range; it takes many sinusoids to represent a sharp object like an impulse. In this way, the impulse probes how the filter behaves to input with rich spectral content. We implement the impulse response in MATLAB as follows:

```
impulse = zeros(1,N);          %Define the input signal,
impulse(N/2) = 1;              %... with an impulse at the midpoint.
impulsef=fft(impulse).*rectangular_filter; %Apply naive filter,
impulse_response = ifft(impulsef);  %...iFT back to time domain.
lag_axis = (-N/2+1:N/2)*dt;         %Define lag axis,
plot(lag_axis, impulse_response)    %...display impulse response.
```

We plot in figure 6.6a the original impulse and the impulse response (i.e., the result of applying the naive rectangular filter to the impulse). Visual inspection suggests how the filter affects the input signal in the time domain; we see that the filtered impulse consists of a large peak (centered at the time of the original impulse). By focusing on a small vertical range, we find that the peak at time 0 s is surrounded by smaller-amplitude fluctuations. Although these fluctuations are small, we notice that they persist across *all* time indices examined (figure 6.6b). By focusing both the vertical and horizontal range (figure 6.6c), we observe that these fluctuations are periodic, with a period of 60 Hz. The impulse response provides important insights into the behavior of the naive rectangular filter: an impulse occurring in the original signal impacts all time points in the resulting filtered signal through small-amplitude 60 Hz fluctuations.

We are concerned about this impulse response for the naive rectangular filter. We find that an impulse, initially localized to a single index in time, becomes broadly distributed in time upon filtering. To further illustrate the impact of the naive rectangular filter, let's consider a more direct method to apply a filter and compute the impulse response. Up

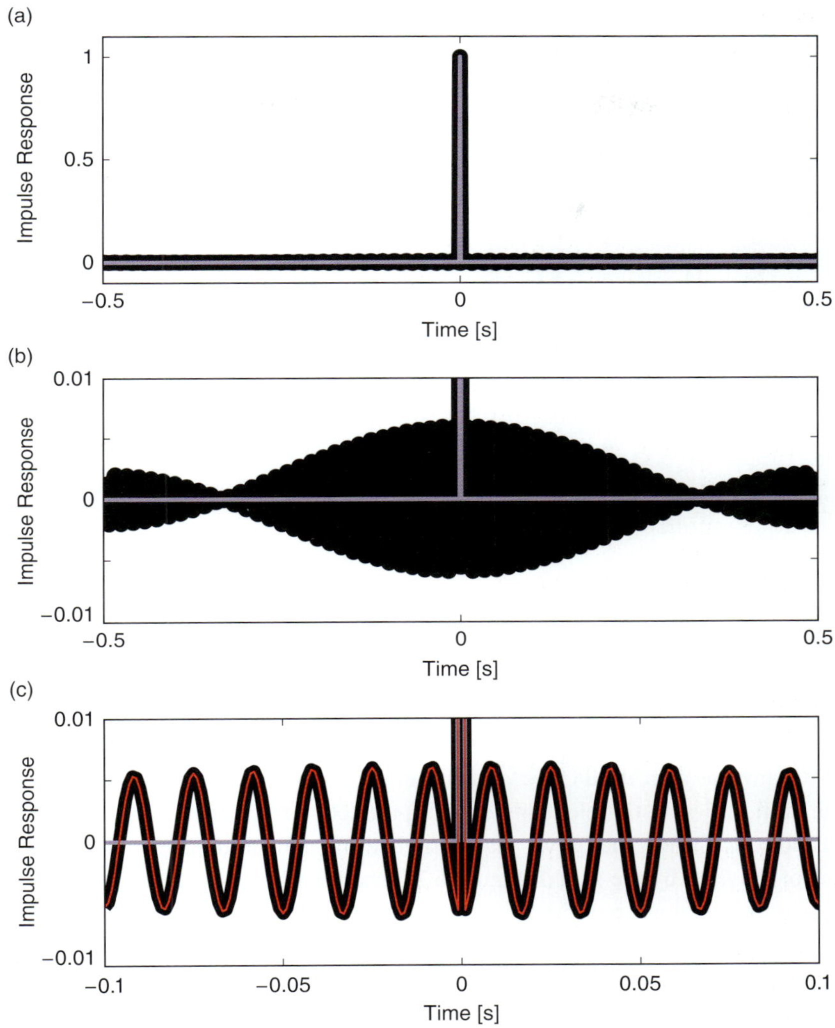

Figure 6.6

Impulse response of naive rectangular filter. Original impulse (*blue*) and impulse response (*black*) at different levels of vertical and horizontal scaling. (*a*) At a wide vertical scale, a peak in the impulse response coincides with original impulse. (*b*) At a narrower vertical scale, impulse response exhibits fluctuations not apparent in original impulse. (*c*) At narrower vertical and horizontal scales, impulse response fluctuates periodically near 60 Hz. Impulse response computed using convolution in the time domain is also shown (*red*).

to this point, we applied the filter by first transforming the input signal to the frequency domain, then performing an element-by-element multiplication of the input signal and filter, and finally transforming the result back to the time domain. We can avoid these transformations by remembering the following important fact.

> Multiplication in the frequency domain is equivalent to convolution in the time domain.

Therefore, as an alternative method for computing the impulse response, we convolve the impulse with the time domain representation of the filter. By doing so, we no longer need to transform to and from the frequency domain. In MATLAB,

```
%Transform the filter to the time domain,
i_rectangular_filter = ifft(rectangular_filter);
%... and define the impulse response,
impulse_response_t = zeros(1,N);
for i=1:N                              %...at each point in time,
    impulse_response_t(i)=...          %...by computing the convolution.
        sum(circshift(i_rectangular_filter, [1,i-1]).*impulse);
end
```

Notice that in this code we compute the convolution by hand. We do so to make explicit the computation performed. At each time index (`i`), we multiply element by element the shifted rectangular filter in the time domain (note the use of `circshift`) by the impulse. At each shift, we sum the elements of this product; the result is the impulse response at time index `i`. Conceptually, we may visualize the convolution as multiplying shifted versions of the filter by the signal (Figure 6.7). The filter begins with a peak at time index 1, and small oscillations are most apparent at lags near the beginning and end of the vector (figure 6.7a). As we (circularly) shift the filter, we move the filter peak to higher time indices (figure 6.7b). When the filter peak reaches the impulse (figure 6.7c), the resulting summed product (i.e., the convolution) is large. Away from this time of large overlap, the resulting convolutions are small. The impulse response function is the result of these summed multiplications performed for all time shifts.

The resulting impulse response function computed in the time domain through convolution (variable name `impulse_response_t`, where `_t` denotes time) is identical to the impulse response function computed in the frequency domain through element-by-element multiplication (variable name `impulse_response`); compare the black (`impulse_response`) and red (`impulse_response_t`) curves in figure 6.6.

We have now examined two equivalent methods of filter application: through multiplication in the frequency domain or through convolution in the time domain. Both approaches

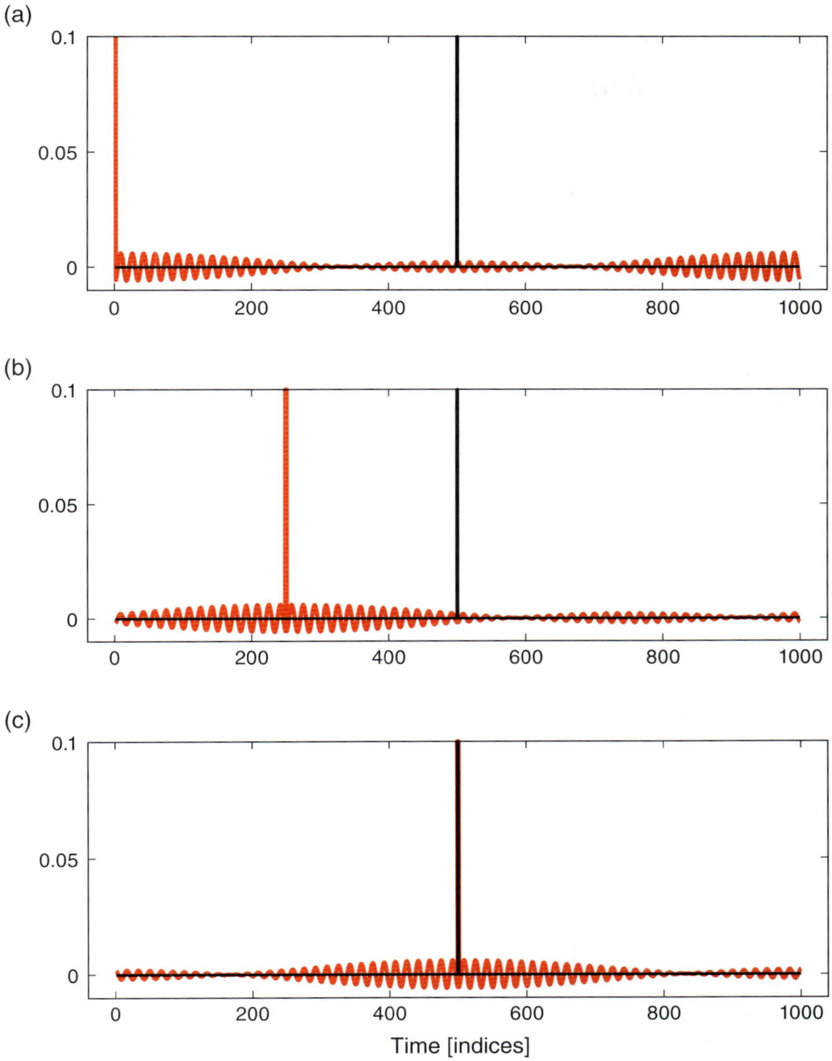

Figure 6.7

Illustration of convolution between impulse and naive rectangular filter. Convolutions computed by element-by-element multiplication of vectors corresponding to impulse (*black*) and naive rectangular filter (*red*) in the time domain. Impulse occurs at time index 500. (*a*) With no shift, filter has a peak at time index 1 and small-amplitude oscillations at both small and large indices. (*b*) With shift of 250 indices, filter peak shifts to index 250, and surrounding small-amplitude oscillations become clear. (*c*) With shift of 500 indices, peaks in impulse function and filter are aligned, and resulting convolution is large.

provide insight into the impact of the naive rectangular filter. The frequency domain representation is easier to interpret. We designed this filter in the frequency domain to eliminate signal components near 60 Hz, which we implemented through an abrupt decrease in the frequency domain representation of the filter (figure 6.4a). It's relatively easy to envision that this filter eliminates signal features near ±60 Hz and preserves features at other frequencies (figure 6.4c).

The time domain representation of the naive rectangular filter is more complicated (figure 6.6). While the impact of the filter in the frequency domain is limited, the impact in the time domain is broad. The impulse response decays at times away from the impulse; however, these contributions do remain. This example illustrates the necessary trade-off between the time and frequency domains. Namely, the sharp transition bands, or roll-off in the frequency domain (i.e., the nearly vertical rectangular-shaped transitions) correspond to a broad impulse response function that extends over many lags in the time domain. There's no escaping this fact: the sharper we make the filter's transitions in the frequency domain, the broader its effects in time. Again, we make this trade-off to implement the sharp roll-off of the naive rectangular filter; the transition band is narrow (a precipitous drop) in frequency and therefore requires many values in time.

Through the interpretation of filtering as convolution in time, we gain additional insight into the naive rectangular filter's impact. In figure 6.5c we first identified suspicious behavior in the filtered signal; we found that after filtering, the large-amplitude discharge was surrounded by small-amplitude oscillations with frequency near 60 Hz. Under the interpretation of filtering as convolution, we expect that a sudden large change in the input signal (i.e., the brief discharge in the EEG) will clearly impact the resulting filtered signal across an extended interval of time. Indeed, because the brief EEG discharge is so large, the small amplitude oscillations surrounding the central peak of the impulse function are apparent in the filtered EEG signal. For these brief large-amplitude peaks in the EEG data, the temporal impact of the filter is obvious in the filtered signal. However, this temporal impact occurs throughout the filtered signal; using the naive rectangular filter, each time point in the original signal impacts every time point in the filtered signal.

We conclude this section by noting that these results are analogous to our discussion of the rectangular window function in chapter 4. In that chapter, we applied a rectangular taper in the time domain and showed that this produced broad effects in the frequency domain. Here, we instead apply an (inverted) rectangular taper in the frequency domain and find broad effects in the time domain. The fundamental concept is that the Fourier transform of a sharp transition in one domain (in this case, the abrupt edge of a rectangular taper) produces broad effects in the other domain.

A Naive Hanning Filter. In the previous section, we developed and applied a naive rectangular filter. We used the word "rectangular" to indicate the rectangular shape of the filter in the frequency domain (figure 6.4). Although well-behaved in the frequency domain, the naive rectangular filter produces undesired effects in the time domain; namely, the filter's sharp transitions in the frequency domain produce wide-ranging effects in the time domain. These long-range temporal effects are an unwanted feature of the filter. To reduce these effects, we propose an alternative filter that softens the sharp transitions in the frequency domain and makes these transitions more gradual. The idea is simple: replace the rectangular function in the original naive filter with a different, smoother function. In what follows we implement many of same procedures employed in the previous section and interpret how these changes affect the filtered EEG data.

Let's begin our filter design in the frequency domain. Our goal is to eliminate the 60 Hz component (i.e., the electrical noise) from the EEG signal without introducing long-lasting temporal effects. To do so, instead of the naive rectangular function employed in our first filter, we'll use a smooth Hanning function (first introduced in our computations of the spectrum in chapter 4). In MATLAB we first load the data and identify the indices corresponding to ± 60 Hz, and then apply a Hanning window centered at each of these indices. Some of the following MATLAB commands are redundant with commands we used previously, but they are repeated here for completeness.

```
load('Ch6-EEG-1.mat')          %Load the EEG data.
x   = EEG(1,:);                %... analyze first trial.
dt  = t(2)-t(1);               %Define sampling interval.
N = length(x);                 %Define no. of points in single trial.
df  = 1/(N*dt);                %Determine the frequency resolution.
fNQ = 1/dt/2;                  %Determine the Nyquist frequency.
faxis = fftshift(-fNQ:df:fNQ-df); %Construct frequency axis.

%Find indices at +/- 60 Hz.
ind   = find((abs(abs(single(faxis))-60)) <= 0.1*df);

hann_filter = ones(1,N);  %Define filter in frequency domain,
win = 15;                      %... set size of the Hann window,
                               %... and apply it:
hann_filter(ind(1)-win:ind(1)+win)  = 1-transpose(hann(2*win+1));
hann_filter(ind(2)-win:ind(2)+win)  = 1-transpose(hann(2*win+1));
```

In this code, we load the EEG from the first trial (variable `x`) and create the frequency axis (variable `faxis`) to find the indices that correspond to frequencies at ± 60 Hz. At these indices, we center a Hanning window, which sets the filter to 0 at ± 60 and gradually returns to 1 away from these values (the parameter `win` sets the width of the window).

> **Q:** Compare the frequency domain representations of the new Hanning filter with the original naive rectangular filter, both plotted in figure 6.8. What differences do you observe?

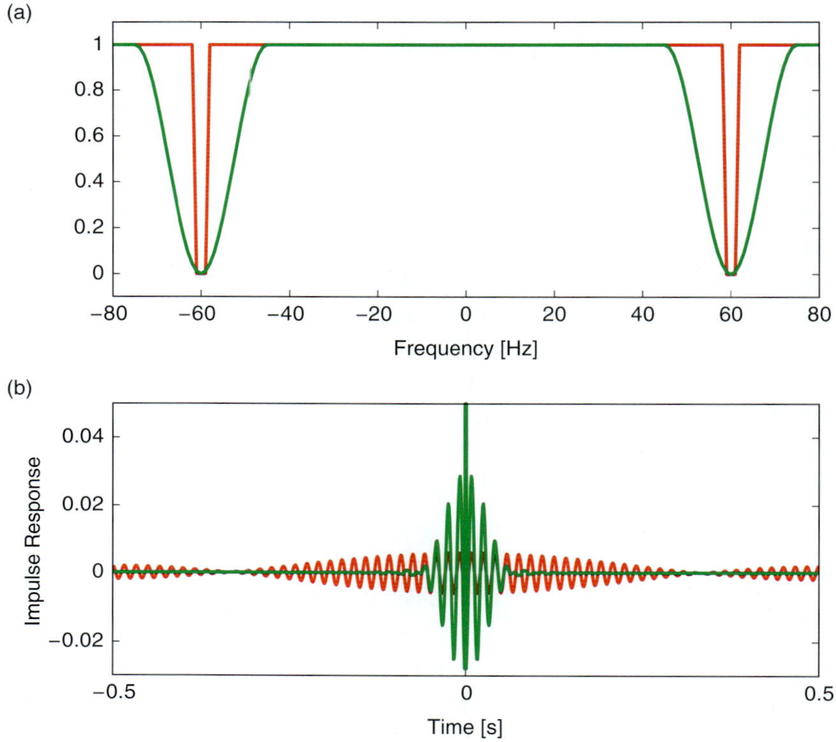

Figure 6.8

(*a*) Frequency domain representations of the Hanning (*green*) and naive rectangular (*red*) filters. (*b*) Impulse response of the Hanning (*green*) and naive rectangular (*red*) filters. Horizontal and vertical axes scaled to emphasize the most relevant features.

A: Although both filters decrease to 0 near ±60 Hz, the Hanning filter returns gradually to 1, while the rectangular filter rapidly changes from 0 to 1. In other words, the roll-off is more gradual in the Hanning filter compared to the square filter.

Let's also examine the time domain representation of the new Hanning filter. To do so, we compute the impulse response. In MATLAB, following the same procedures we employed for the naive rectangular filter,

```
impulse = zeros(1,N);                 %Define the input signal,
impulse(N/2) = 1;                     %...with impulse at the midpoint.
i_hann_filter=ifft(hann_filter);%Transform filter to time domain,
impulse_response_t = zeros(N,1);%...define impulse response
```

```
for i=1:N                              %... vector, compute it with
    impulse_response_t(i) = ... %... convolution.
        sum(circshift(i_hann_filter, [1,i-1]).*impulse);
end
lag_axis = (-N/2+1:N/2)*dt;          %Define lag axis for plotting,
plot(lag_axis, impulse_response_t)%...display impulse response.
```

In this code, we first define the impulse signal and then convolve this impulse with the Hanning filter transformed to the time domain (using `ifft`). The resulting impulse response function is plotted in figure 6.8b.

Q: Compare the time domain representations of the new Hanning filter with the original naive rectangular filter (figure 6.8b). What differences do you observe?

A: Both filters exhibit a large peak at time 0 s and are surrounded by small-amplitude 60 Hz oscillations. Compared to the naive rectangular filter, these oscillations are initially larger for the proposed Hanning filter. However, these oscillations quickly decay for the Hanning filter; after ±50 ms, the Hanning filter remains near zero.

Q: Compute the impulse response function of the Hanning filter *without* using the convolution function. Instead, only use the Fourier transform and the inverse Fourier transform. Do you find the same results as shown in figure 6.8? *Hint*: You should.

The visualizations in figure 6.8 provide insights into the newly designed Hanning filter. The more gradual roll-off in the frequency domain corresponds to a more localized impact in the time domain. Here again we are forced to trade desirable features in the time and frequency domains. In the Hanning filter, we forsake the sharp roll-off of the naive rectangular filter—a desirable property—for the more local temporal impact, also a desirable property.

Now, having visualized the new filter, let's apply it to the EEG data. To do so, we follow the same procedure used to filter the EEG data with the naive rectangular filter. In MATLAB,

```
xf = fft(x);                       %Transform data to frequency domain,
xf_filtered=xf.*hann_filter;%...apply Hanning filter,
xnew = ifft(xf_filtered);    %...transform back to time domain.
```

In these three lines of code, the EEG data are first transformed to the frequency domain using the Fourier transform. Then the Hanning filter is applied through element-by-element

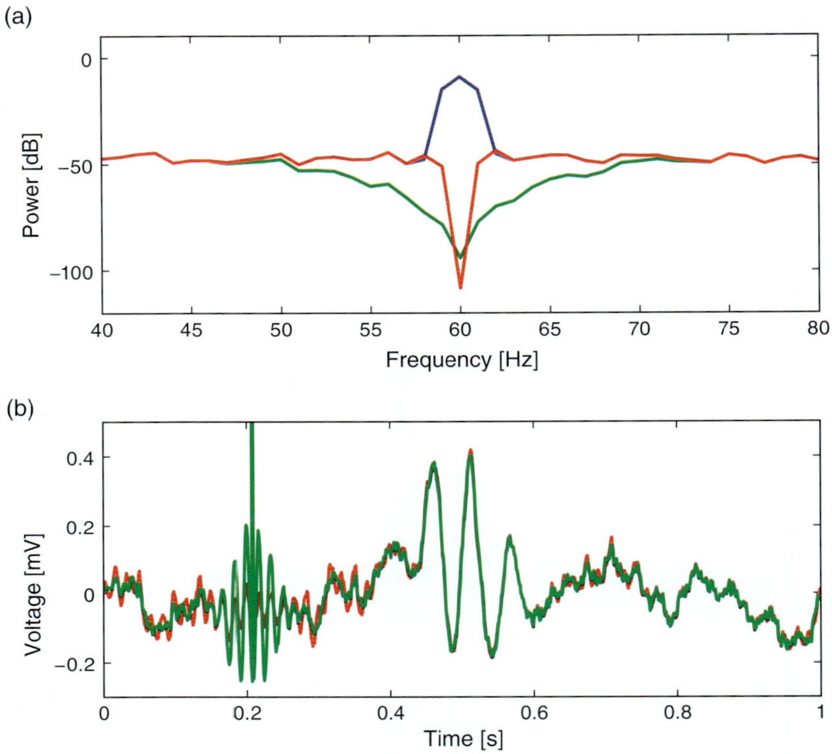

Figure 6.9

(*a*) Spectrum of original EEG data from first trial (*blue*), naive rectangular filtered EEG data (*red*), and Hanning filtered data (*green*). (*b*) Example time evolution of naive rectangular filtered EEG data (*red*) and Hanning filtered data (*green*).

multiplication. Finally, the filtered signal is transformed back to the time domain using the inverse Fourier transform. The results of this filtering procedure are shown in figure 6.9.

Q: Compare the results of filtering the EEG data using the naive rectangular filter and the Hanning filter (figure 6.9). How do the filters produce similar, and different, results?

A: Comparison of the resulting spectra shows a sharper and deeper decrease near the 60 Hz peak for the naive rectangular filtered EEG data. In this way, the naive rectangular filter appears to far outperform the Hanning filter; in figure 6.9a the stop-band of the naive rectangular filter is more focused on eliminating the 60 Hz signal we'd like to remove. The Hanning filter is broader and reduces frequency components extending ±10 Hz around the 60 Hz peak.

However, in the time domain, the Hanning filter is far superior. The example shown in figure 6.9b illustrates the long-range temporal effects imposed by the naive rectangular filter. Near a brief large discharge in the (unfiltered) EEG data, both filters introduce rhythmic activity for an interval of time. For the Hanning filtered data, these oscillations are a bit larger. For the naive rectangular filtered data, these oscillations persist for a much longer duration; notice in figure 6.9b that small-amplitude (60 Hz) oscillations appear from time 0 s to time 0.4 s.

Having filtered the EEG data in two ways and analyzed the results, we may now make an important conclusion: the naive rectangular filter is a poor choice. Although this filter performs admirably in the frequency domain, the results in the time domain are unacceptable. The naive rectangular filter may confound our understanding of the EEG signal through incorporation of new, long-duration temporal effects in the filtered signal. These results suggest the Hanning filter is a better choice. However, we do *not* recommend using this filter. Instead, we recommend the much safer choice of using filters designed in preexisting software functions. We describe one example of this approach in the next section.

6.2.5 More Sophisticated Filtering

In the previous section, we designed two filters. To do so, we started in the frequency domain and applied concepts developed in previous chapters when studying the spectrum (chapters 3 and 4). We undertook this initial approach for one purpose: to build intuition. Developing such intuition is critical; without it, filtering, would be hard to understand. However, in practice, we do not recommend the use of the naive rectangular filter or the Hanning filter. This point is so important, we further emphasize it.

> Do *not* use the naive rectangular filter or the Hanning filter on your data.

Instead, we recommend using preexisting filter design methods provided in MATLAB or other software. In this section, we illustrate the implementation and application of one such method. We continue to use the visualization techniques developed in the previous section to analyze the time and frequency domain representations of the implemented filter. We then apply the filter to the EEG data and examine the results.

The Finite Impulse Response (FIR) Filter: An Example. The most common category of filter applied in neuroscience applications is the *finite impulse response* (FIR) filter. The name for this approach is actually quite informative. In the previous section, we defined the impulse response; it represents the response of the filter to a signal composed of only a single impulse. "Finite impulse response" indicates that the impulse response consists of only a finite number of nonzero terms. The naive rectangular filter was an example of an *infinite* impulse response; to represent the naive rectangular filter in the time domain requires an infinite number of terms.[3] We found that such a broad response in the time domain produces unwanted temporal effects in the filtered signal. Here, we instead implement a filter with only a finite number of nonzero terms in the impulse response.

Before using built-in MATLAB routines to create a FIR filter, we might consider how to design our own. A straightforward approach would be to start with an existing filter we developed (e.g., either the naive rectangular filter or the Hanning filter) and truncate the number of terms in the impulse response. By doing so, we would necessarily create a *finite* impulse response function; all terms beyond a chosen time point would be set to zero. We would therefore eliminate contributions to the filtered signal from inputs far into the past or future. That's the right idea, but we do not pursue this approach. Instead, we use built-in MATLAB functions. In this way, we leverage the expertise already inherent in these preexisting filter design approaches.

We focus specifically on the application of a lowpass FIR filter to the EEG data. We use the `fir1` command in MATLAB to design this filter and visualize the filter in the time and frequency domains. Let's load the data into MATLAB, define useful parameters, and then design the filter:

```
load('Ch6-EEG-1.mat')     %Load the EEG data,
x   = EEG(1,:);           %...and analyze first trial.
dt  = t(2)-t(1);          %Define the sampling interval,
fNQ = 1/dt/2;             %...and the Nyquist frequency.

n=100;                    %Define the filter order,
Wn=30/fNQ;                %... specify the cutoff frequency,
b = fir1(n,Wn,'low');     %... and build the lowpass filter.
```

Notice that to design the filter we call the MATLAB function `fir1` with three inputs. The first input specifies the filter order, which corresponds to the number of nonzero terms in the filter. We specify a filter order of `n=100`. We next specify the upper frequency for the lowpass filter, which we set to 30 Hz. Notice that the frequency is specified as a fraction of the Nyquist frequency. The final input specifies the filter type. In this case, we implement a *lowpass* filter; this filter will pass frequencies below 30 Hz and stop frequencies above 30 Hz. We choose a lowpass filter in this case for physiological reasons; scalp EEG data are often corrupted by muscle artifacts at frequencies above 20–30 Hz. Therefore, in an

3. In practice, all digital filters are finite. We have only a finite amount of computer memory in which to store the filter. In theory, the impulse response of the naive rectangular filter decays to zero as time approaches infinity and therefore contains nonzero contributions for an infinite number of terms.

attempt to better isolate true brain signals and extract an evoked response, we apply a lowpass filter to the EEG data.

The vector `b` contains the coefficients of the lowpass FIR filter. To examine this filter, let's visualize it in the time and frequency domains. In the time domain, we compute the impulse response in MATLAB as follows:

```
N = length(EEG);                      %Define no. of points in data.
bz = [zeros(1,N-n-1), b];             %Amend filter with leading zeros.
impulse = zeros(1,N);                 %Define the test input signal,
impulse(N/2) = 1;                     %... with impulse at midpoint.
impulse_response_t = zeros(N,1);%Define impulse response vector,
for i=1:N                             %...compute it with convolution.
    impulse_response_t(i) = ...
        sum(circshift(bz, [1,i-1]).*impulse);
end
```

Before computing the convolution, we create an augmented vector (variable `bz`). This new vector consists of leading zeros, followed by the filter (variable `b`). The augmented vector has the same length (`N`) as the impulse signal (variable `impulse`), and therefore we can compute the element-by-element multiplication of these two vectors in the same way as we did for the naive rectangular and Hanning filters. Note that the nonzero values of the filter appear at the end of the augmented vector `bz`. To compute the convolution, we shift these nonzero values over the entire length of the augmented vector, and at each shift multiply element by element the two vectors and sum their product.

The resulting impulse response is shown in figure 6.10a. Visual inspection reveals the relation between the original impulse and the filtered impulse following application of the FIR filter. We notice an important difference between the impulse response of the FIR filter and the impulse responses of the naive rectangular and Hanning filters. For the FIR filter, the peak impulse response *follows* the impulse. The reason for this delay is that the FIR filter is *causal*; to compute the convolution at any time requires only past and current values of the input signal. The naive rectangular and Hanning filters required both past and future values of the input signal. Those filters were *noncausal*.

To further explore this idea, let's examine in more detail the computation of the FIR filter. To compute the FIR filter, we first constructed the augmented filter vector. To align the filter peak with the impulse requires that we (circularly) shift the augmented filter vector. In this example, the peak in the augmented filter vector (`bz`) occurs at index $N - n/2$, or equivalently, at time $(N - n/2) * dt = 0.95$ s, where N is the length of the EEG data, and n is the filter order. We must therefore (circularly) shift the augmented filter vector by $N/2 + n/2$, or equivalently, $(N/2 + n/2) * dt = 0.55$ s, to align it with the peak of the impulse (which occurs at index $N/2$, or equivalently, at time $N/2 * dt = 0.5$ s). The largest value in the resulting convolution therefore occurs at an index $n/2$ (or equivalently, 0.05 s)

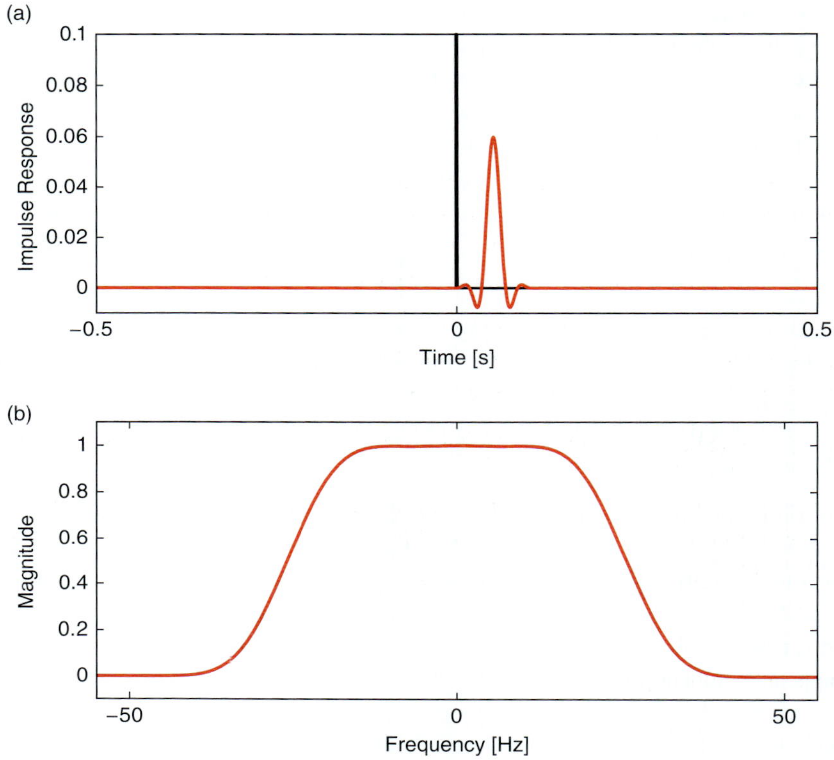

Figure 6.10

(*a*) Impulse response of FIR filter (*red*). An impulse (*black*) is shifted in time upon filtering. (*b*) Magnitude response of FIR filter. The lowpass filter passes frequencies between approximately ±30 Hz.

past the impulse in the original signal. The delay between the impulse and impulse response in figure 6.10a corresponds to $n/2$ indices, or 0.05 s.

> **Q:** What are the implications of the delay induced by the FIR filter? How might this delay impact subsequent analysis?
>
> **A:** We consider this question in more detail later in this chapter.

We may also examine the filter in the frequency domain. To do so in MATLAB, we compute and plot the *magnitude response*:

```
bf = fft(b,N);        %Transform filter to the frequency domain,
Mb = bf.*conj(bf);    %...and compute the mag response.
```

```
df = 1/(N*dt);         %Define the frequency resolution,
faxis = fftshift((-fNQ:df:fNQ-df)); %...create frequency axis,
[~, isorted]=sort(faxis,'ascend');  %...with axes sorted,
plot(faxis(isorted), Mb(isorted))   %...plot mag response.
```

We take the Fourier transform of the filter, with zero padding (see chapter 4) to match the size of the EEG data, and compute the product of the Fourier transform of the filter and its complex conjugate. We then define the frequency axis, and plot the sorted frequency axis to avoid any unwanted lines. The result is shown in figure 6.10b.

Q: Describe the behavior of the FIR filter in the frequency domain. What frequencies are passed? What frequencies are stopped?

A: Inspection of the magnitude response in figure 6.10b shows that frequencies between approximately ±30 Hz are passed; within this interval, the magnitude response of the filter is above zero. Beyond this interval, the magnitude response decreases to zero. Frequencies greater than 30 Hz or less than −30 Hz are removed in the filtered signal.

Having analyzed the filter through inspection of its impulse response and magnitude response, let's now apply this filter to the EEG data:

```
xnew = zeros(N,1);          %Define vector to hold filter results,
for i=1:N                   %...and for each index i,
    xnew(i) = ...           %...compute the convolution.
        sum(circshift(bz, [1,i]).*x);
end
```

In these lines of code we use convolution to compute the lowpass filtered EEG signal and generate the new time series `xnew`.

Inspection of the resulting filtered signal reveals important features of the new time series. Perhaps the most prominent change is the large reduction in the 60 Hz electrical noise (figure 6.11a). Without the contamination of this noise, we now observe a transient oscillatory event near 0.5 s. As expected, the spectrum is now dominated by low-frequency activity, namely, rhythms below 30 Hz (figure 6.11b). Finally, the lowpass filter alters the sharp, brief discharge in the original EEG signal (near 0.2 s in figure 6.11a) in two important ways. First, we directly observe the temporal shift introduced by the FIR filter; the peak in the filtered signal (black curve in figure 6.11a) follows the large voltage deviation in the original signal (blue curve in figure 6.11a) by 0.05 s. Second, the FIR filter acts to reduce and broaden the large voltage deviation in the original EEG signal, consistent with the impulse response for this filter (figure 6.10a).

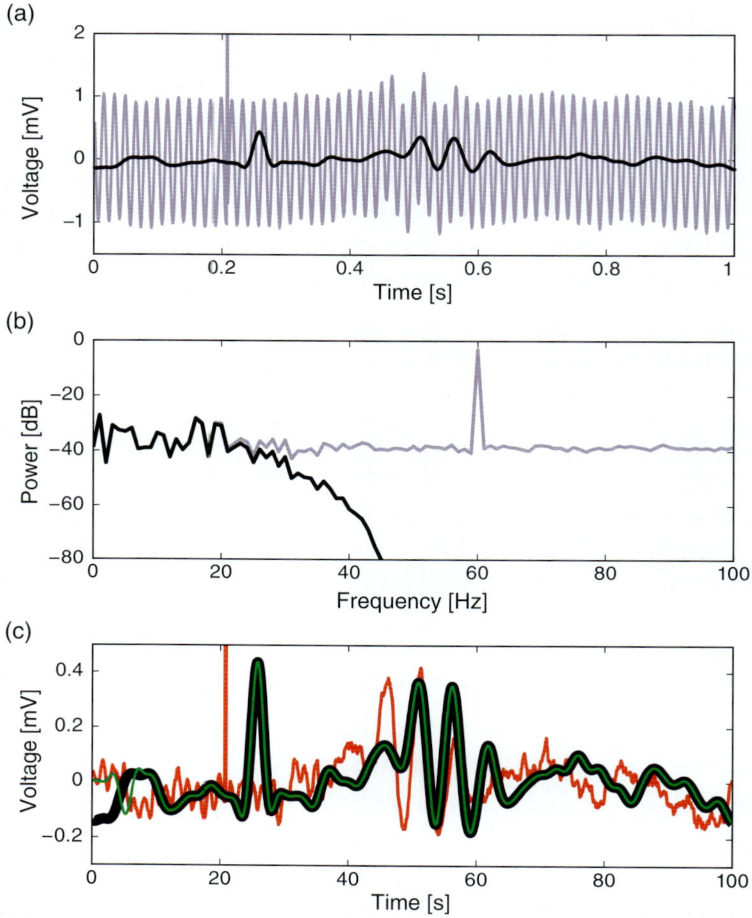

Figure 6.11

Lowpass filter of EEG using the `fir1` filter. Lowpass filtered signal (*black*) and original signal (*blue*) in (*a*) time domain and (*b*) frequency domain. (*c*) Low-amplitude voltage fluctuations in filtered signal introduced by naive rectangular filter (*red*) are not present in FIR filter computed through convolution (*black*) or using a built-in MATLAB routine (*green*).

Q: Why would a lowpass filter act to reduce and broaden a brief large discharge in the EEG?

A: Consider the spectrum of a simpler time series that consists of all zeros except for a single value of 1 at some time index; for example, the simple time series we use to compute the impulse response of a filter. The corresponding spectrum will have nonzero contributions at all frequencies, i.e., representing an impulse requires a combination of sinusoids at many frequencies. Conceptually, an impulse is difficult to represent as a sum of sinusoids. A single sinusoid exists (theoretically) for all time, and we'd somehow like to use these long-duration functions to represent a brief-duration impulse. If we now eliminate some of these sinusoids (e.g., by lowpass filtering the data) we corrupt the representation of the impulse; without these sinusoids, we're no longer able to accurately represent the sharp, brief impulse in time. Instead, we create an impulse that's broader and shorter, just as we've done in figure 6.11a. That's the best we can do to represent the impulse with the sinusoids we're given, i.e., the sinusoids with frequency less than 30 Hz. In this way, the lowpass filter acts to smooth the brief large discharge in time.

Finally, inspection of the filtered signal in a more restricted voltage range (figure 6.11c) reveals an important advantage over the naive rectangular filter. The small-amplitude 60 Hz activity produced by the naive rectangular filter does not appear in the FIR filtered data. Because the FIR filter acts more locally in time (the impulse response is finite), this filter does not produce the long-lasting temporal effects of the naive rectangular filter. We also note the clear delay induced by the FIR filter compared to the naive rectangular filter; compare the black and red curves in figure 6.11c.

In the preceding MATLAB code, we applied the FIR filter by computing the convolution. To conclude this section, we introduce a built-in procedure to apply the FIR filter in MATLAB:

```
xnew_MATLAB=filter(b,1,x);   %Apply filter using MATLAB function.
```

Here, we call the MATLAB function `filter` with three arguments. The first argument is the FIR filter we designed at the beginning of this section using the `fir1` command. The second input (a value of 1) is appropriate for the FIR filter we designed here,[4] and the last input is the EEG signal from trial 1. An example of the resulting filtered signal (variable `xnew_MATLAB`) is plotted in figure 6.11c. We see that after an initial transient, the filtered signal computed using the built-in MATLAB function, and the filtered signal computed explicitly using the convolution produce the same result.

Q: Apply the FIR filter to the EEG data *without* using convolution or the MATLAB function `filter`. Instead, only use the Fourier transform and inverse Fourier transform. Do you find results consistent with the other two computations shown in figure 6.11c? *Hint:* You should.

4. Different values for the second input allow specification of different filter types (e.g., a Butterworth filter) but are not considered here. Consult MATLAB documentation for more details.

To summarize, the design and application of a lowpass FIR filter can be performed in two simple steps:

```
b=transpose(fir1(n,Wn,'low'));  %Design the filter,
xnew_MATLAB = filter(b,1,x);     %...apply it to the signal x.
```

To actually execute these lines of code, we must first define the filter parameters (i.e., the filter order and the cutoff frequency), but the essence of the filtering procedure is captured here. These two lines make the process of filter design and application simple but potentially obfuscate what the filter actually does. We therefore did not immediately implement a filter in this way. Instead, we first examined intuitive ideas for filter design (e.g., the naive rectangular filter and Hanning filter) and visualizations. We expect that these initial examples will provide insight to the built-in MATLAB routines. In practice, application of these routines is typically the best choice when analyzing your own data.

6.2.6 What's Phase Got to Do with It?

We saw in the previous section that the FIR filter (implemented using the `fir1` function in MATLAB) introduced a time shift in the resulting signal; this shift appeared in both the impulse response (figure 6.10a) and in application to the EEG signal (figure 6.11a). In many applications, we're interested in the precise timing of neural events. For example, if we'd like to understand the EEG response following a stimulus presentation, we must carefully preserve the timing of EEG features. We discuss in chapter 7 a specific context in which such timing of features is important to preserve (e.g., cross-frequency coupling). In these contexts and others, shifts in the EEG signal must be either well understood and accounted for, or avoided.

To assess how a filter impacts a signal, we develop another visualization technique. Recall that the Fourier transform of a signal consists of both a real and an imaginary component, or equivalently, a magnitude and phase in the complex plane (see, for example, the discussion of phase in chapter 5). We have already discussed how to visualize the magnitude response of a filter (figure 6.10b). We now consider a second visualization in the frequency domain: the *phase response*. The phase response is similar to the magnitude response in that both are frequency domain visualizations of the filter. The primary difference is that the phase response illustrates the impact of the filter on phase at each frequency.

Let's compute the phase response for the lowpass FIR filter. We first construct this filter (using the same procedure described above), and then compute and display the phase response. In MATLAB, repeating some commands from previous sections for completeness,

```
load('Ch6-EEG-1.mat')     %Load the EEG data.
x   = EEG(1,:);           %...and analyze first trial.
```

```
N   = length(x);             %Define no. of points in trial.
dt  = t(2)-t(1);             %Define the sampling interval.
fNQ = 1/dt/2;                %Define the Nyquist frequency,
df  = 1/(N*dt);              %...and the frequency resolution,
faxis = fftshift((-fNQ:df:fNQ-df)); %...create frequency axis,
[~, isorted]=sort(faxis, 'ascend'); %...with axes sorted.

n=100;                       %Define the filter order,
Wn=30/fNQ;                   %...specify the cutoff frequency,
b = transpose(fir1(n,Wn,'low'));     %...build lowpass filter.
bf  = fft(b,N);              %Transform filter to frequency domain,
plot(faxis(isorted), angle(bf(isorted)))%...plot phase response.
```

We first load the data into MATLAB and define the frequency axis. We then construct the lowpass FIR filter and plot the phase response versus frequency (with axes sorted to avoid spurious features in the plot). Note that in the last line of code we use the MATLAB function `angle` to determine the phase of the vector `bf`. We focus on the passband (from −30 Hz to 30 Hz) because signals outside of this band are greatly reduced by the filter and not relevant in the filtered signal.

Q: Examine the phase response plotted in figure 6.12. How does the phase vary with frequency within the passband (i.e., for frequencies between ±30 Hz)?

A: Visual inspection of figure 6.12a reveals that the phase response varies with frequency. This variation is linear except for discrete jumps occurring at $\pm\pi$. To make this linear variation clear, we can *unwrap* the phase:

```
%Plot the unwrapped phase response.
plot(faxis(isorted), unwrap(angle(bf(isorted))))
```

Notice the application of the `unwrap` function to the angle computed in this line of code. With the phase unwrapped, the smooth and linear variation in the phase response versus frequency becomes clear (figure 6.12b). The wrapped phase response (figure 6.12a) allows us to identify frequencies at which the filter phase advances the signal (e.g., when the phase response is positive, such as at 15 Hz), when the filter phase delays the signal (e.g., when the phase response is negative, such as at 25 Hz), or when the filter leaves the phase unchanged (e.g., when the phase is zero, such as at 20 Hz).

Analysis of the phase response for the FIR filter shows that, consistent with our observations of the impulse response and filtered EEG data, the FIR filter alters the phase of

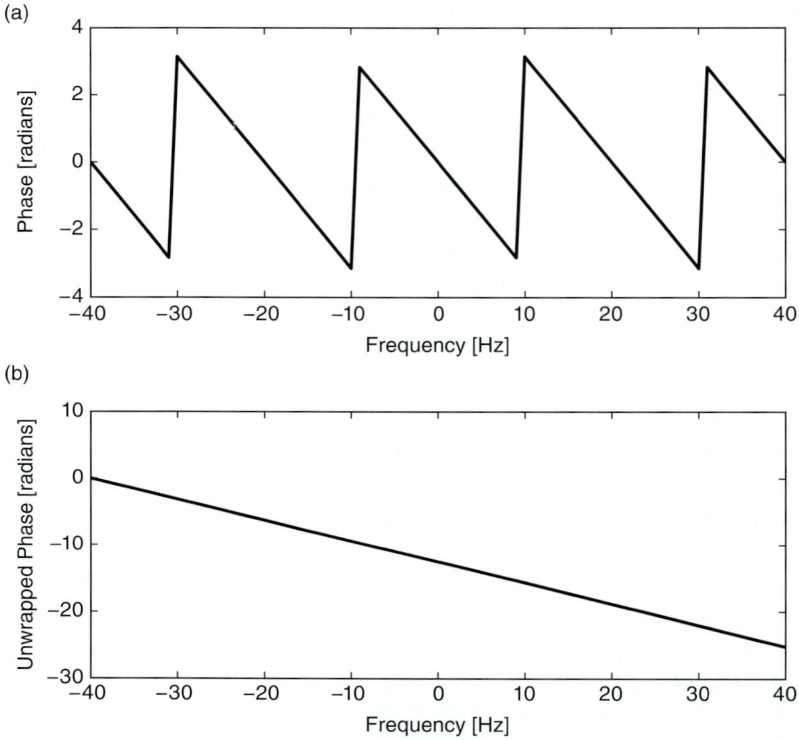

Figure 6.12

(*a*) Phase response and (*b*) unwrapped phase response versus frequency of lowpass FIR filter.

the original signal. To eliminate this shift introduced by the filter, we apply the same filter twice to the data. First, we apply the FIR filter to the original input signal, just as we did to create the lowpass filtered EEG. This filtering operation introduces a shift (of size $n/2$ indices, or 0.05 s for our data) in the resulting EEG (e.g., figure 6.10a). Second, we reverse the filtered signal and then apply this same FIR filter to the reversed sequence. The outcome of this second filtering operation is our desired signal: the filtered data without the phase shift. Let's try this in MATLAB:

```
x1 = filter(b,1,x);        %Apply the filter to the EEG data,
x1 = x1(N:-1:1);           %... reverse the sequence,
x2 = filter(b,1,x1);       %... reapply the filter,
x2 = x2(N:-1:1);           %... and reverse the sequence.
```

In these lines of code, we apply the FIR filter using the MATLAB function `filter`.The first line of code applies the filter once, and the resulting filtered signal is phase shifted relative to the original EEG (figure 6.13, black curve). We then reverse the filtered signal. To do so, we redefine the variable `x1` to begin at the last index `N` and progress to the first index `1` in steps of `-1`; this reverses the sequence. We then filter the reversed sequence, and reverse the result. The resulting double-filtered (and double-reversed) signal no longer exhibits phase distortion relative to the EEG (figure 6.13, green curve); prominent features in the original EEG signal and the *zero-phase* filtered signal, such as the large-amplitude discharge, appear better aligned.

In general, in neuroscience applications, it's often useful to remove phase distortion through zero-phase filtering. MATLAB provides a simple function to perform zero-phase filtering,

```
x3 = filtfilt(b,1,x);          %Perform zero-phase filtering.
```

The `filtfilt` command applies the lowpass FIR filter defined by the parameters `b` to the data in the forward and reverse directions, and (ignoring transients at the beginning and end of the signal) matches our explicit approach to zero-phase filtering (figure 6.13, red curve). In the next section, we apply these commands to complete our analysis of the EEG signal.

Before completing this section, let's briefly consider an intuitive argument to motivate the procedure for performing zero-phase filtering. Consider the impulse response for the lowpass FIR filter of order $n = 100$ we implemented previously (figure 6.10a). An impulse at index k will result in a peak at index $k + n/2$ in the filtered signal; the first filtering operation shifts the index of the peak by $n/2$. Now, consider reversing the filtered signal. For concreteness, let's consider the case where the impulse occurs at index 800 (i.e., $k = 800$)

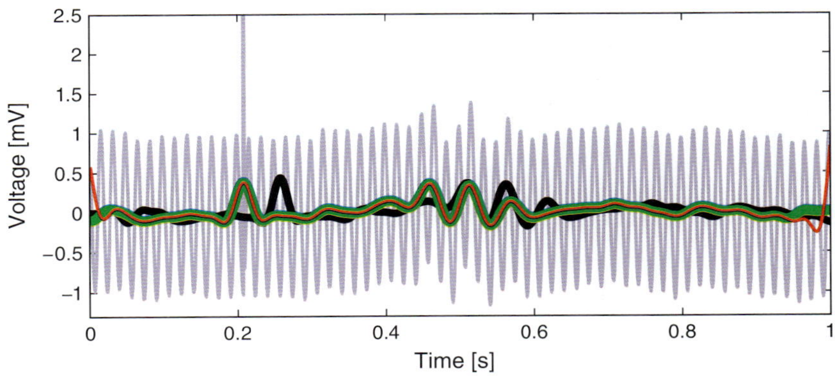

Figure 6.13

Example trace of original EEG signal (*blue*), corresponding lowpass FIR-filtered signal (*black*), and zero-phase lowpass FIR filtered signal (*green*), and zero-phase filtered data computed using the MATLAB function `filtfilt` (*red*).

and the total length of the signal is 2,000 indices. After applying the filter, the peak will occur at index $k + n/2 = 850$. Now, reverse this filtered signal; the peak will then occur at index $2000 - 850 = 1150$. We then filter this new signal, which again shifts the index of the peak by $n/2$; the new peak then occurs at index $1150 + n/2 = 1200$. Finally, we reverse the signal once more; the peak index becomes $2000 - 1200 = 800$. Through this simple example, we gain some intuition for the zero-phase filtering process. By applying the filter twice and reversing the signal appropriately, we maintain the timing of features in the original input signal.

Does applying the same filter twice to the signal impacts the results? Yes. Each time we apply the filter, we convolve the signal with the coefficients `b` determined here for the lowpass FIR filter. Each application changes the resulting signal (see problem 6.8). However, the additional distortion produced by filtering twice is compensated by the elimination of phase distortion.

6.2.7 Analysis of the Filtered EEG Data

Having introduced some basic filtering concepts, let's now return to the EEG data. Our primary scientific goal is to determine whether the provided EEG data exhibit an evoked response. Our initial analysis hinted that an evoked response might occur but was hidden by the large electrical noise—and perhaps other noise—inherent in the EEG data (figure 6.3). To reduce this noise, let's examine the lowpass filtered EEG signal. We choose a lowpass filter to both reduce the 60 Hz electrical noise and reduce other activities associated with nonbrain signals (e.g., muscle artifacts) common in EEG data. In retrospect, the design and application of a lowpass filter with cutoff frequency is now straightforward.

```
load('Ch6-EEG-1.mat')          %Load the EEG data.
dt = t(2)-t(1);                %Define the sampling interval.
fNQ = 1/dt/2;                  %Determine the Nyquist frequency.
K = size(EEG,1);               %Define no. of of trials.

n=100;                         %Define the filter order,
Wn=30/fNQ;                     %...specify the cutoff frequency,
b = transpose(fir1(n,Wn,'low'));    %...build lowpass filter.
EEG_lo = zeros(size(EEG));     %Define matrix to store results,
for k=1:K                      %...zero-phase filter each trial.
    EEG_lo(k,:)=filtfilt(b,1,EEG(k,:));
end
```

Here we use the built-in MATLAB function `fir1` to design the filter and the function `filtfilt` to apply the filter with zero-phase distortion. The design and application of the filter to each trial requires only a few lines of code (including the for-loop). However, we now perform this analysis with a thorough understanding of how the filter behaves; we

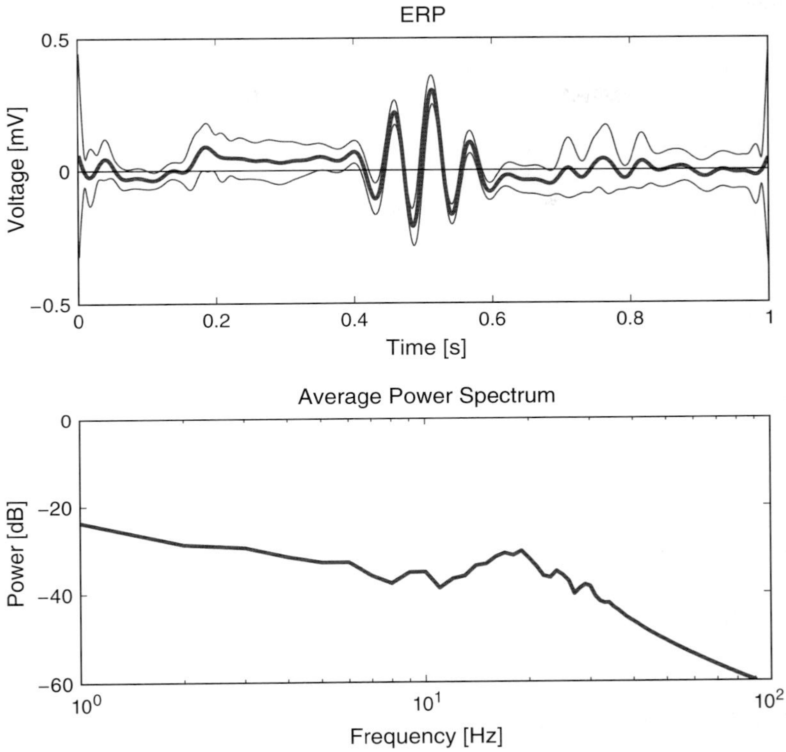

Figure 6.14

ERP and trial-averaged spectrum of lowpass filtered EEG data. (*a*) ERP (*thick curve*) plotted with 95% confidence intervals. (*b*) Trial-averaged spectrum versus frequency.

examined its impulse response (figure 6.10a), magnitude response (figure 6.10b), and phase response (figure 6.12). Let's now analyze the resulting filtered EEG data by computing the evoked response and average spectrum (figure 6.14).

Q: Compare the evoked response and spectrum of the original EEG data (figure 6.3) to the evoked response and spectrum of the filtered data (figure 6.14). What features are similar? What features differ?

A: To compute the evoked response and average spectrum, we apply the same procedures utilized for the original data (see code in section 6.2.3). We now find that between approximately 0.4 s and 0.6 s, the filtered EEG signal exhibits a signifiant ERP; the 95% confidence intervals of the ERP now exclude zero in this range. We are therefore happy to report to our collaborator evidence for a significant ERP in the filtered EEG data. We also note the reduced impact in the ERP of the large, brief discharges that appear in individual trials. In the original EEG data, each discharge from an individual trial was so large that the impact on the ERP was dramatic (e.g., consider times near 0.2 s and near 0.8 s in figure 6.3). In the filtered EEG data, these discharges have been smoothed, and their impact greatly reduced in the ERP.

Inspection of the average spectrum for the filtered EEG data reveals features at low frequencies and perhaps a small peak at 15–25 Hz, as observed in the unfiltered EEG data. Again, we note that the approximately 15–25 Hz peak in the spectrum is consistent with the period of the transient rhythmic discharge in the ERP. The filtered data exhibit much less power spectral density at higher frequencies compared to the original EEG; this is what we expect following application of the lowpass filter.

Summary

We began this chapter with visual analysis of the single-trial data and computation of an ERP and trial-averaged spectrum. The spectrum revealed a large peak at 60 Hz, consistent with visual inspection of the single-trial data. The ERP showed some suggestive evidence for an evoked response; however, we did not find a significant effect. We made an initial conjecture that an interesting evoked response might occur in the data but was hidden by the large-amplitude 60 Hz noise.

To isolate the evoked response, we then focused on reducing the 60 Hz activity in the signal. We introduced the notion of filtering. We put forward two naive approaches, the naive rectangular filter, and the naive Hanning filter and developed these approaches in great detail. We defined the notion of an impulse response and examined how filtering may be equivalently applied in the frequency domain (through multiplication) or in the time domain (through convolution). Through these example filters, we observed the trade-offs that occur in the time and frequency domains. In particular, we observed that the sharp edge in the frequency domain of the naive rectangular filter created long-lasting effects in the time domain, acting to distort the original signal.

We then discussed the application of a finite impulse response (FIR) filter to the data. We showed how this filter may be easily defined and applied in MATLAB using built-in functions. We discussed procedures to visualize a filter's behavior, including the magnitude response and the phase response. Finally, we discussed the importance of zero-phase filtering.

We concluded by reanalyzing the EEG data. To do so, we first lowpass filtered the data and then computed the evoked response and trial-averaged spectrum. After filtering, we found a significant evoked response in the data. Consistent with our initial conjecture, the evoked response was hidden by the high-amplitude 60 Hz noise present in the original signal. Upon filtering to remove this noise, the evoked response became clear.

The design and application of filters is an enormous and rich field of study. The goal of this chapter is not a thorough discussion of filtering. Instead, we introduced only a handful of filtering concepts that motivate a basic understanding of filtering. These concepts extend directly from ideas developed to compute the spectrum in chapters 3 and 4. Through tools such as the Fourier transform and convolution, we are able to visualize and apply filters in the frequency and time domains. These same tools apply and provide context for alternative approaches to filtering. For further details in the design and application of filters see [8, 9]. We apply filters in chapter 7 to assess cross-frequency coupling in neural field data.

Problems

6.1. Load the file `Ch6-EEG-2.mat`, available at

`http://github.com/Mark-Kramer/Case-Studies-Kramer-Eden`

into MATLAB. You will find two variables in your workspace. The variable `EEG` corresponds to an EEG recording. The variable `t` corresponds to the time axis, in units of seconds. Use these data to answer the following questions.

a. Visualize the time series data. What rhythms do you observe?

b. Plot the spectrum versus frequency for these data. Are the dominant rhythms in the spectrum consistent with your visual inspection of the data?

c. Within the EEG data, a low-amplitude sinusoid occurs. Apply the filtering methods developed in this chapter to isolate this sinusoid. Plot both your filtered EEG signal versus time, and the spectrum of your filtered EEG.

6.2. Consider the naive Hanning filter shown in figure 6.8. How do the properties of this filter change as the width of the Hanning window (the variable `win`) is increased or decreased?

6.3. For trial 1, compute the autocovariance (see chapter 3) of the EEG signal, of the naive rectangular filtered EEG signal, and of the lowpass FIR filtered EEG signal. How do the results compare?

6.4. How does changing the order of the lowpass FIR filter impact the resulting EEG signal from the first trial? *Hint:* Consider the following MATLAB code:

```
%Load the EEG data to define useful parameters.
load('Ch6-EEG-1.mat')
x = EEG(1,:);
dt = t(2)-t(1);
N = length(x);
Fs = 1/dt;
fNQ = Fs/2;
```

```
df = 1/(N*dt);
faxis = fftshift((-fNQ:df:fNQ-df));

%Define the filter orders, cutoff frequency, and colors.
n=[1000,500,100,50];
Wn=30/fNQ;
clr = {'r', 'g', 'b', 'm'};

%For each filter, design and visualize the filter.
for k=1:length(n)
    %Design filter.
    b = fir1(n(k),Wn,'low');

    %Visualize filter.
    subplot(2,1,1)
    hold on
    plot(b+0.05*k, 'Color', clr{k})
    hold off
    axis tight
    axis off

    bf = fft(b,N);       %NOTE: zero-pad to same length as x.
    subplot(2,1,2)
    hold on
    plot(faxis,bf.*conj(bf), 'Color', clr{k})
    hold off
    axis tight
    xlim([-55 55])
    xlabel('Frequency [Hz]')

end
subplot(2,1,1)
legend({'n=1000'; 'n=500'; 'n=100'; 'n=50'})
```

6.5. Use the MATLAB function `fir1` to develop a *bandstop* filter that reduces the 60 Hz line noise in the EEG signal. See MATLAB Help for the function `fir1` to assign the function inputs correctly. Use figures to show that your filter performs as expected.

6.6. Compute the phase response of the naive rectangular filter. How does the phase response differ from the phase response computed for the lowpass FIR filter examined in this chapter?

6.7. Use the MATLAB function `fir1` to develop a lowpass FIR filter with cutoff frequency of 30 Hz. Apply this filter to three synthetic signals:

   a. A sinusoid at frequency 15 Hz.
   b. A sinusoid at frequency 20 Hz.
   c. A sinusoid at frequency 25 Hz.

   For each synthetic signal, compare the phase difference between the original (sinusoidal) signal and filtered signal. How do these phase differences compare with the phase response plotted in figure 6.12?

6.8. Your colleague proposes the following new filter in two steps: (1) Apply the lowpass FIR filter with cutoff frequency of 30 Hz we developed in this chapter, and save the result. (2) To the result from (1) apply the same lowpass FIR filter with cutoff frequency of 30 Hz. What is the impact of this double-filtering operation? Consider the impulse response, magnitude response, and phase response.

6.9. In what scenarios would zero-phase filtering be inappropriate? In other words, in what scenarios would a causal filter be appropriate? As a concrete example, consider an experiment consisting of a brief large-amplitude electrical stimulus delivered to the brain. During the stimulus, the recording device becomes saturated by large-amplitude fluctuations; these are artifacts of the stimulus, which we cannot remove from the signal. We would like to analyze the EEG data immediately following the stimulus, when the artifacts have ceased. We would also like to filter these data. What impact would a noncausal filter have on the EEG data immediately following the stimulus? How would a causal filter help address this?

7 Investigation of Cross-Frequency Coupling in a Local Field Potential

Synopsis

Data 100 s of local field potential data sampled at 1000 Hz.

Goal Characterize the coupling between rhythms of different frequency.

Tools Hilbert transform, analytic signal, instantaneous phase, cross-frequency coupling.

7.1 Introduction

7.1.1 Background

In Chapter 5, we focused on the coherence between voltage activity recorded at two electrodes. The coherence is a measure of association between rhythms at the same frequency. In this chapter, we continue our study of field data but now focus on local field potential (LFP) recordings. The LFP is a measure of local population neural activity, produced from small aggregates of neurons [15]. In these data, we examine the association between rhythms of *different* frequencies.

In general, lower-frequency rhythms have been observed to engage larger brain areas and modulate spatially localized fast oscillations [16–20]. This cross-frequency coupling (CFC) between the power (or amplitude) of high-frequency rhythms and the phase of low-frequency rhythms has been observed in many brain regions, has been shown to change in time with task demands, and has been proposed to serve a functional role in working memory, neuronal computation, communication, and learning [21]. Although the cellular and dynamic mechanisms of specific rhythms associated with CFC are relatively well understood, the mechanisms governing interactions between different frequency rhythms and the appropriate techniques for measuring CFC remain active research areas. Although we consider only a single electrode recording here, note that these techniques can be extended to association measures between electrodes as well.

7.1.2 Case Study Data

We are approached by a collaborator recording the local field potential (LFP) from rat hippocampus. She has implanted a small bundle of electrodes, which remain (chronically) implanted as the rat explores a large circular arena. She is interested in assessing the association between different frequency rhythms of the LFP, and more specifically whether an association between different frequency rhythms exists as the rat explores the arena. To address this question, she has provided us with 100 s of LFP data recorded during the experiment (i.e., while the rat spontaneously explored the arena).

7.1.3 Goal

Our goal is to assess the associations between different frequency rhythms recorded in the LFP. To do so, we analyze the LFP data by computing the cross-frequency coupling of the time series. We construct two CFC measures that characterize how the phase of a low-frequency signal modulates the amplitude envelope of a high-frequency signal. This chapter assumes some knowledge of neural rhythms and their assessment. If the material seems overly dense, consult the earlier case studies in chapters 3–6.

7.1.4 Tools

In this chapter, we develop two CFC measures. We introduce the concepts of the Hilbert transform, analytic signal, instantaneous phase, and amplitude envelope.

7.2 Data Analysis

7.2.1 Visual Inspection

To access the data for this chapter, visit

`http://github.com/Mark-Kramer/Case-Studies-Kramer-Eden`

and download the file `Ch7-LFP-1.mat`. Let's begin with visual inspection of the LFP data. To do so, we first load the LFP data into MATLAB and plot it:

```
load('Ch7-LFP-1.mat')          %Load the LFP data,
plot(t,LFP)                    %... and plot it,
xlabel('Time [s]');            %... with axes labeled.
ylabel('Voltage [mV]')
```

Within an example 1 s interval, rhythmic activity in the LFP is apparent (figure 7.1). Visual inspection immediately suggests a dominant low-frequency rhythm interspersed with smaller-amplitude blasts of high-frequency activity.

> **Q:** Approximate the rhythmic activity by visual inspection of the LFP data plotted in figure 7.1. What is the frequency of the large-amplitude rhythm? Do you observe high-frequency activity? If so, where in time, and at what approximate frequency? What is the sampling frequency of these data? If you were to compute the spectrum of the entire dataset (100 s of LFP), what would be the Nyquist frequency and the frequency resolution? *Hint:* In figure 7.1, consider the times near 4.35 s and 4.5 s. Do you see the transient fast oscillations?

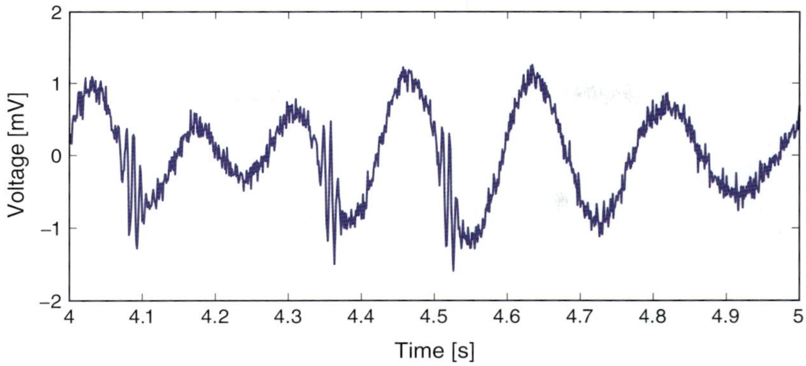

Figure 7.1
Example trace of LFP data.

> *Note:* If you have no idea how to address these questions, or if the terminology seems completely unfamiliar, consider reviewing the case studies in chapters 3 and 4. In those chapters, the notions of rhythms are developed in much greater detail, as well as the spectrum, Nyquist frequency and frequency resolution.

7.2.2 Spectral Analysis

Visual inspection of the LFP data suggests that multiple rhythms appear. To further characterize this observation, we compute the spectrum of the LFP data.[1] We analyze the entire 100 s of data and compute the spectrum with a Hanning taper (see chapter 4). In MATLAB,

```
load('Ch7-LFP-1.mat')               %Load the LFP data.
dt = t(2)-t(1);                     %Define the sampling interval.
T = t(end);                         %Define the duration of data.
N  = length(LFP);                   %Define no. of points in data.

x = hann(N).*transpose(LFP);        %Multiply data by Hanning taper.
xf = fft(x-mean(x));                %Compute Fourier transform of x.
Sxx = 2*dt^2/T *(xf.*conj(xf));     %Compute the spectrum.
Sxx = Sxx(1:N/2+1);                 %Ignore negative frequencies.

df = 1/max(T);                      %Define frequency resolution.
fNQ = 1/dt/2;                       %Define Nyquist frequency.
```

1. We could instead write the *sample* spectrum because we use the observed data to estimate the theoretical spectrum that we would see if we kept repeating this experiment. However this distinction is not essential to the discussion here.

```
faxis = (0:df:fNQ);                    %Construct frequency axis.
plot(faxis, 10*log10(Sxx))             %Plot spectrum vs frequency.
xlim([0 200]); ylim([-80 0])           %Set frequency & decibel range.
xlabel('Frequency [Hz]')               %Label axes.
ylabel('Power [ mV^2/Hz]')
```

Q: Does the use of the Fourier transform and Hanning taper make sense? Do the expressions for the frequency resolution (`df`), Nyquist frequency (`fNQ`), and spectrum (`Sxx`) make sense?

A: If you answered yes in all cases, you're right. If not, consider reviewing the case studies in chapters 3 and 4.

The resulting spectrum, shown in figure 7.2, reveals two intervals of increased power spectral density. The lowest-frequency peak at 6 Hz is also the largest and corresponds to the slow rhythm we observe dominating the signal through visual inspection of figure 7.1. At higher frequencies, we find an additional broadband peak at approximately 80–120 Hz. These spectral results support our initial visual inspection of the signal; there exist both low- and high-frequency activities in the LFP data. We now consider the primary question of interest: Do these different frequency rhythms exhibit associations?

7.2.3 Cross-Frequency Coupling

To assess whether different frequency rhythms interact in the LFP recording, we implement a measure to calculate CFC. The idea of CFC analysis is to determine whether a

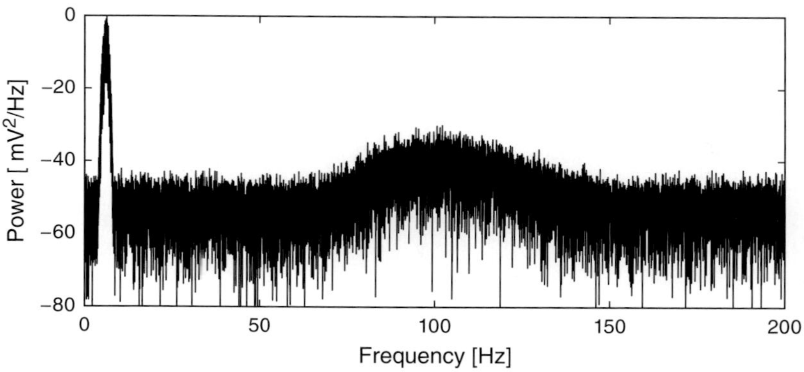

Figure 7.2
Spectrum of LFP data.

relation exists between the phase of a low-frequency signal and the envelope or amplitude of a high-frequency signal. In general, computing CFC involves three steps. Each step contains important questions and encompasses entire fields of study. Our goal in this section is to move quickly forward and produce a procedure we can employ, investigate, and criticize. Continued study of CFC—and the associated nuances of each step—is an active area of ongoing research.

CFC analysis steps

1. Filter the data into high- and low-frequency bands.
2. Extract the amplitude and phase from the filtered signals.
3. Determine if the phase and amplitude are related.

1. Filter the Data into High- and Low-Frequency Bands. The first step in the CFC analysis is to filter the data into two frequency bands of interest. The choice is not arbitrary: the separate frequency bands are motivated by initial spectral analysis of the LFP data. In this case, we choose the low-frequency band as 5–7 Hz, consistent with the largest peak in the spectrum, and the high-frequency band as 80–120 Hz, consistent with the second-largest broadband peak (figure 7.2). To consider alternative frequency bands, the same analysis steps would apply.

There are many options to perform the filtering. To do so requires us to design a filter that ideally extracts the frequency bands of interest without distorting the results. Here, we apply a finite impulse response (FIR) filter (see chapter 6). In MATLAB,

```
dt  = t(2)-t(1);                 %Define the sampling interval.
Fs  = 1/dt;                      %Define the sampling frequency.
fNQ = Fs/2;                      %Define the Nyquist frequency.
                                 %For low-frequency interval,
Wn  = [5,7]/fNQ;                 %...set the passband,
n   = 100;                       %...and filter order,
b   = fir1(n,Wn);                %...build bandpass filter.
Vlo = filtfilt(b,1,LFP);         %...and apply filter.
                                 %For high-frequency interval,
Wn  = [80,120]/fNQ;              %...set the passband,
n   = 100;                       %...and filter order,
b   = fir1(n,Wn);                %...build bandpass filter.
Vhi = filtfilt(b,1,LFP);         %...and apply filter.
```

For each frequency band, we specify a frequency interval of interest by defining the low- and high-cutoff frequencies in the vector `Wn`. This vector contains two elements, and we divide this vector by the Nyquist frequency (`fNQ`). In this way, we specify the passband of the filter on the interval between 0 and 1, where 1 represents the Nyquist frequency. We then set the filter order (`n`) and design the filter using the MATLAB function `fir1`. Finally, we apply the filter using the MATLAB function `filtfilt`, which performs zero-phase filtering by applying the filter in both the forward and reverse directions (see chapter 6). We note that the filtering procedure is nearly the same in both frequency bands; the only change is the specification of the frequency interval of interest.

> If the filtering procedure seems puzzling consider completing the case study in chapter 6. That chapter describes filtering in detail.

To understand the impact of this filtering operation on the LFP, let's plot the results (figure 7.3). As expected, the low-frequency band captures the large-amplitude rhythm dominating the LFP signal, while the higher-frequency band isolates the brief bursts of faster activity.

2. Extract the Amplitude and Phase from Filtered Signals. The next step in the CFC procedure is to extract the phase of the low-frequency signal and the amplitude envelope (or simply, amplitude) of the high-frequency signal. To gain some intuition for this operation, let's consider the amplitude and phase for the example signals in figure 7.3. We plot the low-frequency signal and its phase in figure 7.4a. As time progresses, the phase increases nearly linearly from $-\pi$ to π. At π, the phase jumps suddenly to $-\pi$. This apparent discontinuity is imposed by the space on which the phase evolves: a circle.

As another example of such a space, consider the time on a 24-hour clock changing suddenly from 23:59 to 00:00. In this case, the measurement has suddenly shifted discontinuously, but time has not; this discontinuity appears because we choose to measure time in 24-hour intervals (for good reason). The same is true of the phase that we choose to

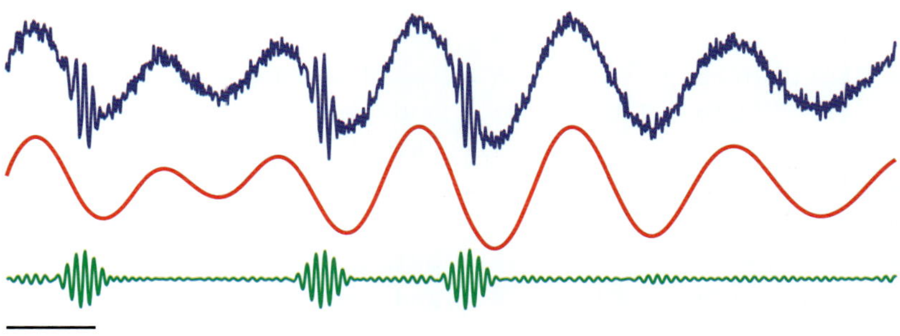

Figure 7.3

Examples of filtering the LFP data. Original LFP signal (*blue*) is filtered into a low-frequency band (*red*) and a high-frequency band (*green*). Scale bar indicates 0.1 s.

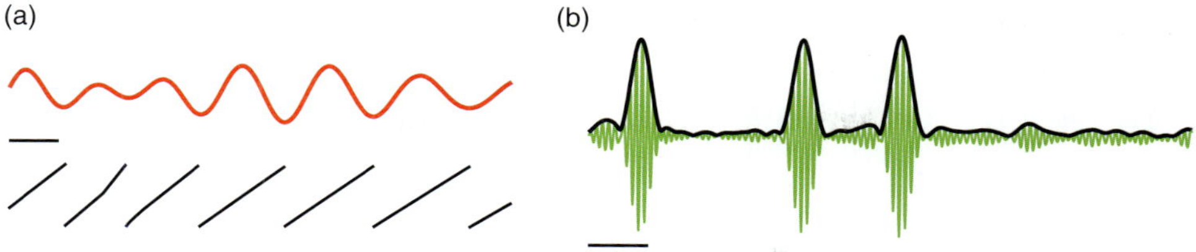

Figure 7.4

Examples of phase and amplitude of filtered LFP data. (*a*) Phase (*black*) of low-frequency signal (*red*) increases from $-\pi$ to π over time. (*b*) Amplitude envelope (*black*) outlines deviations of high-frequency signal (*green*). Scale bar indicates 0.1 s.

measure, from $-\pi$ to π. The two endpoints ($-\pi$ and π) actually touch on the space of a circle but appear to discontinuously jump when plotted on the plane.

The amplitude envelope (figure 7.4b) outlines the extent of deviations of the high-frequency signal. Notice that the envelope fluctuates much less rapidly than the underlying high-frequency signal.

To compute CFC, we compare the two signals we've extracted from the data, the phase of the low-frequency activity and the amplitude envelope of the high-frequency activity (figure 7.4). How do we actually extract the phase and amplitude signals from the data? There are a variety of options to do so, and we choose here to employ the *analytic signal* approach, which allows us to estimate the instantaneous phase and amplitude envelope of the LFP.

The first step in computing the analytic signal is to compute the *Hilbert transform*. We begin with some notation. Define x as a narrowband signal (i.e., a signal with most of its energy concentrated in a narrow frequency range,[2] e.g., the low- or high-frequency band filtered signals in figure 7.3). Then the Hilbert transform of x, let's call it y, is

$$y = H(x).$$

It's perhaps more intuitive to consider the effect of the Hilbert Transform on the frequencies f of x,

$$H(x) = \begin{cases} -\pi/2 \text{ phase shift} & \text{if } f > 0, \\ 0 \text{ phase shift} & \text{if } f = 0, \\ \pi/2 \text{ phase shift} & \text{if } f < 0. \end{cases}$$

> The Hilbert transform $H(x)$ of the signal x produces a phase shift of ± 90 degrees for $\mp$ frequencies of x.

2. The impact of this narrowband assumption on CFC estimates remains an open research topic. One might consider, for example, the meaning and implications of estimating phase from a broadband signal, and the impact on subsequent CFC results.

The Hilbert Transform can also be described in the time domain, although its representation is hardly intuitive (see the appendix at the end of this chapter). Then the analytic signal z is

$$z = x + iy = x + iH(x). \tag{7.1}$$

The effect of the Hilbert transform is to remove negative frequencies from z. As it stands, this is not obvious. To get a sense for why this is so, let's consider a simple example.

What Does the Hilbert Transform Do? Let x_0 be a sinusoid at frequency f_0,

$$x_0 = 2\cos(2\pi f_o t) = 2\cos(\omega_0 t), \tag{7.2}$$

where to simplify notation we have defined $\omega_0 = 2\pi f_o$. We know from Euler's formula that

$$x_0 = e^{i\omega_0 t} + e^{-i\omega_0 t}. \tag{7.3}$$

The real variable x_0 possesses both a positive and a negative frequency component (i.e., ω_0 and $-\omega_0$). So, the spectrum has two peaks (figure 7.5). For real signals, which include nearly all recordings of brain activity, the negative frequency component is redundant, and we usually ignore it. However, the negative frequency component still remains.

By definition, the effect of the Hilbert transform is to induce a phase shift. For positive frequencies, the phase shift is $-\pi/2$. We can produce this phase shift by multiplying the positive frequency part of the signal by $-i$.

Q: Why does a phase shift of $-\pi/2$ correspond to multiplication by $-i$?

A: Consider the complex exponential $e^{i\omega_0 t}$, which consists of only a positive frequency component (ω_0). This signal shifted in phase by $-\pi/2$ corresponds to the new signal $e^{i(\omega_0 t - \pi/2)}$, which simplifies to

$$e^{i(\omega_0 t-\pi/2)} = e^{i\omega_0 t} e^{-i\pi/2} = e^{i\omega_0 t}\Big(\cos(\pi/2) - i\sin(\pi/2)\Big) = e^{i\omega_0 t}(-i).$$

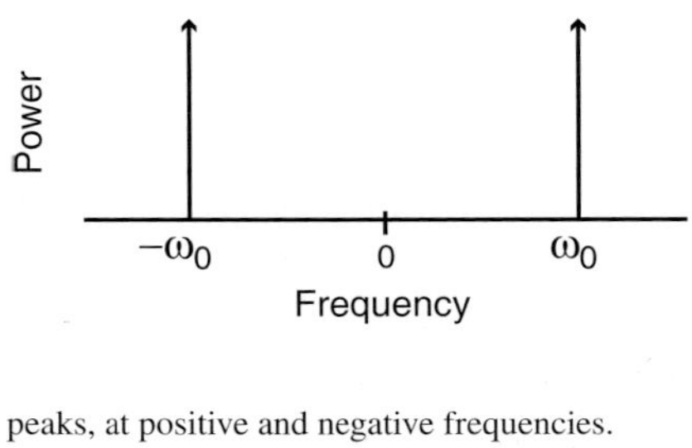

Figure 7.5
Spectrum of a sinusoid has two peaks, at positive and negative frequencies.

Notice the result simplifies to the original complex exponential $e^{i\omega_0 t}$ multiplied by $-i$. This shows that the $-\pi/2$ phase shift corresponds to multiplication of the positive frequency component ($e^{i\omega_0 t}$) by $-i$.

Q: Can you show that a $\pi/2$ phase shift corresponds to multiplication by i?

This analysis shows that we can represent the Hilbert Transform of x at frequency f as

$$H(x) = \begin{cases} -ix & \text{if } f > 0, \\ 0 & \text{if } f = 0, \\ ix & \text{if } f < 0. \end{cases}$$

Therefore, the Hilbert transform of $x_0 = e^{i\omega_0 t} + e^{-i\omega_0 t}$ becomes

$$y_0 = H(x_0) = -ie^{i\omega_0 t} + ie^{-i\omega_0 t}.$$

In this equation, we multiply the positive frequency part of x_0 (i.e., $e^{i\omega_0 t}$) by $-i$, and the negative frequency part of x_0 (i.e., $e^{-i\omega_0 t}$) by i. Simplifying this expression using Euler's formula, we find

$$y_0 = 2\sin(\omega_0 t). \tag{7.4}$$

Q: Can you perform this simplification? In other words, can you show that $-ie^{i\omega_0 t} + ie^{-i\omega_0 t} = 2\sin(\omega_0 t)$?

The Hilbert Transform of x_0 (a cosine function) is a sine function. We could perhaps have guessed this: sine is a 90-degree ($\pi/2$) phase shift of cosine.

We are now ready to define the analytic signal (z_0) for this example. Using the expressions for x_0 and y_0 and Euler's formula, we find

$$z_0 = x_0 + iy_0 = 2\cos(\omega_0 t) + i2\sin(\omega_0 t) = 2e^{i\omega_0 t}.$$

Notice that this analytic signal z_0 contains no negative frequencies; as mentioned, the effect of the Hilbert Transform is to eliminate the negative frequencies from x. The spectrum of this signal consists of a single peak at ω_0, compared to the two peaks at $\pm\omega_0$ in the original signal x (figure 7.5). In this sense, the analytic signal (z_0) is simpler than the original signal x_0. To express the original signal x_0 required two complex exponential functions—see (7.3)—compared to only one complex exponential required to express the corresponding

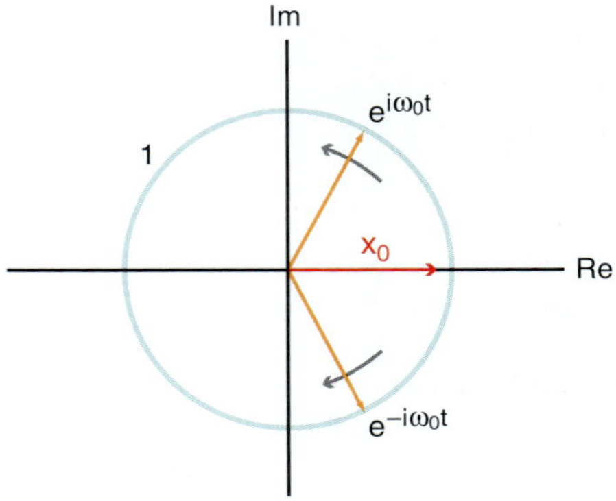

Figure 7.6
Cartoon of real time series x_0 (*red*) plotted in the complex plane. The two complex exponentials (*orange*) cancel in their imaginary parts to produce x_0. Over time, the two complex exponentials rotate (*gray*) so that their imaginary parts cancel at each moment in time and x_0 remains on the real axis. Blue circle indicates radius 1.

analytic signal z_0. There's an interesting geometrical interpretation of this. Consider plotting x_0 in the complex plane (figure 7.6). Because x_0 is real, this quantity evolves in time along the real axis. To keep x_0 on the real axis, the two complex exponentials that define x_0 (i.e., $e^{i\omega_0 t}$ and $e^{-i\omega_0 t}$) rotate in opposite directions along the unit circle. By doing so, the imaginary components of these two vectors cancel, and we're left with a purely real quantity x_0.

> **Q:** The phase of a complex quantity is the angle with respect to the real axis in the complex plane. What is the angle of x_0 in figure 7.6?

2. Extract the Amplitude and Phase from Filtered Signal (Continued). Having developed some understanding of the Hilbert Transform, let's now return to the LFP data of interest here. It's relatively easy to compute the analytic signal and extract the phase and amplitude in MATLAB:

```
phi=angle(hilbert(Vlo));%Compute phase of low-freq signal.
amp=abs(hilbert(Vhi));  %Compute amplitude of high-freq signal.
```

These operations require just two lines of code. But beware of the following.

> **Alert!** The command `hilbert(x)` returns the analytic signal of x, not the Hilbert transform of x.

To extract the phase, we apply the MATLAB function `angle` to the analytic signal of the data filtered in the low-frequency band (variable `Vlo`). We then extract the amplitude of the analytic signal of the data filtered in the high-frequency band (variable `Vhi`) by computing the absolute value (MATLAB function `abs`).

To summarize, in this step we apply the Hilbert transform to create the analytic signal and get the phase or amplitude of the bandpass-filtered data.

3. Determine if the Phase and Amplitude are Related. As with the previous steps, we have at our disposal a variety of procedures to assess the relation between the phase (of the low-frequency signal) and amplitude (of the high-frequency signal). We do so here in two ways.

Method 1: Phase-amplitude plot. To start, define the two-column phase-amplitude vector,

$$\begin{pmatrix} \phi(1) & A(1) \\ \phi(2) & A(2) \\ \phi(3) & A(3) \\ \vdots & \vdots \end{pmatrix},$$

where $\phi(i)$ is the phase of the low-frequency band activity at time index i, and $A(i)$ is the amplitude of the high-frequency band activity at time index i. In other words, each row defines the instantaneous phase and amplitude of the low- and high-frequency filtered data, respectively.

We now use this two-column vector to assess whether the phase and amplitude envelope are related. Let's begin by plotting the average amplitude versus phase. We divide the phase interval into bins of size 0.1 beginning at $-\pi$ and ending at π. The choice of bin size is somewhat arbitrary; this choice will work fine, but you might consider the impact of other choices.

For a chosen phase bin, determine the time indices where the phase ϕ falls within this phase bin; we can think of these times as indexing specific rows of the two-column phase-amplitude vector. Then compute the average amplitude at these same time indices. The result is the average amplitude for phases that lie within the chosen phase bin. Finally, repeat this procedure for all phase bins. We implement these steps in MATLAB and plot the results as follows:

```
p_bins = (-pi:0.1:pi);                 %Define the phase bins.
a_mean = zeros(length(p_bins)-1,1);  %Vector for average amps.
p_mean = zeros(length(p_bins)-1,1);  %Vector for phase bins.
```

```
for k=1:length(p_bins)-1                %For each phase bin,
    pL = p_bins(k);                     %... lower phase limit,
    pR = p_bins(k+1);                   %... upper phase limit.
    indices=find(phi>=pL & phi<pR);     %Find phases falling in bin,
    a_mean(k)  = mean(amp(indices));    %... compute mean amplitude,
    p_mean(k)  = mean([pL, pR]);        %... save center phase.
end
```

Q: Consider the plot of average amplitude versus phase (figure 7.7a). Does this result suggest CFC occurs in these data?

A: We plot in figure 7.7a the phase bins (variable `p_mean`) versus the mean amplitude in each bin (variable `a_mean`). Visual inspection of this phase-amplitude plot suggests that the amplitude of the high-frequency signal depends on the phase of the low-frequency signal. In particular, we note that when the phase is near a value of 2 radians, the amplitude tends to be large, while at other phases the amplitude is smaller. This conclusion suggests that CFC does occur in the data; the high-frequency amplitude depends on the low-frequency phase.

Q: If no CFC occurred in the data, what would you expect to find in the plot of average amplitude versus phase?

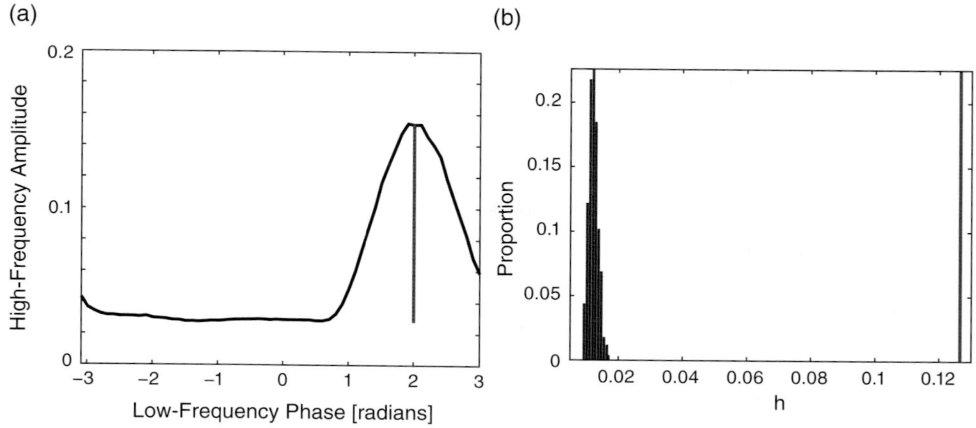

Figure 7.7

Phase-amplitude plot between low-frequency phase and high-frequency amplitude envelope. (*a*) Average amplitude of the high-frequency band versus phase of low-frequency band. Value of statistic h is length of the blue vertical line. (*b*) Distribution of h values for surrogate data (*bars*) versus observed value of h (*red vertical line*).

As a basic statistic to capture the extent of this relation, we compute the difference between the maximum and minimum of the average amplitude envelope over phases (blue vertical line in figure 7.7a). Let's assign this difference the label h. In MATLAB,

```
%Difference between max and min modulation.
h = max(a_mean)-min(a_mean);
```

We find a value of $h = 0.1265$. This value, on its own, is difficult to interpret. Is it bigger or smaller than we expect? To assess the significance of h, let's generate a surrogate phase-amplitude vector by resampling without replacement the amplitude time series (i.e., the second column of the phase-amplitude vector). Resampling is a powerful technique that we have applied in our analysis of other case study data.[3] By performing this resampling, we reassign each phase an amplitude chosen randomly from the entire 100 s LFP recording. We expect that if CFC does exist in these data, then the timing of the phase and amplitude vectors will be important; for CFC to occur, the amplitude and phase must coordinate in time. By disrupting this timing in the resampling procedure, we expect to eliminate the coordination between amplitude and phase necessary to produce CFC.

For each surrogate phase-amplitude vector, we compute the statistic h. To generate a distribution of h values, let's repeat 1,000 times this process of creating surrogate data through resampling and computing h. In MATLAB,

```
n_surrogates = 1000;                    %Define no. of surrogates.
hS = zeros(n_surrogates,1);             %Vector to hold h results.
for ns=1:n_surrogates;                  %For each surrogate,
    ampS = amp(randperm(length(amp)));  %Resample amplitude,
    p_bins = (-pi:0.1:pi);               %Define the phase bins.
    a_mean = zeros(length(p_bins)-1,1); %Vector for average amps.
    p_mean = zeros(length(p_bins)-1,1); %Vector for phase bins.
    for k=1:length(p_bins)-1             %For each phase bin,
      pL = p_bins(k);                    %...lower phase limit,
      pR = p_bins(k+1);                  %...upper phase limit.
      indices=find(phi>=pL & phi<pR);    %Find phases in bin,
      a_mean(k) = mean(ampS(indices));   %...compute mean amp,
      p_mean(k) = mean([pL, pR]);        %...save center phase.
    end
    hS(ns) = max(a_mean)-min(a_mean);    %Store surrogate h.
end
```

In this code, we first define the number of surrogates (variable `n_surrogates`) and a vector to store the statistic h computed for each surrogate phase-amplitude vector (variable `hS`). Then, for each surrogate, we use the MATLAB function `randperm` to randomly

3. For a review of resampling, applied in a different case study, see chapter 2.

permute the amplitude vector without replacement. We then use this permuted amplitude vector (variable `ampS`) and the original phase vector (variable `phi`) to compute h; this last step utilizes the MATLAB code developed earlier to compute h for the original (unpermuted) data.

> **Q:** Do you notice any inefficiencies in this code? If so, how would you modify the code to avoid these inefficiencies (e.g., repeated calculations of unchanging quantities)?

Figure 7.7b shows the results of this resampling procedure as a histogram of the variable `hS`. The value of h computed from the original data lies far outside the distribution of surrogate h values. To compute a p-value, we determine the proportion of surrogate h values greater than the observed h value:

```
p = length(find(hS > h))/length(hS);     %Compute p-value.
```

We find a p-value that is very small; there are no surrogate values of h that exceed the observed value of h. We therefore conclude that in this case the observed value of h is significant. As a statistician would say, we reject the null hypothesis of *no* CFC between the phase of the 5–7 Hz low-frequency band and the amplitude of the 80–120 Hz high frequency band.

Method 2 (advanced): A generalized linear model approach. The method described to assess the relation between the phase (of the low-frequency activity) and amplitude (of the high-frequency activity) is common in the neuroscience literature (as reviewed in [22]). However, alternatives exist. Here, we outline a method based on the concept of a *generalized linear model* (GLM). This method involves more advanced statistical concepts than the phase-amplitude plot described in method 1. We discuss generalized linear models in more detail in chapter 9. Here we only outline the highlights of a procedure to create a GLM-CFC statistic. A more detailed description of this procedure, which includes many simulated examples, may be found in [23].

The idea of the GLM-CFC statistic is to build a statistical model that describes the amplitude (A) of the high-frequency signal as a function of the phase (ϕ) of the low-frequency signal. To create this model, we first assume a distribution for the amplitudes. We choose a gamma distribution because the amplitudes are real and positive, and typically confined to a limited interval with infrequent large events. We must then choose a function that links the amplitudes to the phases. Here we choose the log link,

$$\log(\mu) = \beta X,$$

where μ is the expected value of the amplitudes, X is a function of the phases, and β are the unknown model coefficients to determine. The number of coefficients depends on the model choice. The log link function is common for GLMs using a gamma distribution and leads to models where predictors have multiplicative effects on the response.

We construct two GLMs to fit the amplitude A as a function of the phase ϕ. In the first, we assume that the amplitude does not depend on the phase; we label this the *null model* because this model represents the null hypothesis of no CFC. In this case, X is a constant and does not depend on the phase. We set $X = 1$, and there is a single unknown coefficient to estimate, which we label β_0. Conceptually, the null model estimates the average amplitude across all phases.

In the second model, which we label the *spline model*, we use cardinal splines to fit a smooth function for the expected amplitude as a function of the phase. Cardinal splines are smooth, piecewise-connected third-order polynomial functions that are defined by a set of control points. The advantage of the cardinal spline is that it is capable of approximating any continuous functional relation between phase and amplitude with a small number of parameters [24]. These parameters, the control points, are directly interpretable as the expected amplitude at a specific set of phase values. The spline fit then smoothly interpolates between the estimated control points.

To fit a spline model, we generate X by applying a set of cardinal spline basis functions to the observed phase values, ϕ, at each time step. We select a number of control points, n, and space these evenly between 0 and 2π. The value of the spline estimate at any phase is determined by the two nearest control points to the left and the two nearest control points to the right, where the control point values below zero or above 2π are taken modulo 2π. In this way, the spline function is defined over the circular topology of phase values. We must also select the tension parameter for the spline, which controls the curviness of the function at the control points. Here we choose a tension parameter of 0.5, a standard choice. In this case, X is a matrix consisting of n independent variables (the control points), and therefore n unknown model coefficients to estimate, which we label β_S. Exponentiating these estimates represents the multiplicative effect of the phase on the expected value of the amplitude envelope at each control point.

Principled methods exist to determine the number of control points (or knots) n in the spline model (e.g., the Akaike information criterion (AIC); see [23]). An alternative method for selecting the number of knots is to employ prior physical or observational knowledge about the system. For example, if we believe that the amplitude increases at one or two broad phase intervals, then we may choose to utilize 4 or 10 knots, respectively, in the spline model. However, if instead we believe that the amplitude increases in a sharp phase interval, then we may choose to utilize more knots (e.g., more than 10). By selecting too few knots, we may fail to detect amplitude increases restricted to narrow phase intervals, and by selecting too many knots, we may lose statistical power. In general, by selecting the number of knots, we impose a class of models in the GLM-CFC procedure, and we may do so using either quantitative techniques (e.g., AIC) or prior physical knowledge about the system.

After estimating the unknown coefficients β_0 and β_S of the two models, we then compute the predicted values for the amplitude using the spline model (A_S) and the null model (A_0) at 100 phase values evenly spaced between $-\pi$ and π. As a scalar statistic to characterize the CFC we compute,

$$r = \max\big[\text{abs}\big[1 - A_S/A_0\big]\big], \tag{7.5}$$

which is the maximum absolute fractional change between the spline and null models. This statistic is therefore simply interpreted as the largest proportional (or percentage) change between the null and spline models. A large value of r is indicative of CFC; when r is large, the amplitude at some phase in the spline model differs from the constant amplitude of the null model.

An advantage of this modeling approach is that we can use the GLM to generate confidence intervalss for the statistic r. To do so, we use the estimated coefficients β_S to generate 10,000 normally distributed samples of the coefficients (β^j_S, where $j = \{0, 1, 2, \ldots, 10{,}000\}$). For each j, we then compute the predicted values for the amplitude using the spline model (A^j_S) and reestimate the amplitude for the null model (A^j_0) as the mean of A^j_S. Finally, we compute the measure r^j_S for each j. From the resulting distribution of r^j_S, we determine the 0.025 and 0.975 quantiles. In this way, we use the surrogate distribution to define the 95% confidence intervals for the statistic r.

The statistic r provides a single scalar value representative of CFC between the phase and amplitude time series, and the associated confidence intervals provides a range of certainty in the statistic. To visualize CFC, we plot the predicted amplitude as a function of phase for the null and spline GLMs, and the pointwise 95% confidence bounds of the predicted amplitude values for both models. This provides a graphical representation of the differences between the two models; strong CFC at some phase results in large differences between the two models.

To summarize, we measure CFC using the single quantity r in (7.5) with a corresponding 95% confidence intervals determined from the GLM. This measure represents the largest deviation between the null model (which allows no variation in amplitude with phase) and the spline model (which permits variation of amplitude with phase). To visualize CFC as a function of phase, we plot both models with corresponding pointwise 95% confidence intervalss.

To compute the GLM-CFC statistic in MATLAB, download the file `GLM_CFC.m` from

`http://github.com/Mark-Kramer/Case-Studies-Kramer-Eden`

This file defines a MATLAB function to compute the GLM-CFC statistic. We provide as input to this function the low-frequency signal, high-frequency signal, and the number of control points to use in the spline model. The function returns the statistic r and its 95% confidence intervals. In MATLAB,

```
nCtlPts = 8;                          %Define no. of control points,
[r,r_CI]=GLM_CFC(Vlo,Vhi,nCtlPts);%...and compute r.
```

Here we have chosen to use eight control points, and analyzed the low- (variable `Vlo`) and high- (variable `Vhi`) frequency signals, as defined, previously. The function returns for the GLM-CFC statistic $R = 1.73$, with 95% confidence intervals $[1.71, 1.76]$.

In addition to returning the statistic r and its 95% confidence intervals, the function also produces the plot in figure 7.8. Graphically, the approximately 173% maximal deviation between the two models (corresponding to $r = 1.73$) is represented by the height of the blue vertical line at a phase near 2 radians. The statistic r is quite large, with a 95% confidence intervals that exceeds zero, in support of the conclusion that CFC occurs between the chosen low- and high-frequency bands. We note that these results are consistent with the conclusions from method 1 (the phase-amplitude plot) as well, which inspires further confidence.

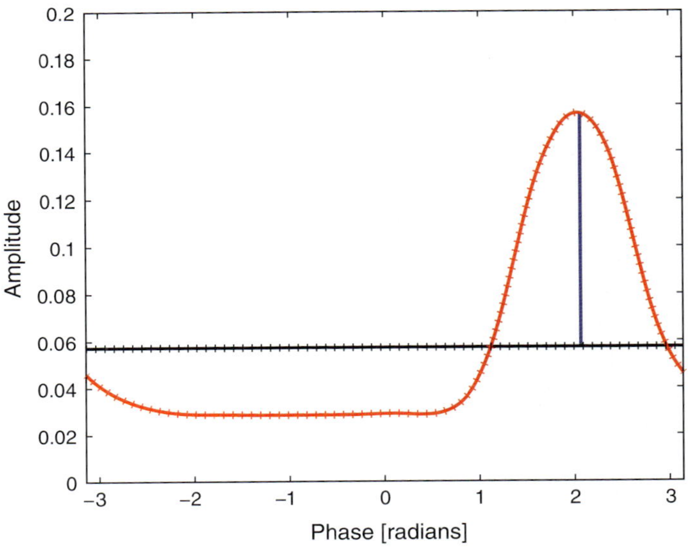

Figure 7.8

GLM fits to high-frequency band amplitude as a function of low-frequency band phase. Fits of the null model (*black*) and spline model (*red*). Means are solid curves; 95% confidence intervals are dotted curves, which remain close to the means. Blue vertical line indicates largest difference between the two models.

Summary

In this chapter, we considered techniques to characterize cross-frequency coupling (CFC), associations between rhythmic activity observed in two different frequency bands. To do so, we introduced the Hilbert transform, which can be used to compute the instantaneous phase and amplitude of a signal. We focused on characterizing the relation between the phase of low-frequency band activity (5–7 Hz) and the amplitude of high-frequency band activity (100–140 Hz) using two approaches. In one approach, we computed the average amplitude at each phase and determined the extent of the variability (or wiggliness). In the other approach, we utilized the GLM framework to develop a statistical model of the data.

For the LFP data of interest here, we found evidence for CFC between the two frequency bands using both methods. Importantly, these results also appear consistent with visual inspection of the unfiltered data. Careful inspection of the example in figure 7.1 suggests that CFC does in fact occur in these data. In general, such strong CFC, visible to the naked eye in the unprocessed LFP data, is unusual. Instead, data analysis techniques are required to detect features not obvious through visual inspection alone. Developing techniques to assess CFC and understanding the biophysical mechanisms of CFC and implications for function, remain active research areas.

In developing these approaches, we utilized expertise and procedures developed in other chapters. In particular, we relied on the notions of frequency and power, amplitude and phase, filtering, resampling, and generalized linear models. Such a multifaceted approach is typical in the analysis of neural data, where we leverage the skills developed in analyzing other datasets.

Problems

7.1. Load the file `Ch7-LFP-2.mat` available at

```
http://github.com/Mark-Kramer/Case-Studies-Kramer-Eden
```

into MATLAB. You will find two variables. The variable `LFP` corresponds to an LFP recording. The variable `t` corresponds to the time axis, in units of seconds. Use these data to answer the following questions.

a. Visualize the time series data. What rhythms do you observe? Do you detect evidence for CFC in your visualizations?

b. Plot the spectrum versus frequency for these data. Are the dominant rhythms in the spectrum consistent with your visual inspection of the data?

c. Apply the two CFC methods developed in this chapter to these data. In doing so, you must choose the low-frequency and high-frequency bands. What choices will you make, and why? What do you find using the two CFC methods?

d. Describe (in a few sentences) your spectrum and CFC results, as you would to a colleague or collaborator.

7.2. Load the dataset `Ch7-LFP-3.mat` available at

`http://github.com/Mark-Kramer/Case-Studies-Kramer-Eden`

into MATLAB. You will find two variables. The variable `LFP` corresponds to an LFP recording. The variable `t` corresponds to the time axis, in units of seconds. Use these data to answer the following questions.

a. Visualize the time series data. What rhythms do you observe? Do you detect evidence for CFC in your visualizations? Note: Look carefully at these traces.

b. Plot the spectrum versus frequency for these data. Are the dominant rhythms in the spectrum consistent with your visual inspection of the data?

c. Apply the two CFC methods developed in this chapter to these data. In doing so, you must choose the low-frequency and high-frequency bands. What choices will you make, and why? What do you find using the two CFC methods?

d. Describe your spectrum and CFC results, as you would to a colleague or collaborator.

7.3. Generate synthetic data consisting of Gaussian noise. More specifically, generate 100 s of artificial noise data sampled at 1000 Hz. Then compute CFC of these data. To do so, use the low-frequency band of 5–7 Hz and the high-frequency band of 80–120 Hz. What do you expect to find (i.e., will this noisy signal exhibit CFC)? What do you find?

7.4. In this chapter, we examined the relation between the phase of the low-frequency band (5–7 Hz) and the amplitude of a selected high-frequency band (80–120 Hz) and found strong evidence in support of CFC. Perhaps CFC occurs between other frequency bands in these data. Repeat both methods of CFC analysis described, but now consider the relation between the low-frequency band (5–7 Hz) and a second high-frequency band of 40–60 Hz. What evidence do you find for CFC between these two frequency bands?

7.5. In our computation of the GLM-CFC statistic for the LFP data analyzed in this chapter, we fixed at 8 the number of control points in the spline model. Repeat the analysis of the LFP data using different choices for the number of control points. How are the results affected?

7.6. In our analysis of CFC, we focused on distinct choices of high- and low-frequency bands. However, sometimes we would like to explore a broader range of potential cross-frequency interactions. To do so, we need a comodulagram. Use the code developed in this chapter to define a new function that computes a comodulogram. Your comodulogram should have two axes:

a. x-axis: the phase frequency (e.g., 3 Hz to 12 Hz in 1 Hz steps)
b. y-axis: the amplitude envelope frequency (e.g., 50 Hz to 200 Hz in 10 Hz steps)

For each pair of (x-axis, y-axis) values, determine the statistic h (as defined in method 1) and plot the three-dimensional results. For reference and motivation, consider the comodulograms in [25].

Appendix: Hilbert Transform in the Time Domain

We have presented the Hilbert Transform in the frequency domain: it produces a quarter cycle phase shift. It's reasonable to consider as well the *time domain* representation of the Hilbert Transform. To do so, let's write the Hilbert Transform as

$$H(x) = \left\{ \begin{array}{rl} -ix & \text{for positive frequencies of } x \\ 0 & \text{for 0 frequency of } x \\ ix & \text{for negative frequencies of } x \end{array} \right\} = x(f)\big(-i\,\text{sgn}(f)\big),$$

where we have written $x(f)$ to make the frequency dependence of x explicit, and the sgn (pronounced "sign") function is defined as

$$\text{sgn}(f) = \begin{cases} 1 & \text{if } f > 0, \\ 0 & \text{if } f = 0, \\ -1 & \text{if } f < 0. \end{cases}$$

In the frequency domain, we perform the Hilbert Transform by multiplying the signal $x(f)$ by a constant (either i or $-i$ depending on the frequency f of x).

To convert the Hilbert Transform in the frequency domain to the Hilbert Transform in the time domain, we take the inverse Fourier transform. Looking up the inverse Fourier transform of $-i\,\text{sgn}(f)$, we find

$$\text{Inverse Fourier transform}\{-i\,\text{sgn}(f)\} = \frac{1}{\pi t}.$$

Let's represent the inverse Fourier transform of $x(f)$ as $x(t)$.

Now, let's make use of an important fact. Multiplication of two quantities in the frequency domain corresponds to convolution of these two quantities in the time domain (see chapter 4). The convolution of two signals x and y is

$$x \star y = \int_{-\infty}^{\infty} x(\tau)y(t-\tau)d\tau.$$

So, in the time domain, the Hilbert Transform becomes

$$H(x) = x(t) \star \frac{1}{\pi t} = \frac{1}{\pi} \int_{-\infty}^{\infty} \frac{x(\tau)}{t - \tau} d\tau. \tag{7.6}$$

This time domain representation of the Hilbert Transform is equivalent to the frequency domain representation. However, the time domain representation is much less intuitive. Compare (7.6) to the statement, "The Hilbert Transform is a 90-degree phase shift in the frequency domain." The latter, we propose, is much more intuitive.

8 Basic Visualizations and Descriptive Statistics of Spike Train Data

Synopsis

Data Spontaneous spiking activity from a retinal neuron in culture, exposed to low-light and high-light environments.

Goal Visualize spike trains, compute and interpret descriptive statistics, and build simple models of interspike interval distributions as a function of the ambient light level.

Tools Raster plots, interspike interval histograms, firing rate, autocorrelograms, maximum likelihood estimation, Kolmogorov-Smirnov plots.

8.1 Introduction

8.1.1 Background

Neurons in the retina typically respond to patterns of light displayed over small sections of the visual field. However, when retinal neurons are grown in culture and held under constant light and environmental conditions, they will still spontaneously fire action potentials. In a fully functioning retina, this spontaneous activity is sometimes described as background firing activity, which is modulated as a function of visual stimuli. It is useful to understand the properties of this background activity in order to determine in future experiments how these firing properties are affected by specific stimuli.

8.1.2 Case Study Data

A researcher examining the background firing properties of one of these neurons contacts you to discuss his data. He records the spiking activity in one of two states, with the room lights off (low ambient light levels) or with the room lights on (high ambient light levels). He would like to collaborate with you to determine whether there is a difference in background firing between these two conditions, and whether one environment is more conducive to future experimental analyses. He records the spiking activity for 30 seconds in each condition.

8.1.3 Goal

Typically the first step in any data analysis involves visualizing and using simple descriptive statistics to characterize pertinent features of the data. For time series data that take on a continuous value at each time point, like the field potentials analyzed in earlier chapters, we typically start by simply plotting each data value as a function of time. For spike train data, things can become a bit more complicated. One reason for this is that there are multiple equivalent ways to describe the same spike train data. The data could be stored as a sequence of spike times; as a sequence of waiting times between spikes, or *interspike intervals*; or as a discrete time series indicating the number of spikes in discrete time bins. Knowing how to manipulate and visualize spike train data using all these different representations is the first step to understanding the structure present in the data and is the primary goal of this chapter.

8.1.4 Tools

We develop tools in this chapter to visualize spike train data and to provide basic statistical methods appropriate for analyzing spike trains.

8.2 Data Analysis

8.2.1 Visual Inspection

To access the data for this chapter, download the MATLAB file `Ch8-spikes-1.mat` from

`http://github.com/Mark-Kramer/Case-Studies-Kramer-Eden`

Our data analysis begins with visual inspection. We load the spike train data into MATLAB:

```
load('Ch8-spikes-1.mat')            %Load the spike train data.
```

where the file `Ch8-spikes-1.mat` resides in your current MATLAB directory.

> **Q:** What variables are now accessible to you in MATLAB? *Hint:* Consider the function `who`.

You should find two variables now accessible in MATLAB:

`SpikesLow`: spike times over 30 s in the low ambient light condition
`SpikesHigh`: spike times over 30 s in the high ambient light condition

Each variable is a single vector that gives a set of increasing spike times for the associated condition. The two vectors are of different sizes because the neuron fired a different number of spikes in each condition.

Q: What is the size of the vector `SpikesLow`?

A: To answer this in MATLAB, we use the command

```
size(SpikesLow)
```

MATLAB returns the answer

```
1 750
```

which reveals that `SpikesLow` is a vector with dimensions 1 × 750 (i.e., a vector with 1 row and 750 columns). Our collaborator who collected the data told us that each column holds a single spike time, and we continue to consider the implications of this statement.

Q: What is the size of the vector `SpikesHigh`?

Inspection of the sizes of the vectors `SpikesLow` and `SpikesHigh` reveals an important fact: the neuron fires more in the high-light condition. To make this observation more concrete, let's compute the *firing rate* (f), defined mathematically as

$$f = \frac{n}{T}, \tag{8.1}$$

where n is the number of spikes over the time interval T.

Q: What is the firing rate f of the neuron recorded in the low ambient light condition?

A: To answer this question, we must first define two quantities of interest: n and T. We consider here the entire duration of the recording, so $T = 30$ (our collaborator recorded the spiking activity for 30 s in each condition). During this interval, we found that the vector `SpikesLow` contains 750 spikes. With these two pieces of information, we may compute the firing rate.

```
T = 30;       %The duration of the recording in seconds,
n = size(SpikesLow,2);%...and # spikes in low light condition,
f = n/T;      %... to compute the firing rate.
```

These commands reveal that `f` = 25 spikes/s = 25 Hz. The firing rate computed in this way represents the firing rate for the entire 30 s interval. Is this single number a good representation of the spike train data? What if the spiking changes dramatically throughout the recording? These are important questions to keep in mind as we continue the analysis.

Q: What is the firing rate of the neuron recorded in the high ambient light condition?

These calculations allow us to compute a simple number representative of one aspect of the data: the firing rate over the entire duration of the recording. Do the two datasets exhibit a statistically significant change in the firing structure between conditions? Or, does the difference in firing rates lie within the range of expected fluctuations between any two trials of random spiking data? To answer these types of questions, we need to develop statistical methods that are appropriate for analyzing spike trains. Let's look at the data more carefully and visualize the structure of the spiking in the low ambient light condition. Motivated by the results of the previous chapters, it may be tempting to visualize the spike train by simply plotting the `SpikesLow` variable,

```
plot(SpikesLow)
```

However, as shown in figure 8.1, the resulting plot does not produce the typical picture of a spike train, and the spiking nature of these data is not obvious.

Q: What went wrong here? How do we interpret figure 8.1?

A: These data are stored as a sequence of *N* time stamps representing an increasing sequence of times at which the neuron spiked. When we run the MATLAB `plot`

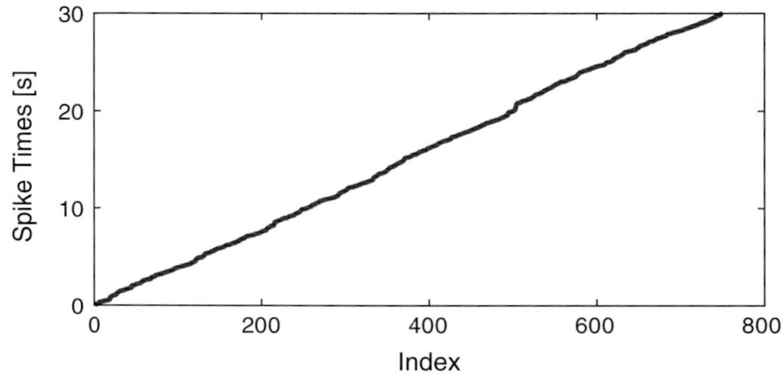

Figure 8.1
Result of plotting the variable `SpikesLow`.

command with only one input vector, we plot an index that runs from 1 to n (the length of the vector) on the x-axis against the numerical values in that vector on the y-axis. Therefore the plot shows an increasing line where the x-axis represents the spike number and the y-axis represents the spike time. Notice that the vector ends at an x-axis value of 750, which corresponds to the length of the vector. Also, the values on the y-axis range from 0 to 30; these correspond to times starting near 0 s and ending near 30 s, as we expect for the 30 s recording. Although this plot is not immediately useful, the results are consistent with our expectations for the data.

Instead of the data representation in figure 8.1, we would like to plot the spike train data as a set of points in a single row with x-coordinates that occur at the spike times. One way to produce such a plot in MATLAB is the following:

```
plot(SpikesLow,ones(size(SpikesLow)),'.');%Plot spikes as a row,
xlim([0 5]);                           %...display times (0,5)s,
xlabel('Time [s]')                     %...label the x-axis.
```

Q: What's happening in the first line of this code?

A: The `plot` function receives three inputs. The first input defines the x-axis values for the plot, which here are the spike times. The second input is itself a function:

```
ones(size(SpikesLow))
```

The function `ones` produces an array filled entirely with 1s. Notice that the input to the function `ones` is the size of the vector `SpikesLow`. So, this command acts to create a vector filled with 1s with the same dimensions as the vector `SpikesLow`. The last input to `plot` instructs MATLAB to display the data using the dot symbol. To summarize, we're calling the `plot` command to display

x-axis values: spike times in the *low* ambient light condition

y-axis values: 1

as blue dots. The last two commands set the range of the x-axis (in this case from 0 s to 5 s) and provide an x-axis label.

The plot of the spike train is shown in figure 8.2a. Each spike time corresponds to a blue dot at a y-axis value of 1. The value on the y-axis is arbitrary. We could have chosen to use a y-axis value of 2 or -100 or 412. What matters in figure 8.2a is the x-axis, which indicates the time at which each spike occurs in the 5 s interval.

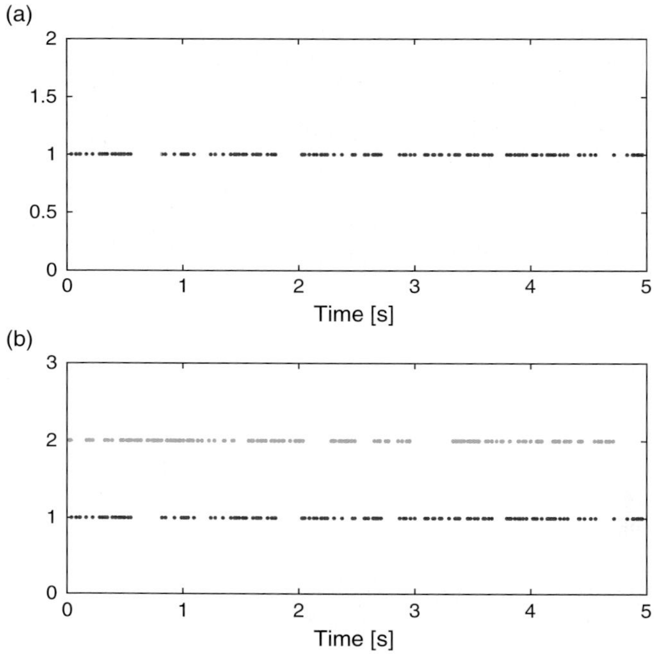

Figure 8.2

(*a*) Spike train for first 5 s of data from low-light condition. (*b*) Spike train data from low-light condition (*blue*) and high-light condition (*green*).

To compare the spiking in the low- and high-light conditions, we can plot both in the same figure (see figure 8.2b):

```
%Plot low-light condition spikes:
plot(SpikesLow, ones(size(SpikesLow)), '.');
hold on                  %... freeze the graphics window.
%Plot high light condition spikes:
plot(SpikesHigh, 2*ones(size(SpikesHigh)), '.g');
hold off                 %... release the graphics window,
xlim([0 5]);             %... and display times 0 to 5 s,
ylim([0 3]);             %... set y-axis range,
xlabel('Time [s]')       %... and label the x-axis.
```

Q: What's happening in the fifth line of this code segment?

A: The fifth line of this code segment is similar to the second line, but it plots the data for the high ambient light condition. The first input to the `plot` function is the variable `SpikesHigh`. The second input to `plot` is a function; here we're creating an array of 1s, this time with dimensions to match the variable `SpikesHigh`. Notice that we multiply this array by a scalar value of 2; this command acts to create a vector of 2s with the same dimensions as the vector `SpikesHigh`. The last input to `plot` indicates to display the data using a green dot symbol. To summarize, here we're calling the `plot` command to display

x-axis values: spike times in the high ambient light condition
y-axis values: 2

as green dots.

Q: Explain why the following MATLAB command fails to execute:

```
plot(SpikesLow,ones(size(SpikesHigh)), '.');
```

With the data visualized in this way, we're now able to ask an interesting question: What structure do you notice in the two spike trains plotted in figure 8.2b? At first glance, your answer might be "not much." Spikes occur fairly regularly throughout the 5 s interval under both conditions. Perhaps a careful visual inspection suggests there are fewer spikes in the low-light than in the high-light condition. But the spike times themselves do not seem to be directly comparable between these conditions. Often, when we examine data from a stimulus response experiment, we expect to see regions where spiking activity increases or decreases as a function of a changing stimulus. In this case, the stimulus is the ambient light level, which remains constant over the entire experiment. How else can we analyze these data and identify differences in the spiking activity (if any) between the two conditions?

Q: We've plotted in figure 8.2 a 5 s interval of data that begins at time 0 s. Through visual inspection, do you find similar conclusions for other 5 s intervals chosen from the data?

8.2.2 Examining the Interspike Intervals

So far, we have examined the long-term structure of the spiking over multiple seconds. Let's now focus on the short-term structure that occurs within a single second or less.

Q: Plot the spike train data for an interval of 1 s or less. What do you find? *Hint:* Use the MATLAB code from the previous section, and update the inputs to `xlim`.

Inspecting smaller time intervals, you might notice bursts of spikes that cluster near each other in time, interspersed with longer periods that contain less spiking (see figure 8.3). These patterns of bursts and quiescence look different between the low- and high-light stimuli. Visual inspection is an important tool, but we would like a quantitative result. How might we compare this fine temporal structure in the two conditions?

One approach to further characterizing the differences in spiking between the two conditions is to transform the data. One of the most useful transformations focuses on the waiting times between the spikes, or *interspike intervals* (*ISIs*), instead of the spike times themselves. We can compute the ISIs for the two conditions in MATLAB as follows:

```
ISIsLow=diff(SpikesLow);  %Compute ISIs in low-light condition.
ISIsHigh=diff(SpikesHigh);%Compute ISIs in high-light condition.
```

Q: How do these MATLAB commands represent the ISIs of the data?

A: Let's focus on the first command, which defines the variable `ISIsLow`. Here, we use the function `diff` with input `SpikesLow`. If you have not seen the command `diff` before, look it up in MATLAB Help. Briefly, the `diff` command computes the difference between adjacent elements of the input. In this case, the vector `SpikesLow` represents the times at which spikes occur. Therefore, the difference between adjacent elements of `SpikesLow` produces the time interval or waiting time between successive spikes. To further explore the concept of an ISI, let's write

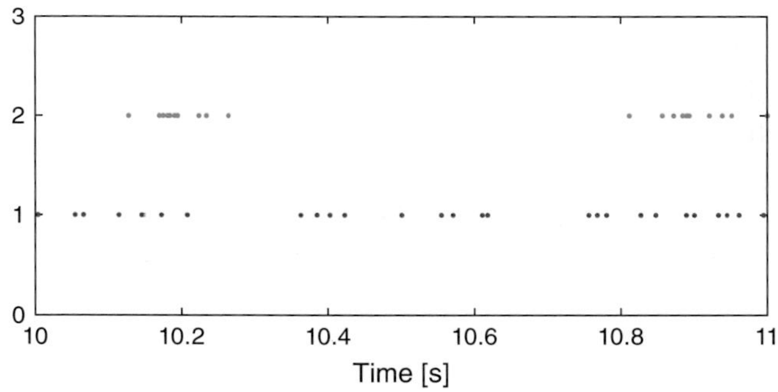

Figure 8.3

Example spike trains from 1 s of data in both conditions.

the spike times as a vector,

$$v = \{t_1, t_2, t_3, \ldots, t_N\}, \tag{8.2}$$

where t_i is the time of the ith spike. The difference between the first two adjacent elements of v is

$$t_2 - t_1 \,. \tag{8.3}$$

By convention, MATLAB subtracts the first element from the second. In words, this difference represents the time of the second spike (t_2) minus the time of the first spike (t_1), or the first interspike interval in the data. The difference between the next two adjacent elements of v is

$$t_3 - t_2 \,, \tag{8.4}$$

which is the second ISI, and so on. In this way, the command `diff` converts the spike times in the vector `SpikesLow` into interspike intervals saved in the variable `ISIsLow`.

Q: Consider the variables `ISIsLow` and `ISIsHigh`. How do the sizes of these variables compare to the sizes of the corresponding spike trains `SpikesLow` and `SpikesHigh`, respectively? *Hint:* Given N spikes, how many ISIs must occur?

The variables `ISIsLow` and `ISIsHigh` are vectors, and we can visualize these vectors using the same tools we've applied to visualize vectors in other scenarios. For example, we may simply plot these vectors:

```
plot(ISIsLow);
```

Figure 8.4 is a plot of the vector `ISIsLow`. The x-axis is the vector index, which ranges from 1 to the length of the vector `ISIsLow`. The y-axis is the value of the ISI at each index. We see that the ISI values range from small times (less than 0.05 s) to large times (over 0.4 s). In this way, the visualization provides some insight into the ISI values for the low-light condition.

Q: Plot the ISI vector for the high-light condition and compare it to figure 8.4. What similarities and differences do you notice in the ISIs from the two conditions?

Q: What is the smallest ISI you expect to observe for a neuron? Would you be surprised to find an ISI of less than 1 second? of less than 1 millisecond? of less than 1 nanosecond?

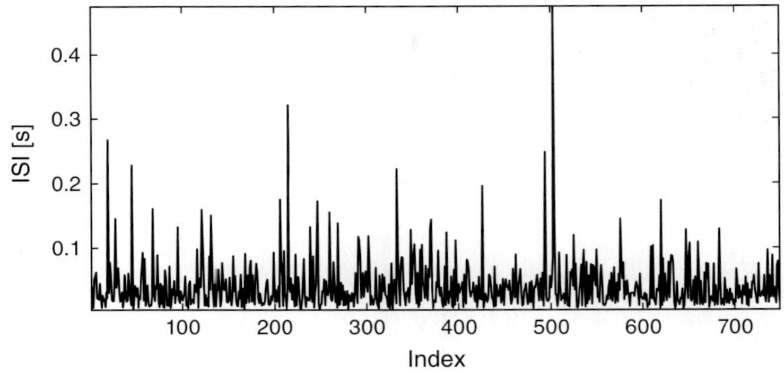

Figure 8.4
Plot of ISI vector `ISIsLow`.

Plots of the ISI vectors provide some information about the data (e.g., the approximate range of the ISI values), but there's more insight to be gained. To that end, let's now implement another approach to visualizing these types of data: the histogram. The idea of a histogram is to count the number of occurrences of each value in the data. In this case, we count the number of times we observe an ISI value in different bins of time. Let's define the time bins for the histogram. Inspection of the ISI data for the low-light condition (figure 8.4) reveals values that range from near 0 s to over 0.4 s. Therefore, we choose the following time bins:

Bin 1 [0.00 0.01]

Bin 2 [0.01 0.02]

Bin 3 [0.02 0.03]

Bin 4 [0.03 0.04]

Bin 5 [0.04 0.05]

Bin 6 [0.05 0.06]

Bin 7 [0.06 0.07]

. . .

Bin *N* [0.49 0.50]

The bins begin at time 0 s and end at time 0.5 s, with a bin size of 0.01 s. The purpose of the histogram is to count the number of times the data values fall into each bin. Notice that we've chosen the range of bins to extend beyond the observed range of data `ISIsLow`; that's fine, and we expect to count no values in the bins near 0.5 s. To further explore this counting process, let's examine the first eight values of the data `ISIsLow`:

```
0.0410 0.0290 0.0075 0.0521 0.0555 0.0620 0.0227 0.0213
```

We see that the first value of `ISIsLow` is `0.0410`. In which bin does this value belong? Examining the list of bins, we find that `ISIsLow[1]` lies in bin 5, so we increment the number of counts in bin 5 by 1. The second value of the vector `ISIsLow[2] = 0.0290` lies in bin 3, so we increment the number of counts in bin 3 by 1. The third value of `ISIsLow[3] = 0.0075` lies in bin 1, so we increment the number of counts in bin 1 by 1. The fourth value of `ISIsLow[4] = 0.0521` lies in bin 6, so we increment the number of counts in bin 6 by 1. And the fifth value of `ISIsLow[5] = 0.0555` also lies in bin 6, so we again increment the number of counts in bin 6 by 1.

Q: At this point, for the first five entries of the vector `ISIsLow`, how many counts are there in each bin?

A: We find for the first five ISIs in the low-light condition, zero counts in all bins except

Bin 1: 1 count
Bin 3: 1 count
Bin 5: 1 count
Bin 6: 2 counts

Notice that only four bins have counts and that bin 6 has two counts; for the first five ISIs in the low-light condition, we observe two ISIs in the interval $(0.05, 0.06)$.

Q: Repeat the binning procedure for the first eight ISIs observed in the high-light condition. Which bins have 1 or more counts for the first eight ISI values? What is the number of counts in each of these bins?

Q: Consider the first eight ISI values. In the previous question, you placed each of these values in a bin. Sum the counts across all bins. What do you find? Think about the value you compute; does the result make sense?

Of course, we're free to choose any interval of bins for the histogram. In the preceding examples, we chose a bin size of 0.01 s = 10 ms. Based on our understanding of a neuron, we might instead choose to examine a smaller bin size, 0.001 s = 1 ms. Let's do so now, and examine the histogram of *all* ISIs for the low-light condition. Of course, with enough patience, we could examine by hand each ISI value from the low-light condition and place each value in the correct 1 ms bin. However, this process would be time consuming and extremely error prone. Instead, the process of binning the ISI data is better done in

MATLAB. To create a histogram of all the ISI data in the low-light condition in MATLAB is straightforward:

```
bins = (0:0.001:0.5);      %Define the bins for the histogram.
hist(ISIsLow, bins)        %Plot the histogram of the ISI data,
xlim([0 0.15])             %... focus on ISIs from 0 to 150 ms,
xlabel('ISI [s]')          %... label the x-axis,
ylabel('Counts')           %... and the y-axis.
```

In the first line of this code segment, we define the bins. These bins start at time 0 s and end at time 0.5 s, and the size of each bin is 0.001 s. We then call the MATLAB function `hist` with two inputs: the first input is the variable we'd like to examine (here, `ISIsLow`, the ISIs in the low-light condition), and the second input is the bins. The function `hist` computes the histogram and displays the result (figure 8.5a). By setting the x-axis limit with `xlim`, we've chosen to examine the ISI values from 0 ms to 150 ms. We've also labeled the axes in the resulting figure. Notice that the x-axis indicates the binned ISI intervals, while the y-axis indicates the number of counts in each bin.

Q: Repeat this procedure to create a histogram of the ISI data in the high-light condition. Use the same bins we applied in the low-light condition (i.e., a bin size of 1 ms, extending from 0 s to 0.5 s). What do you find? *Hint:* Compare your answer to figure 8.5b.

Now, we have visualized the distributions of ISIs in both conditions by creating two histograms. Here, we chose to focus on the 1 ms bins extending from 0 ms to 150 ms. The histograms for both conditions are plotted in figure 8.5.

Q: Describe the features of the two histograms plotted in figure 8.5. What features of the ISI distributions are similar for the two conditions? What features are most strikingly different?

In the ISI histogram of the low-light condition data (figure 8.5a), very few counts occur at small ISI values (near 0 ms) and high ISI values (beyond approximately 100 ms). Instead, the distribution of counts is broadly peaked in the approximate interval 5–20 ms.

Q: What does the ISI distribution reveal about the spiking activity in the low-light condition? From this ISI distribution in figure 8.5a, could you sketch a spike train consistent with these data?

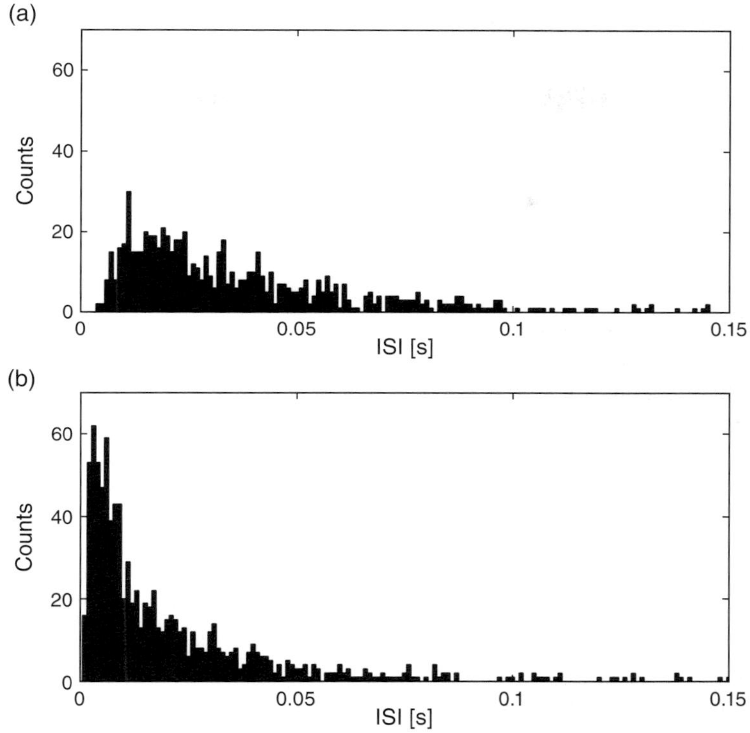

Figure 8.5
ISI histograms for (*a*) low- and high- (*b*) light conditions.

A: We conclude from the ISI distribution that many spikes are separated by time intervals 5–20 ms. So, we might be tempted to imagine near-periodic spiking with a period 5–20 ms (see figure 8.6a). However, the histogram contains additional structure beyond the single broad peak. Indeed, the histogram has a long tail, with counts extending up to 150 ms. Therefore, the intervals between spikes are varied. We often see ISIs in the 5–20 ms interval, but we also find much longer ISIs (e.g., from 50 to 150 ms). The structure of the ISI histogram is consistent with *bursting* activity, which consists of intervals of rapid spiking interspersed with quiescence (see an example in figure 8.6b). The intervals of rapid spiking produce many shorter ISIs, and the longer intervals produce (typically fewer) longer ISIs. We may conceptualize a bursting neuron as having two time scales: fast and slow. From the shape of the histogram in figure 8.5a, we conclude that the spike train data in the low-light condition are more consistent with bursting activity than with periodic, metronome-like spiking activity.

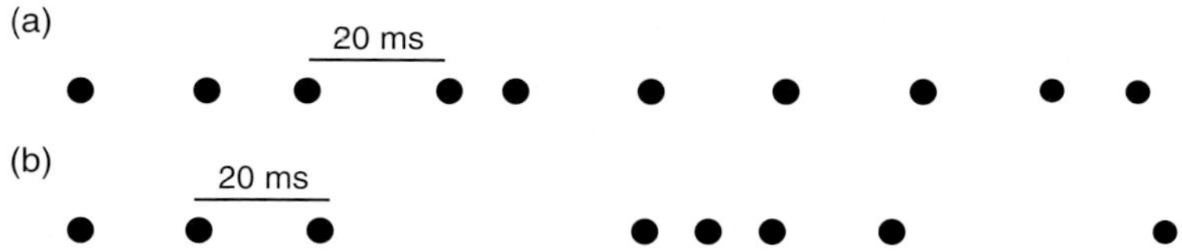

Figure 8.6

Cartoon representations of spiking activity. Black circles indicate time of a spike. (*a*) Near-periodic spiking, with period 5–20 ms. (**b**) Bursting activity, with both short and long intervals between spikes.

Q: What does the ISI distribution reveal about the spiking activity in the high-light condition? From this ISI distribution (figure 8.5b), could you sketch a spike train consistent with these data?

Let's now consider the two ISI histograms representing the spiking activity from the two conditions (figure 8.5). We note that both histograms show no ISI values below 0 ms. This is to be expected; the intervals between spikes cannot be negative. Both histograms also show broad peaks at 5–20 ms, indicating that a large number of short ISIs appear in the spike trains. In addition, both histograms possess long tails (i.e., counts of ISIs at larger bins, beyond 50 ms). A reasonable conclusion is that both neurons exhibit bursting activity, intervals of rapid spiking separated by periods of quiescence. However, a prominent difference exists between the ISI histograms in the two conditions. In the high-light condition, the proportion of small ISIs is much larger. The visualizations of the ISI histograms provide additional evidence of the similarities and differences in the spiking activity from the two conditions. We continue to investigate these two datasets—and build our scientific conclusions—in the next sections.

Q: We claim that the neuron exhibits bursting activity in both histograms. But clearly the two ISI histograms are different. How does the bursting activity differ in the low- and high-light conditions? *Hint*: Consider the impact of the large proportion of small ISIs in the high-light condition.

Q: So far, we've investigated two bin sizes: 10 ms and 1 ms. How do the shapes of the histograms in figure 8.5 depend on the bin size used? Was 1 ms a good choice in this case? Why, or why not?

8.2.3 Examining Binned Spike Increments

Another common approach to analyzing spiking data is to discretize time into bins of fixed width and count the number of events that occur in each time bin. The sequence of spike counts across all the bins is sometimes called the *increment process* for the spike train. When the time bins are sufficiently small, say, 1 ms for typical spike train data, the resulting increment process is just a sequence of zeros and ones. In this case, the time bins are so small that the probability of more than one spike occurring in each bin is zero or negligibly small.[1] Each tiny time bin then contains a spike (and we assign that bin a value of 1) or does not (and we assign that bin a value of 0). This idea of representing the spike train as a sequence of zeros and ones for small bin increments will be important when we build statistical models of the spike trains (chapter 9). In this section, we compute the increment process with multiple bin sizes in order to characterize the amount of variability in the spiking data and to examine temporal dependencies between spikes.

Let's bin the spike train data of the low-light condition into time bins of size 50 ms. To do so, we make use of the function `hist`:

```
time_bins = (0:0.05:30);     %Define time bins, 50 ms increments,
%... compute a histogram of spike time data,
IncrementsLow50 = hist(SpikesLow,time_bins);
%... and plot the resulting counts versus time,
plot(time_bins,IncrementsLow50,'.')
ylabel('Number of Spikes')  %... with x-axis labeled,
xlabel('Time [s]')          %... and y-axis labeled.
```

Notice that, in this case, we use the function `hist` in a new way. Instead of simply generating a plot, we call the function `hist` with the output variable `IncrementsLow50`. This variable `IncrementsLow50` is a vector containing the number of counts (i.e., the number of spikes) in each 50 ms increment. The time bins (`time_bins`, the second input to the function `hist`) is a vector containing the bin locations in time. In this case, the time bins start at time 0 s and end at time 30 s, with 0.05 s between bins. The variable name `IncrementsLow50` is quite descriptive. It reminds us that the variable represents the increments process in the low-light condition with a time bin of 50 ms.

> A descriptive choice of variable name is often useful.

1. The biophysical mechanisms that produce a spike support this statement. After generating a spike, the neuron experiences a refractory period typically lasting a few milliseconds, in which generating a subsequent spike is very unlikely.

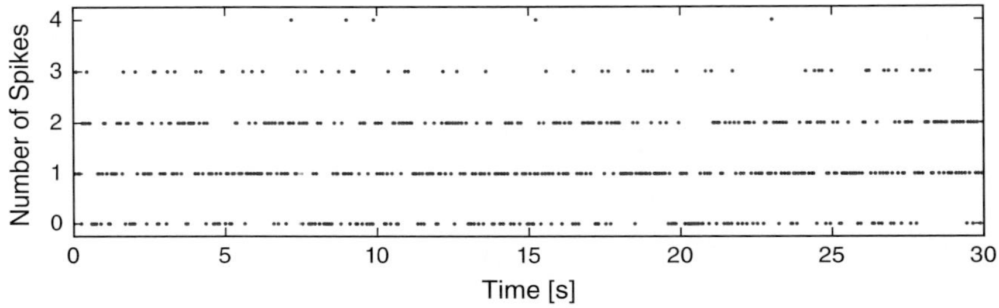

Figure 8.7
Plot of increment process for spike train data in low-light condition. Time bins specify a 50 ms resolution.

> **Q:** What can you say about the spike train data based on the increment process, plotted in figure 8.7? Approximately how often do you observe a 50 ms increment with zero spikes? With four spikes?

One question that arises quite often is how variable these binned counts are. To illustrate this variability, let's consider two scenarios. In the first, consider a neuron that fires perfectly regularly, like a metronome. In this case, we expect the number of spikes in each time bin to be nearly identical. On the other hand, consider the scenario of a neuron that fires in irregular bursts. In this case, we expect much more variability in the number of spikes in each time bin, depending on whether a time bin contained a burst of spikes or a quiet period. To characterize this variability, a standard measure to compute is the sample *Fano factor* (FF). It's easy to define the Fano factor: FF is the sample variance of the increment process divided by the sample mean of the increment process. The implementation of FF in MATLAB is also relatively simple:

```
FF50Low = var(IncrementsLow50)/mean(IncrementsLow50)
```

We use the MATLAB functions `var` and `mean` to compute the sample variance and sample mean, respectively, of the increment process `IncrementsLow50`. We store the result using a descriptive variable name, `FF`, that represents the Fano factor for an increment process with time bin 50 ms from the low-light condition data. We find a value of the Fano factor in this case of 0.72.

> **Q:** How do we interpret this FF value?

To answer that question, we need to introduce the concept of a *Poisson process*. A Poisson process is a model for a spiking process for which each spike occurrence is independent of every other spike occurrence. In other words, the probability of a neuron spiking at any instant does not depend on when the neuron fired (or did not fire) previously. A useful way to conceptualize this process is as a coin flip. For example, consider the following outcome of 20 coin flips:

H T H T T T H T T T T T H H H H H H T H

where H indicates heads, and T indicates tails.

Q: Based on your intuitive knowledge of a coin flip, does the result of a chosen coin flip depend on any other coin flip?

A: No. Consider, for example, the fifth coin flip. In the example outcome, the fifth coin flip resulted in T (tails). Does this result depend on the previous coin flip? on the next coin flip? on the first coin flip? on a future one-hundredth coin flip? In all cases, intuition suggests that it does not. Each coin flip is independent of every other coin flip. That's the assumption we make in assuming a Poisson process as a model for spiking activity: each spike occurrence is independent of every other spike occurrence.

The Poisson process is rarely an accurate model for spike train data. Our biological knowledge reveals that the occurrence of a spike *does* depend on the occurrence of previous spikes (e.g., because of the refractory period of a neuron, we do not expect a spike to occur immediately after another spike). However, the Poisson process has many nice theoretical properties, that make it a good model against which to compare the data. For example, for any Poisson process, the number of spikes in any time interval has a Poisson probability distribution for which the theoretical variance and mean are equal (see the appendix at the end of the chapter).

The theoretical Fano factor for a Poisson process is exactly equal to 1.

When measuring the variability of the increments of a spike train, we typically compare it to the variability of a Poisson process. If we compute a Fano factor well below the value 1 for a particular set of increments, this suggests that the spiking is more regular than a Poisson process for the time scale at which the increments were binned. In this case, spiking activity in the past is influencing the neuron to spike in a more predictable manner in subsequent bins. If we compute a Fano factor well above the value 1, this suggests that the spiking is more variable than a Poisson process for the time scale at which the increments were binned.

For the 50 ms binned spikes in the low-light condition, we obtained a sample Fano factor value of 0.72, well below 1. We might therefore conclude that the spiking data in the low-light condition are more regular (i.e., more like a metronome) than we expect for a Poisson process.

Q: What can we conclude about the variability of the counts for the spiking data in the high-light condition?

A: Repeating the analysis, we find that the sample Fano factor in the high-light condition is 1.78, well above 1. We might therefore conclude that the spiking data in the high-light condition are less regular than we expect for a Poisson process.

Does the observed Fano factor differ from 1? The preceding results are somewhat unsatisfying. We claimed that in the low-light condition, the calculated Fano factor of 0.72 was well below 1. What if, instead, we calculated a Fano factor of 0.8; is that well below 1? Is a Fano factor of 0.9 well below 1? These questions highlight an important issue when drawing a conclusion from a Fano factor calculation: How far above or below the value of 1 does the calculated Fano factor have to be before we are confident that there is really a statistically significant difference in the variability from a Poisson process? After all, even if we had spiking from a true Poisson process, from one experiment to the next we would expect to find different values for the increments, and values for the sample Fano factor that fluctuate slightly above and below 1. Fortunately, a bit of statistical theory can help us out. It can be shown that the distribution of Fano factors that we might compute from a Poisson process follows a gamma distribution with shape parameter $(N-1)/2$ and scale parameter $2/(N-1)$, where N is the number of time bins used in the Fano factor calculation [26].

Q: What is the correct value of N for the increment process computed in section 8.2.3 and saved in the variable `IncrementsLow50`?

A: We find in MATLAB that

```
length(IncrementsLow50)
```

is 601. Therefore, $N = 601$. This makes sense; 601 time bins of 50 ms duration fit between 0 s and 30 s.

With the value of $N = 601$ now determined, let's plot the gamma distribution in MATLAB and investigate its shape:

```
N =length(IncrementsLow50); %Determine number of time bins.
FF=(0.5:0.001:1.5);          %Define possible FF values for plot,
Y=gampdf(FF,(N-1)/2,2/(N-1));%... compute gamma distribution,
plot(FF,Y);                  %... and plot it.
```

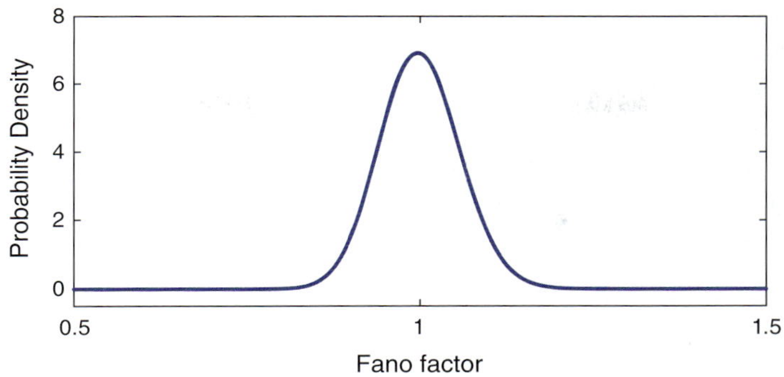

Figure 8.8
Gamma distribution of Fano factor values with $N = 601$.

Notice that we're evaluating the MATLAB function `gampdf`, which returns as output the gamma probability density function. We provide three inputs to this function. The first specifies a range of Fano factors to investigate (here we choose Fano factors ranging from 0.5 to 1.5; we choose a small interval between Fano factor values to make a smooth plot of the gamma distribution). The second and third inputs to the function specify the shape and scale parameters of the gamma distribution. The resulting gamma distribution is plotted in figure 8.8.

When N is large, as it is here, the gamma distribution looks like a normal distribution (i.e., like a bell-shaped curve). We can use this distribution to construct an interval where we would expect the Fano factor to lie if the data were generated by a Poisson process. More specifically, if the data were generated by a Poisson process, then we would expect the Fano factor to lie in the 95% confidence interval around the value of 1. Let's use MATLAB to construct this 95% confidence interval:

```
gaminv([.025,.975],(N-1)/2,2/(N-1))
```

In this case, we use the MATLAB function `gaminv` to compute the gamma inverse cumulative distribution function. The first input specifies the boundaries of the probabilities of interest, which range from 2.5% to 97.5%, to encompass 95% of the probability mass around the value of 1. The next two inputs specify the shape and scale parameters of the gamma distribution. Evaluating this line of MATLAB code, we find the interval (0.890, 1.116). Therefore, if we observed a true Poisson process with $N = 601$ bins and computed the Fano factor, we would not be surprised to find values between 0.890 and 1.116. However, the observed Fano factor for an increment process with 50 ms bins in the low-light condition is well outside of this range; this result suggests that it is very unlikely that these data were generated by a Poisson process. Instead, it's more likely that these data were generated by a process with less variability in the low-light condition (FF = 0.72).

Q: Consider the Fano factor for the high-light condition. Do you believe these data were generated by a Poisson process? Why, or why not?

A: We found for the high-light condition a Fano factor of 1.78 (again using an increment process with 50 ms bins). This calculated value is well above the interval expected for a Poisson process. We therefore reject the hypothesis that these data were generated by a Poisson process. Instead, we observe more variability in the high-light condition than expected for a Poisson process (with $N = 601$).

Q: How do the results for the Fano factor change in each condition with different choices for the bin size of the increment process (e.g., 25 ms, 100 ms, 500 ms)? Note that by changing the bin size, you also change N.

8.2.4 Computing Autocorrelations for the Increments

Another way to characterize the history dependence structure of a spike train is with the *autocorrelation* function of the increments. A correlation coefficient describes the degree of linear dependence between any two variables. The value of the correlation ranges from -1 to 1. A correlation value of -1 indicates a perfect linear relation between the two variables with a negative slope. A value of 0 indicates no linear relation between the two variables. And a value of 1 indicates a perfect linear relation between the two variables with a positive slope. Any other value indicates that one variable can be predicted using a linear function of the other, but that prediction will be imperfect; the closer the value is to ± 1, the better the prediction will be. The sign of the coefficient indicates the slope of the linear relation. Figure 8.9 shows scatterplots for a variety of possible relations between two variables, and the values of the correlation coefficients.

Figure 8.9

Correlation values for example relations between two variables. Variables are plotted, one against the other, on the x-axis and y-axis. Numbers indicate values of the correlation between the two variables.

Mathematically, the formula for the sample autocorrelation at a lag L is

$$\rho_{xx}\big[L\big] = \frac{\sum_{i=1}^{N-L} (x_i - \bar{x})(x_{i+L} - \bar{x})}{\sum_{i=1}^{N} (x_i - \bar{x})^2}, \tag{8.5}$$

where x_i is the ith data point, and $\bar{x}$ is the sample mean of the data over index i.

Q: This formula is rather complicated, so let's consider a simple case: $L = 0$. What is $\rho_{xx}\big[0\big]$ at lag 0?

A: To answer this, let's substitute $L = 0$ into the mathematical expression for the autocorrelation,

$$\begin{aligned}
\rho_{xx}\big[0\big] &= \Big(\sum_{i=1}^{N-0} (x_i - \bar{x})(x_{i+0} - \bar{x})\Big) \Big/ \Big(\sum_{i=1}^{N} (x_i - \bar{x})^2\Big) \\
&= \Big(\sum_{i=1}^{N} (x_i - \bar{x})(x_i - \bar{x})\Big) \Big/ \Big(\sum_{i=1}^{N} (x_i - \bar{x})^2\Big) \\
&= \Big(\sum_{i=1}^{N} (x_i - \bar{x})^2\Big) \Big/ \Big(\sum_{i=1}^{N} (x_i - \bar{x})^2\Big) \\
&= 1\,.
\end{aligned}$$

At $L = 0$, the autocorrelation is by definition equal to 1. In words, an individual dataset x is perfectly correlated with itself at zero lag.

To gain some intuition for the autocorrelation, let's consider the relatively simple increment process shown in figure 8.10a. Our goal is to understand the autocorrelation of these data, labeled x. From visual inspection of figure 8.10a, we conclude that the data x have mean 0, so that $\bar{x} = 0$; note that the data appear to oscillate between equal and opposite values over time. For simplicity, let's focus on the numerator of equation (8.5). At lag zero (i.e., $L = 0$), the numerator of (8.5) tells us to multiply the data by themselves at each time index, and then sum the product. To visualize this multiplication and sum, consider multiplying the data in figure 8.10a by the data in figure 8.10b at each index in time and then summing the result.

Q: What is the value of this sum?

A: We can't answer this question without knowing the values of x. However, we can deduce the sign and relative size of the sum. At a time index where x is positive, the product is positive; and at a time index where x is negative, the product is still positive. So, the product at each time index will always be positive or zero. Therefore, we expect the sum to be a positive number. And if we add many terms, we expect this sum to be large. So, we may conclude that the value of this sum will be a large, positive number. Note that this number is the numerator of the autocorrelation. At $L = 0$, the numerator equals the denominator in (8.5) and the autocorrelation at lag zero is 1.

Let's consider the numerator of the autocorrelation (8.5) for a small positive lag (i.e., $L > 0$). We can think of the small positive lag as shifting the data x a little bit to the right; this is illustrated in figure 8.10c. Now, to compute the numerator of the autocorrelation, we multiply x by a shifted version of x at each time index and then sum the result. In this case, we find some indices where the product of x_i and x_{i+L} is positive, and some indices where the product of x_i and x_{i+L} is negative; an example of an index where the product is negative is shown in figure 8.10c. Visual comparison of figures 8.10a and 8.10c suggests that the product will be positive more often then negative. Therefore, we expect the sum of these products to still be positive, although not as large as we found for the $L = 0$ case. You may have noticed some points where the data in figures 8.10a and 8.10c do not overlap, at the far right and far left of the figure. At these locations, the product is zero.

Q: Consider the autocorrelation of the data x at the positive lag L shown in figure 8.10d. What is the sign (positive or negative) and relative size of the numerator of the autocorrelation at this lag?

Let's return to the spike train data of interest here, recorded in the low- and high-light conditions. For the corresponding spike train increments, the autocorrelation at a particular lag describes the relation between the spike counts in different bins separated by that lag. The autocorrelation function describes the autocorrelation across a range of lags over which we are interested. Given our visualizations of the ISI histograms (figure 8.5), we might expect relations between spiking events to extend up to 200 ms.

Let's compute the autocorrelation for increment processes deduced from the spike train data in the low-light condition. We compute the autocorrelation of the 50 ms increment process for lags ranging from 0 to 200 ms. We need only three lags to cover this range; lag 1 covers 50–100 ms, lag 2 covers 100–150 ms, and lag 3 covers 150–200 ms. In MATLAB, we can compute the autocorrelation using the function `xcorr`:

```
%Compute autocorrelation of increment process for low light.
xcorr(IncrementsLow50-mean(IncrementsLow50),3,'coeff')
```

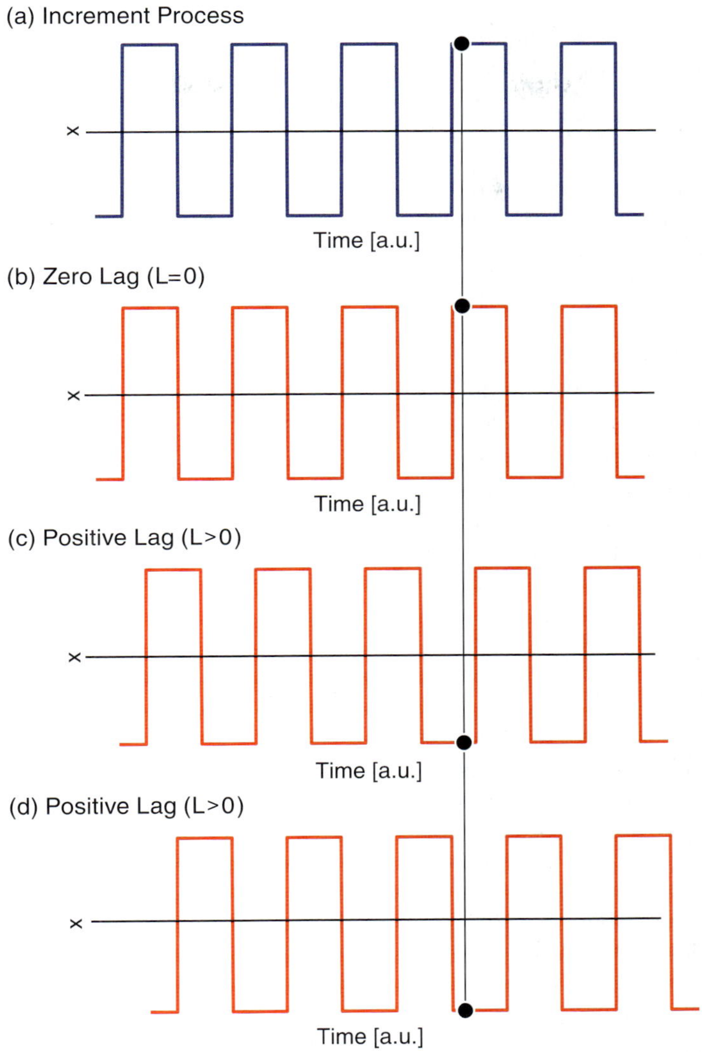

Figure 8.10

Example data for visual inspection of autocorrelation of an increment process. (*a*) Data x as a function of time (*arbitrary units*). (*b*) Data x at lag $L = 0$. (*c*, *d*) Data x at positive lags ($L > 0$). Black dots indicate a single time index in the different traces. We multiply the data at time index i (*black dot in* (*a*)), by the data at time index $i + L$ (*black dot in* (*b*), in (*c*), and in (*d*)).

The function `xcorr` takes three inputs. The first input is the data for which we want to compute the autocorrelation, in this case, the increment process for the low-light condition with 50 ms time bins. The second input is the number of lags to compute, and the `'coeff'` term in the third input indicates that we want to output correlation coefficients, which range from -1 to 1. Notice that we subtract from `IncrementsLow50` the mean of this variable before computing the autocorrelation. This command outputs a vector with seven numerical values corresponding to the autocorrelation at lag indices -3 through 3. The autocorrelation at negative lags is mathematically identical to the autocorrelation at the equivalent positive lags.

Q: Examine the numerical values returned by `xcorr`. What do you find?

A: As expected, the autocorrelation at zero lag is exactly equal to 1; the data matches itself at lag 0. At lag 1, corresponding to 50–100 ms, the autocorrelation value is 0.10. This positive correlation value indicates that when the number of spikes in one bin is higher than expected, the number of spikes in the next bin tends to be higher than expected. Similarly, when the number of spikes in one bin is lower than expected, the number of spikes in the next bin will also tend to be lower than expected. At lag 2, corresponding to 100–150 ms, the autocorrelation value of 0.017 is again positive but much smaller. At lag 3, corresponding to 150–200 ms, the autocorrelation value remains small and positive at a value of 0.013.

Q: How do we know whether these autocorrelation values are statistically significant?

A: This can be a difficult question when N is small (less than 30), but for larger N we can approximate a confidence interval about the correlation coefficient using a normal approximation with standard deviation $1/\sqrt{N}$. In this case, any correlation value exceeding $\pm 2/\sqrt{N}$ is unlikely to be generated by chance and likely reflects real dependence structure. For the increment process considered here, $N = 601$, and the significance bound is $\pm$ 0.08. We conclude that only the 50–100 ms lag has significant autocorrelation.

If we are particularly interested in the fine-scale temporal dependence structure of the spikes, we would do better to compute the autocorrelation function for more finely binned intervals. To that end, let's repeat the autocorrelation analysis for an increment process that uses 1 ms bins. We first compute a new increment process and then apply the `xcorr` function to this process. In MATLAB,

```
%Define time bins, 1 ms increments,
time_bins = (0:0.001:30);
% ... compute histogram to create increment process,
```

```
IncrementsLow1 = hist(SpikesLow,time_bins);
%... and then compute the autocorrelation function.
ACFLow = xcorr(IncrementsLow1-mean(IncrementsLow1),100,'coeff');
```

Q: What is the size of the output variable `ACFLow`? How does this size correspond to the lags?

A: `ACFLow` is a vector with dimensions 1 x 201. The 201 values correspond to lag indices -100 to 100 (or lag times -100 ms to 100 ms).

In order to examine history dependence going back 100 ms, we need 100 lags (because each lag index corresponds to 1 ms). There are now too many values to examine them printed one by one at the MATLAB command line, so instead we construct a plot of the autocorrelation function with lag on the x-axis and correlation on the y-axis. Let's also include in this figure two approximate significance lines at $\pm 2/\sqrt{N}$. In MATLAB,

```
plot(-100:100,ACFLow,'.')          %Plot autocorrelation vs lags,
N1 = length(IncrementsLow1);       %... compute the sample size,
%... and plot the upper and lower significance lines,
line([-100 100], [2/sqrt(N1) 2/sqrt(N1)])
line([-100 100],-[2/sqrt(N1) 2/sqrt(N1)])
xlim([-100, 100])                  %... set x-limits,
ylim([-0.1, 0.1])                  %... and y-limits.
```

We see in figure 8.11 that, as expected, the autocorrelation function is symmetric for positive and negative lags. The two approximate significance lines at $\pm 2/\sqrt{N}$ suggest significant negative correlation structure is present up to about ± 6 ms. This reflects the refractory period of the neuron: if you observed a spike in the previous 6 ms, you are less likely to observe a spike in the next few milliseconds. Beyond this point, the values of the autocorrelation mostly remain between the two significance bounds.

Alert! We are using these significance bounds here for exploratory purposes. We are not performing a rigorous statistical test for significance of the autocorrelation at every lag.

If we choose to perform a rigorous statistical test for the significance of the autocorrelation in figure 8.11, we would face the multiple comparisons problem. Briefly, if we perform many independent tests, and each has a 5% chance of reaching significance by chance, then the probability that any of these tests reaches significance by chance can be very large. If we wanted to perform many tests, we would need to control for multiple comparisons by adjusting the significance level so that the probability of any test being significant by chance is small. In figure 8.11, the significance lines are not corrected for multiple comparisons. Therefore, we accept that some of the correlation values that exceed these bounds may occur by chance. However, it is still very unlikely that all the significant correlations we observed from 1 to 6 ms are occurring purely by chance.

Now that we've computed and interpreted the autocorrelation function for the low-light condition, let's compare it to the autocorrelation in the high-light condition. We repeat our previous MATLAB commands using the `SpikesHigh` data values and choosing a time bin of 1 ms:

```
%Define time bins (1 ms increments), compute increments & ACF.
time_bins = (0:0.001:30);
IncrementsHigh1 = hist(SpikesHigh,time_bins);
ACFHigh=xcorr(IncrementsHigh1-mean(IncrementsHigh1),100,'coeff');
%Plot the autocorrelation vs lags with significance lines,
plot(-100:100,ACFHigh,'.')
N2 = length(IncrementsHigh1);
line([-100 100], [2/sqrt(N2) 2/sqrt(N2)])
line([-100 100],-[2/sqrt(N2) 2/sqrt(N2)])
xlim([-100, 100])                    %... set x-limits,
ylim([-0.1, 0.1])                    %... and y-limits.
```

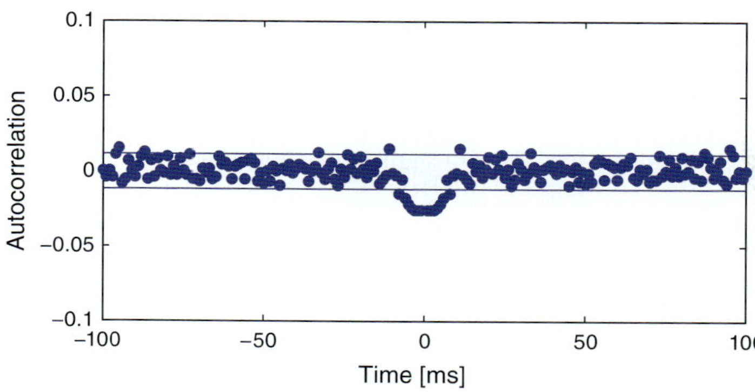

Figure 8.11

Autocorrelation of spike train data in low-light condition, with 1 ms increments. Horizontal lines indicate approximate significance.

Q: Consider the autocorrelation of the spike train data in the high-light condition shown in figure 8.12. What do you observe? How do the autocorrelations differ in the two conditions?

A: We find in the high-light condition significant correlation structure going all the way out to about 50 ms. Once again, we see refractoriness reflected in the negative autocorrelation at a lag of 1 ms, but this lasts much less time than in the low-light condition. Instead, there is now significant positive correlation at intermediate lags (approximately 2–50 ms). This positive correlation at short time lags reflects the tendency of the neuron to fire in bursts with small ISIs in the high-light condition; after a spike, another spike is more likely to occur in the next 2–50 ms than in the subsequent 50–100 ms.

Now that we've visualized the autocorrelations in the two light conditions, we can ask an important related question: Are the differences in the autocorrelations between these two conditions real? To answer this, we compute the difference in the autocorrelation functions between the low- and high-light conditions at every lag. If we assume that the firing in each condition is independent, the significance bounds for this difference can be computed by adding the variance of the autocorrelation from each condition. The standard deviation of the autocorrelation for the low-light condition is `1/sqrt(N1)`, so the variance of the autocorrelation for the low-light condition is `1/N1`. For the high-light condition, the variance of the autocorrelation is `1/N2`. Then, in MATLAB, we plot the differenced autocorrelations and the significance bounds:

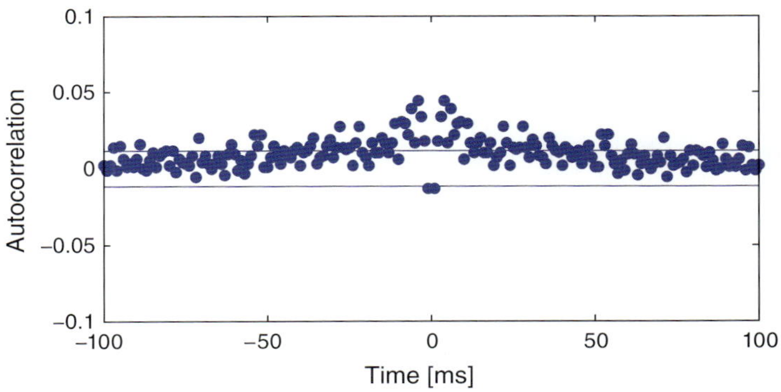

Figure 8.12

Autocorrelation of spike train data in high-light condition with 1 ms increments. Horizontal lines indicate approximate significance.

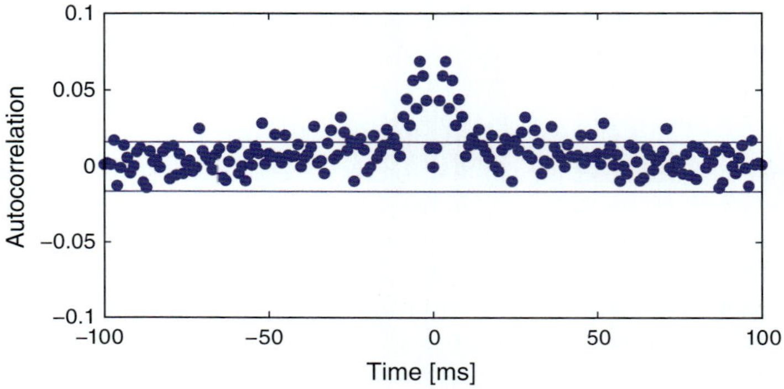

Figure 8.13

Difference in autocorrelation between high- and low-light conditions. Horizontal lines indicate approximate significance.

```
ACFDiff=ACFHigh-ACFLow; %Compute difference of autocorrelations,
% ... and plot it with upper and lower significance lines,
plot(-100:100,ACFDiff,'.')
line([-100 100], [2*sqrt(1/N1+1/N2) 2*sqrt(1/N1+1/N2)])
line([-100 100],-[2*sqrt(1/N1+1/N2) 2*sqrt(1/N1+1/N2)])
xlim([-100, 100])                   %... set x-limits,
ylim([-0.1, 0.1])                   %... and y-limits.
```

The results in figure 8.13 suggest significant differences in the autocorrelation between the two conditions at intermediate time lags (at approximately 2–50 ms). These are the same time lags we identified with bursting activity in the high-light condition. This suggests that the neuron fires with more intermediate ISIs in the bursting range in the high-light condition.

8.2.5 Computing Autocorrelations of the ISIs

The autocorrelation of the increments indicates the amount of time for which there are dependencies in the spiking data. In the high-light condition, we found large correlation values extending out to approximately 50 ms. This could be a consequence of the influence of patterns of many spikes with shorter ISIs or of single spikes with longer ISIs. We can distinguish between these possibilities by looking at the autocorrelation of the *sequence of ISIs*. In this case, the lag represents the number of spikes in the past rather than the amount of time in the past. If the dependence is only due to the last spike, we expect the ISIs to be uncorrelated at any nonzero lag. This would necessarily be true for data from a Poisson process. If we see correlation between ISIs, this suggests that the data do not come from a Poisson process and that the past spiking has an influence over multiple spikes. To investigate this, let's compute the autocorrelation of the sequence of ISIs for the low-light condition:

```
%Compute and plot the autocorrelation of the low-light ISIs,
ISI_ACF_Low = xcorr(ISIsLow-mean(ISIsLow),20,'coeff');
plot(-20:20, ISI_ACF_Low, '.')
%... with upper and lower significance lines,
N3 = length(ISIsLow);
line([-20 20], [2/sqrt(N3) 2/sqrt(N3)])
line([-20 20],-[2/sqrt(N3) 2/sqrt(N3)])
xlim([-20, 20])            %... set x-limits,
ylim([-0.2, 0.2])          %... and y-limits.
```

Notice that we include confidence bounds determined by the size of the sequence of interest (in this case, the length of the ISIs).

We see in figure 8.14a that the autocorrelation function has just a few isolated lags that are outside of the significance bounds. This could indicate a weak relation at particular lags or could be due to chance. Assuming the latter suggests that the data may come from a *renewal process*, a spiking process with independent ISIs for which the probability of a spike at any time only depends on the time of the most recent spike. One advantage of working with renewal processes is that it is fairly easy to write down and fit statistical models to the data. That is our next step.

> **Q:** Compute the correlation between ISIs for the data in the high-light condition. What do you find? Are these data consistent with a renewal process? *Hint:* Compare your answer to figure 8.14b.

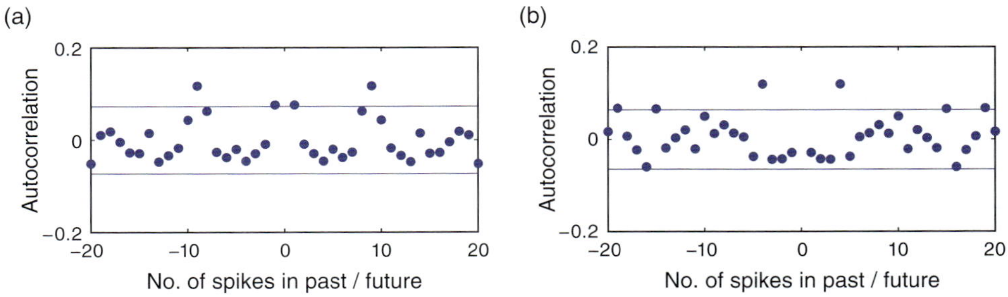

Figure 8.14

Autocorrelations of ISI sequences in (*a*) low- and (*b*) high-light conditions. Horizontal lines indicate approximate significance.

8.2.6 Building Statistical Models of the ISIs

In the previous sections, we constructed autocorrelation functions of the increment processes and autocorrelation functions of the sequences of ISIs. The former suggested dependence going back up to ≈ 50 ms (figure 8.12), while the latter suggested that the spiking at any time depends only on the timing of the most recent spike (figure 8.14). We now consider another powerful technique to understand these data: building a model. More specifically, we construct a *statistical model* of these data. This model captures important features of the data but does not consist of explicit biophysical components (an example of a biologically explicit model is the Hodgkin-Huxley equations [27]). The notion of a model can be confusing and is audience dependent, so we clarify here.

To construct a statistical model for these data we assume that the ISIs are independent samples from some unknown distribution. We typically posit some class of distributions from which the data might arise, and identify the one distribution in that class that maximizes the chance of observing the actual data.

What class of distributions should we use to build an ISI model? Previously, we discussed a Poisson process as a basic model for a spiking process, consistent with the conceptual idea of spikes as coin flips. Let's fit a Poisson process with a constant firing rate to the observed data. In other words, we begin with a model where the number of spikes in any time bin is independent of all previous (and future) spiking and has a Poisson distribution with a fixed but unknown rate parameter λ. The probability P of k spikes in any time bin is given by the Poisson distribution,

$$P(k) = \frac{\lambda^k e^{-\lambda}}{k!}, \tag{8.6}$$

where $k!$ is the factorial of k. Under this model, the distribution for the number of spikes in a bin is Poisson, but what is the distribution of the waiting time between spikes (i.e., what is the distribution of the ISIs)? It can be shown that for any Poisson process with constant firing rate the ISIs have an exponential distribution [6]. Mathematically, the probability density function for any ISI taking on a value x is

$$f(x) = \lambda \exp(-\lambda x), \tag{8.7}$$

where λ is the rate parameter for the Poisson process.

Alert! This is a common point of confusion. The increments of a Poisson process have a Poisson distribution, and the ISIs have an exponential distribution. The Poisson distribution takes on non-negative integer values $\{0, 1, \ldots, \infty\}$, which make it appropriate for counting the number of spikes in an interval. The Poisson distribution does not make sense to describe the waiting time between spikes, since this typically takes on a continuous value in $[0,\infty]$.

Our goal is to find a good value of λ so that our statistical model (8.7) matches the observed ISI distributions. Let's guess some values for λ, evaluate the model (8.7), and see how well the model matches the data. We plot the *probability* of observing ISI values in 1 ms bins for the low-light condition. This is similar to the ISI histogram we plotted previously except that the y-axis should represent probability instead of counts. To do so, we simply divide each count value by the total number of ISIs in the low-light condition:

```
bins = (0:0.001:0.5);                %Define 1 ms bins for histogram,
counts = hist(ISIsLow, bins);        %...compute histogram of ISIs,
prob = counts/length(ISIsLow);       %...convert to probability,
bar(bins, prob);                     %...and plot it,
xlim([0 0.15])                       %...with fixed x-limits,
xlabel('ISI [s]')                    %...and with x-axis labeled,
ylabel('Probability')                %...and y-axis labeled.
```

This empirical probability distribution is plotted in figure 8.15. Now, on this same figure, let's choose a value for λ and plot the statistical model (8.7):

```
lambda = 5;                          %Chocse a value for lambda,
model = lambda*exp(-lambda*bins)*0.001; %...and create model,
hold on                              %...freeze the graphics window,
plot(bins,model, 'r');               %...plot the model in red,
hold off                             %...and release the window.
```

In this code, we have chosen $\lambda = 5$ Hz and evaluated the statistical model at each time bin. We've also scaled the statistical model by a factor of 0.001 to match the 1 ms bin size, and plotted the model on top of the empirical ISI probability distribution. We repeat this procedure for two additional values of λ and plot the results in figure 8.15.

Q: Consider the model fits shown in figure 8.15. By visual inspection, what value of λ provides a good fit to the empirical distribution of ISI values?

The process of guessing values of λ and comparing the model (8.7) to the empirical ISI distribution is not satisfying. How do we identify the parameter λ that *best* fits the observed ISI distribution? We now consider a procedure to do so. Our goal is to find the value of λ that maximizes the likelihood of the data given the statistical model (8.7); this value of λ will be the best fit of the model to the data. To implement this procedure, let's consider the probability density of observing a sequence of ISIs, $x_1, x_2, \ldots, x_n$. If we assume that the

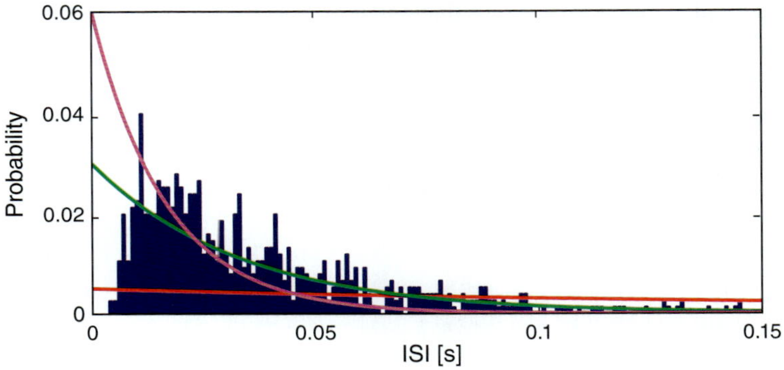

Figure 8.15

Exponential Poisson process models with $\lambda = 5$ (*red*), $\lambda = 30$ (*green*), and $\lambda = 60$ (*magenta*) compared to empirical distribution from observed ISI data (*bar plot*).

ISIs are independent, then the probability density is

$$\begin{aligned} f(x_1, x_2, \ldots, x_n) &= f(x_1) f(x_2) \ldots f(x_n) \\ &= \lambda \exp(-\lambda x_1) \lambda \exp(-\lambda x_2) \ldots \lambda \exp(-\lambda x_n) \\ &= \lambda^n \exp(-\lambda \sum_{i=1}^{n} x_i) \,. \end{aligned} \tag{8.8}$$

We call this expression the joint probability distribution of the observed data. In the first equality, we separate the joint probability distribution $f(x_1, x_2, \ldots, x_n)$ into a product of probability distributions of each event (i.e., $f(x_1)$, the probability of the first ISI equaling x_1, multiplied by $f(x_2)$, the probability of the second ISI equaling x_2, multiplied by $f(x_3)$, the probability of the third ISI equaling x_3, and so on). This partitioning of the joint probability is valid here because we assume the ISIs are independent. In the second equality, we replace each probability distribution with the exponential distribution we expect for the ISIs of a Poisson process. In the last equality, we rewrite the expression as a single exponential. Notice that this last expression is a function of the unknown rate parameter, λ.

When considered as a function of the unknown parameters, the joint distribution of the data (8.8) is also called the *likelihood*. In this case, we write

$$L(\lambda) = \lambda^n e^{-\lambda(x_1 + x_2 + \ldots + x_n)} \,, \tag{8.9}$$

to indicate that the likelihood L is a function of λ. To understand what the likelihood function $L(\lambda)$ looks like, let's plot it in MATLAB. We do so for the data from the low-light condition, and consider a range of possible λ values.

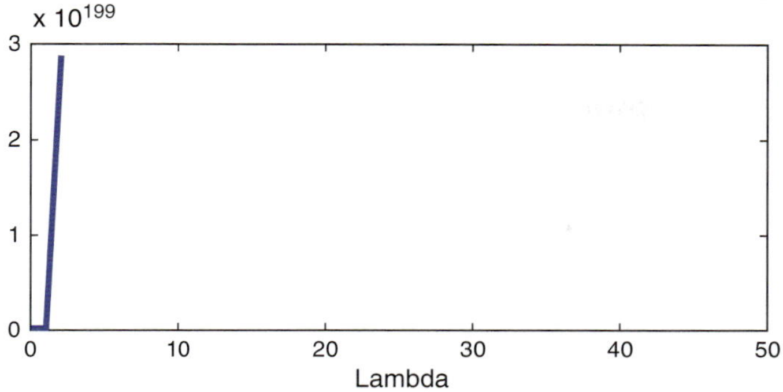

Figure 8.16
Likelihood of ISIs as a function of rate parameter λ for Poisson process model.

```
lambdas = 0:1:50;                %Range of lambda values.
N3 = length(ISIsLow);            %Number of low-light ISIs observed.
L = lambdas.^N3.*exp(-lambdas*sum(ISIsLow));%Compute likelihood,
plot(lambdas,L)                  %...and plot it.
```

Q: Consider the plot of the likelihood in figure 8.16. Does this answer seem okay?

A: Something went wrong here. The plot gives part of a line that rises toward 3×10^{199} and then vanishes. Why does this happen, and how can we fix it? To answer this, consider the first term in the likelihood function, λ^n. In this case we are raising λ to the power of `length(ISIsLow)` = 749. This is beyond the numerical precision limits of standard MATLAB computations.

So, we can't easily plot the likelihood directly. Instead, we plot the *log* of the likelihood. In this case, computing the log is useful because extremely large values are reduced to a more manageable range. To plot the log likelihood in MATLAB,

```
lambdas = 0:1:50;                %Range of lambda values.
N3 = length(ISIsLow);            %Number of low-light ISIs observed.
l = N3*log(lambdas)-lambdas*sum(ISIsLow);%Compute log likelihood,
plot(lambdas,l)                  %...and plot it.
```

Q: Consider the third line of code. Does the definition for `l` correspond to the $\log\left[L(\lambda)\right]$ in (8.9)? *Hint*: It should. Remember $\log(x^a), = a\log x$, and $\log(e^b) = b$.

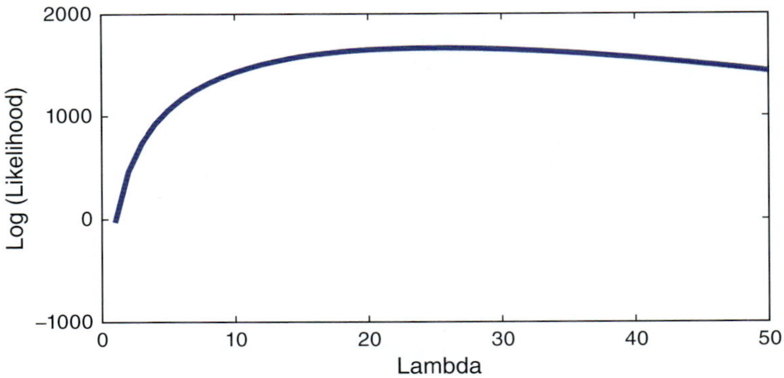

Figure 8.17
Log likelihood of ISIs as a function of rate parameter λ for an exponential model.

We see in figure 8.17 that the log likelihood is low for small λ, rises quickly as λ increases, and then starts to fall off once λ becomes larger than ≈ 25. The point $\lambda = 25$, where the log likelihood is maximized, is called the *maximum likelihood estimate* of λ. We use the symbol $\hat{\lambda}$ to denote the maximum likelihood estimate of λ.

We observe that although the values of the likelihood go beyond the precision range in MATLAB, the peak in the log likelihood stands out very clearly in figure 8.17. Note that the likelihood (figure 8.16) is maximized at the same point as the log likelihood (figure 8.17). This is always true.

Q: Can you explain why?

We could also have computed the maximum likelihood estimator theoretically, by differentiating the log likelihood with respect to λ, setting that equal to zero, and solving for λ. This gives $\frac{n}{\hat{\lambda}} - \sum_{i=1}^{n} x_i = 0$, which can be solved to find $\hat{\lambda} = n(\sum_{i=1}^{n} x_i)^{-1} = 1/\bar{x} = 25.0$ spikes/s. Remember that x_i is the ith ISI value, so $\bar{x}$ is the average ISI value. This computation shows that the maximum likelihood estimate for the rate parameter of a Poisson process is just 1 divided by the average ISI value. For some statistical models, it is convenient to compute maximum likelihood estimates theoretically in this manner, but sometimes no closed-form solution exists. In these cases, we typically use numerical methods to solve for the maximum likelihood estimates.

Q: What is the maximum likelihood estimate for the Poisson rate parameter in the high-light condition?

A: Repeating the analysis for the high-light condition, the maximum likelihood estimate for a Poisson rate parameter is $\hat{\lambda} = n(\sum_{i=1}^{n} x_i)^{-1} = 1/\bar{x} = 32.3$ spikes/s. The difference in the Poisson rate parameter of $32.3 - 25.0 = 7.3$ spikes/s reflects the difference in the overall firing rate of the neuron between the low- and high-light conditions.

Q: Is the difference in the Poisson rate parameter between the low- and high-light conditions statistically significant?

To address this last question, let's use a bootstrap analysis (see chapter 2). We combine all the ISIs from both conditions into one pool, sample many new datasets with replacement from that pool, and compare the actual difference in rate parameters to the distribution of differences across the samples. In MATLAB,

```
%Compute observed difference in lambdas,
MLDiff = 1/mean(ISIsHigh)-1/mean(ISIsLow);
%and then perform the bootstrap analysis.
ISIs  = [ISIsLow ISIsHigh]; %Merge all ISIs.
Nall = length(ISIs);        %Save length all ISIs.
Nlo = length(ISIsLow);      %Save length low-light condition.
Nhi = length(ISIsHigh);     %Save length high-light condition.
for i = 1:1000              %For each bootstrap sample,
 sampLo=ISIs(randsample(Nall,Nlo,1));%...resample low-light ISIs,
 sampHi=ISIs(randsample(Nall,Nhi,1));%...resample high-light ISIs,
 SampDiff(i)=1/mean(sampHi)-1/mean(sampLo);%...and difference.
end
hist(SampDiff,30)           %Plot resampled ISIs distribution,
line([MLDiff MLDiff],[0 100]);%... and the empirical ISIs.
```

Based on this test (figure 8.18), we might conclude that if the Poisson model is good in both conditions, there is a significant difference in the rate parameter between these conditions.[2] Notice that the actual difference in rate parameters lies well outside the distribution of differences from the bootstrap samples.

2. The distribution you generate will probably look a bit different from the distribution in figure 8.18 as it depends on random resampling.

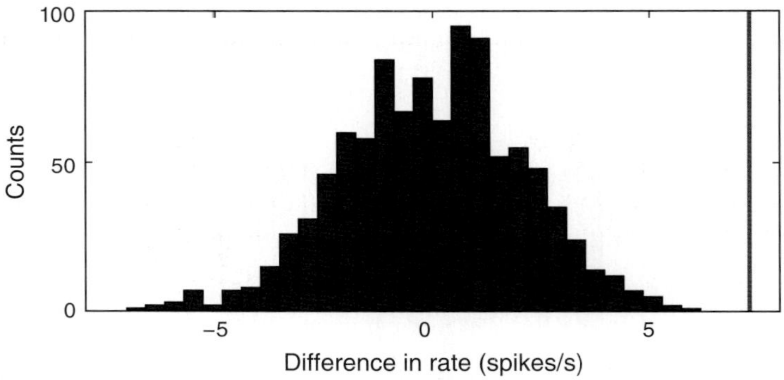

Figure 8.18

Bootstrap distribution of differences in Poisson rate parameter (*bars*) compared to observed value of 7.3 spikes/s (*vertical line*).

> There are more powerful tests we could use to compare the Poisson rate parameters. By more powerful, we mean that the tests are more likely to show a significant difference when one is actually present. However, the fact that the bootstrap test gives a significant result suggests that these more powerful tests would also be significant.

But, is the Poisson model good? To answer this, let's visualize the model fits compared to the data. There are a number of ways to do this. We start by comparing the expected proportion of ISIs for a Poisson process to the ISI histograms we actually observe in each condition. Let's do so first for the low-light condition:

```
bins = (0:0.001:0.5);             %Define 1 ms bins for histogram.
counts = hist(ISIsLow, bins);     %Compute histogram,
prob = counts/length(ISIsLow);    %... convert to probability,
bar(bins, prob);                  %... and plot probability.
lambda = 1/mean(ISIsLow);         %Compute best guess for lambda,
model = lambda*exp(-lambda*bins)*.001; %... build the model,
hold on                                %... and plot it.
plot(bins, model, 'r')
hold off
xlim([0 0.15])                    %... xlim from 0 to 150 ms,
xlabel('ISI [s]')                 %... label the x-axis,
ylabel('Probability')             %... and label the y-axis.
```

Q: Compare the model fit to the empirical ISI distribution for the low-light condition in figure 8.19. Does the model fit the data?

A: No, the model does not provide a very good fit to the data. Since Poisson processes have spikes that are independent of past activity, they do not capture either the refractoriness (i.e., the few spikes observed at short times) or the bursting (i.e., the increased spiking at times 5–20 ms) that we observe in the data.

Q: Repeat the analysis and compare the empirical ISI histogram to the best-fit model in the high-light condition. Does the model fit the data?

To go beyond visual inspection of the model fits and quantify the goodness of fit, we compare the cumulative distributions computed from the data and model. The *cumulative distribution function* (CDF), $F(x)$, is the probability that a random variable will take on a value less than or equal to x. For the exponential ISI model with rate parameter λ, the model CDF is

$$\begin{aligned} F_{\text{mod}}(x) &= \Pr(\text{ISI} \le x) \\ &= \int_0^x \lambda e^{-\lambda t} dt \\ &= 1 - e^{-\lambda x}. \end{aligned}$$

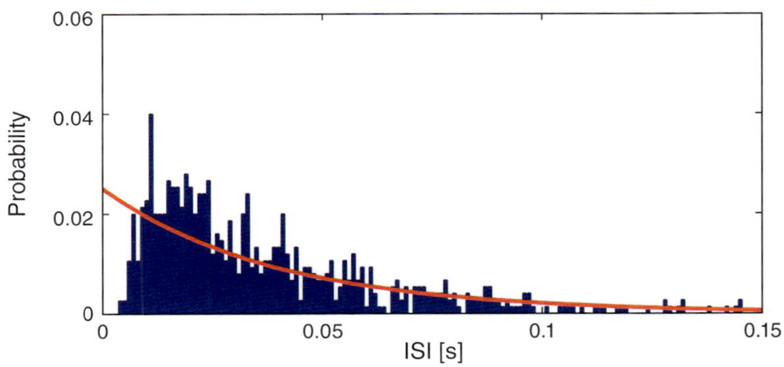

Figure 8.19
Scaled ISI histogram versus model fit (*red*) for low-light condition.

We compare this to the empirical CDF of the data, $F_{\text{emp}}(x)$, which is defined as the proportion of observations less than or equal to x. The code to compute and plot these CDFs for the low light-condition is as follows:

```
bins = (0:0.001:0.5);              %Define 1 ms bins for histogram.
lambda = 1/mean(ISIsLow);          %Compute best guess for lambda,
FmodLow = 1-exp(-lambda*bins);     %... and define model CDF.
FempLow = cumsum(prob);            %Define empirical CDF.
plot(bins,FmodLow)                 %Plot the model CDF,
hold on
plot(bins,FempLow, 'r')            %... and the empirical CDF,
hold off
xlim([0 0.2])                      %... with specified x-limits.
```

Q: Have you used the function `cumsum` before? If not, look it up in MATLAB Help.

Q: Compare the model and empirical CDFs in figure 8.20. What do you think?

A: If the model were a perfect fit, the red and blue curves would align. However, that's not what we find here. We conclude that the model may not provide a good fit to the data.

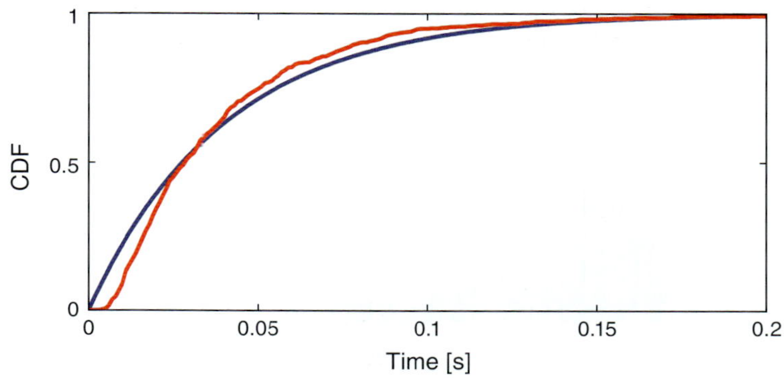

Figure 8.20

Comparison of model CDF (*blue*) and empirical CDF (*red*) in low-light condition.

Q: Compare the model and empirical CDF for the data in the high-light condition. What do you find? Is the model a good fit to the data?

Another common way to visualize the difference between the model and empirical distributions is a *Kolmogorov-Smirnov* (KS) *plot*. This is just a plot of the empirical CDF against the model CDF directly. In MATLAB,

```
plot(FmodLow,FempLow)     %Plot model vs empirical CDFs.
axis([0 1 0 1])           %Set the axes ranges.
```

Since the KS plot compares CDFs (figure 8.21), both the x-axis and y-axis range from 0 to1. A perfect fit between the model and empirical CDFs would look like a straight, 45-degree line between the points (0,0) and (1,1). Any deviation from this line represents deviation between the observed and model distributions. One nice result for comparing CDFs is that with enough data, the maximum difference between the model and empirical CDFs has a known asymptotic distribution, which can be used to put confidence bounds about the KS plot [6]. For 95% confidence bounds, a well-fit model should stay within $\pm 1.36/\sqrt{N}$ of the 45-degree line, where N is the number of ISIs observed. Let's place these confidence bounds on the KS plot (figure 8.22):

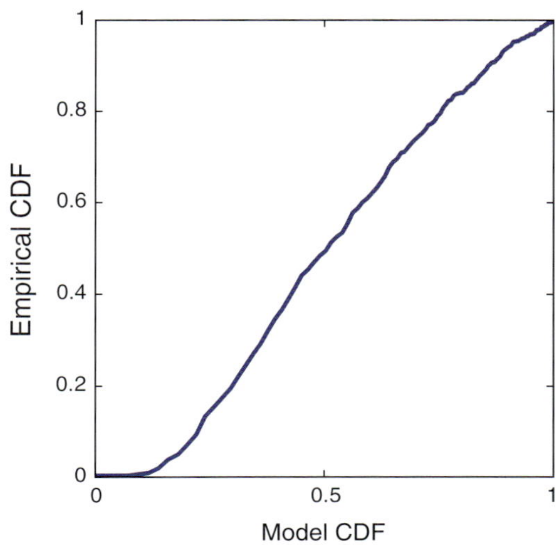

Figure 8.21
KS plot for exponential ISI model in low-light condition.

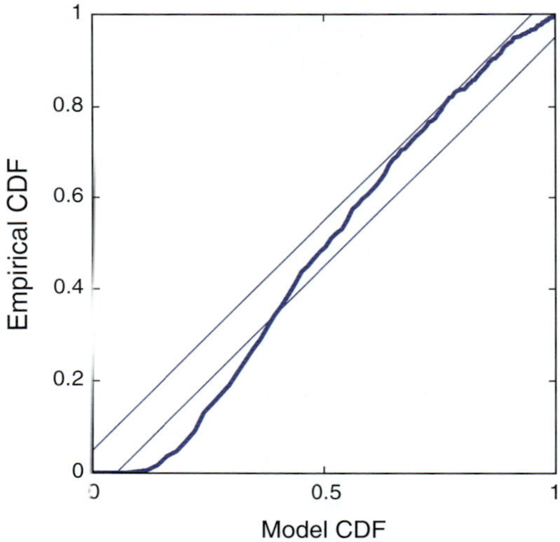

Figure 8.22
KS plot with 95% significance bounds (*thin lines*) for low-light condition.

```
plot(FmodLow,FempLow)          %Plot model vs empirical CDFs.
Nlow = length(ISIsLow);        %Length of low-light condition.
%Plot the upper and lower confidence bounds,
line([0 1], [0 1]+1.36/sqrt(Nlow));
line([0 1], [0 1]-1.36/sqrt(Nlow));
axis([0 1 0 1])                %... with fixed axes.
```

A well-fit model should stay entirely within these bounds. In this case, the KS plot for the low-light condition extends well outside these bounds. The exponential ISI model—or equivalently, the Poisson process model—does not fit the data in the low-light condition well. This suggests that we need a better model if we want to make meaningful comparisons about differences in the structure of the data between the two conditions.

> **Q:** Compute the KS plot with 95% significance bounds for the high-light condition. Does the exponential ISI model fit the data well?

A More Advanced Statistical Model. We've now investigated one class of models, the exponential distribution, to fit the observed ISI distributions. However, through analysis, we've found that this statistical model is not sufficient to mimic the observed data. There are many other choices for statistical models; let's try one other class of models.

The inverse Gaussian probability model has already been used successfully to describe ISI structure in this system [28]. The mathematical expression for the inverse Gaussian probability density is

$$f(x) = \sqrt{\frac{\lambda}{2\pi x^3}} \exp\left(\frac{-\lambda(x-\mu)^2}{2x\mu^2}\right). \tag{8.10}$$

The inverse Gaussian distribution has two parameters that determine its shape: μ, which determines the mean of the distribution, and λ, which is called the shape parameter. At $x = 0$, the inverse Gaussian has a probability density equal to zero, which suggests it could capture some of the refractoriness seen in the data.

If we again assume that the ISIs are independent of each other, then the likelihood of observing the sequence of ISIs, $x_1, x_2, \ldots, x_n$, is the product of the probabilities of each ISI,

$$L(\mu,\lambda) = f(x_1,x_2,\ldots x_n) = \prod_{i=1}^{N} \sqrt{\frac{\lambda}{2\pi x_i^3}} \exp\left(\frac{-\lambda(x_i-\mu)^2}{2x_i\mu^2}\right). \tag{8.11}$$

The log likelihood is then

$$\log\big(L(\mu,\lambda)\big) = \frac{N}{2}\log\frac{\lambda}{2\pi} - \frac{3}{2}\sum_{i=1}^{N}\log x_i - \sum_{i=1}^{N}\frac{\lambda(x_i-\mu)^2}{2x_i\mu^2}. \tag{8.12}$$

Since this distribution has two parameters, the maximum likelihood solution for this model is the pair of parameter estimates $\hat{\mu}$, $\hat{\lambda}$ that maximizes the likelihood of the data. We can solve for the maximum likelihood estimate analytically by taking the derivative with respect to both parameters, setting these equal to zero, and solving the resulting set of equations. In this case, the maximum likelihood estimators are

$$\hat{\mu} = \frac{1}{N}\sum_{i=1}^{N} x_i \tag{8.13}$$

and

$$\hat{\lambda} = \left(\frac{1}{N}\sum_{i=1}^{N}\left(\frac{1}{x_i} - \frac{1}{\hat{\mu}}\right)\right)^{-1}; \tag{8.14}$$

see problem 8.2.

Using this expression, we can fit an inverse Gaussian model to the data in each condition and evaluate the goodness-of-fit of the model. Let's do so now for the low-light condition. Here we present all of the MATLAB code to implement and evaluate the model fit:

```
bins = (0:0.001:0.5);                    %Define 1 ms bins.
Nlow = length(ISIsLow);                  %Length low-light condition.
```

```
mu=mean(ISIsLow);                    %Mean of inverse Gaussian.
lambda=1/mean(1./ISIsLow-1/mu);      %... and shape parameter,
model=sqrt(lambda/2/pi./bins.^3).*...    %... to create model.
  exp(-lambda.*(bins-mu).^2/2/mu^2./bins)*.001;
model(1) = 0;              %Numerator to 0 faster than denominator.

%Plot the data and the model,
counts = hist(ISIsLow, bins);    %Compute histogram,
prob = counts/length(ISIsLow);   %... convert to probability,
bar(bins, prob);                 %... and plot probability.
hold on
plot(bins,model, 'r');           %Plot the model.
hold off
xlim([0 0.2])                    %xlim from 0 to 200 ms.
xlabel('ISI [s]')                %Label the x-axis,
ylabel('Probability')            %... and the y-axis.

%Plot the KS plot.
FmodLow = cumsum(model);      %Define the model CDF,
FempLow = cumsum(prob);       %...and define empirical CDF,
plot(FmodLow,FempLow)         %...plot model vs empirical CDF,
line([0 1], [0 1]+1.36/sqrt(Nlow)); %...upper confidence bound,
line([0 1], [0 1]-1.36/sqrt(Nlow)); %...lower confidence bound,
axis([0 1 0 1])               %... set the axes ranges,
xlabel('Model CDF')           %... and label the axes.
ylabel('Empirical CDF')
```

From the computations, we find maximum likelihood estimates $\mu = 40.0$ ms and $\lambda = 49.3$ ms in the low-light condition. The results of the model fit to the data are shown in figure 8.23.

Q: Consider the fit of the inverse Gaussian model to the data in the low-light condition shown in (figure 8.23). Does the inverse Gaussian model provide a good fit to the ISIs?

A: This model provides a much better fit to the data; the KS plot is contained within the 95% confidence bounds.

Q: Consider the fit of the inverse Gaussian model to the data in the high-light condition. Does the inverse Gaussian model provide a good fit to these ISIs?

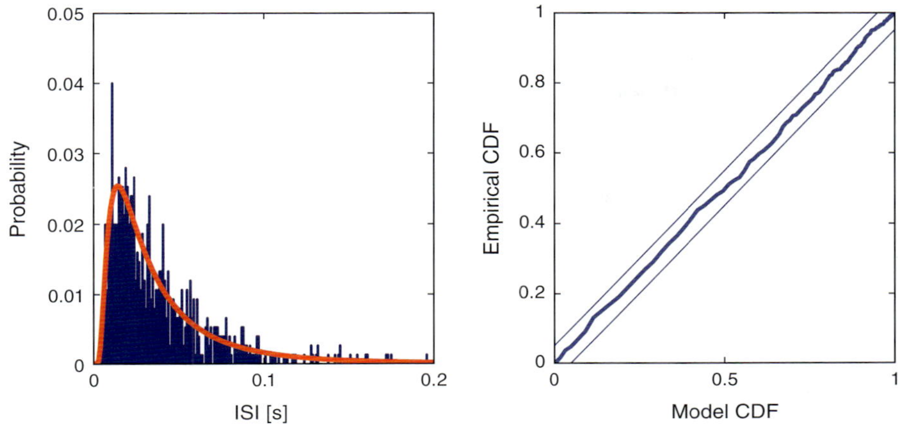

Figure 8.23
Model fit (*left*) and KS plot (*right*) for inverse Gaussian ISI model in low-light condition.

> **Q:** Compare the estimates of the two parameters μ and λ of the inverse Gaussian model in the two conditions. What do these reveal about the differences between the low- and high-light conditions?

Summary

In this chapter, we considered the spiking activity recorded in two conditions. We began with visualizations of the spiking data, and construction and visualization of the increment process (i.e., binned spike counts). We then assessed the variability in the increments through computation of the Fano factor, and showed that the low- and high-light conditions had less and more variability, respectively, than expected for a Poisson process. We also assessed the autocorrelation of the increment processes and observed the impact of refractoriness and bursting activity. In addition, we created and visualized the ISIs for each condition. Inspection of the ISI histograms suggested bursting activity in both conditions, and more small ISIs in the high-light condition. Analysis of the ISI autocorrelations revealed no compelling evidence for dependence and supported the hypothesis of a renewal process. Finally, we built two statistical models of the observed ISIs. We discussed how to fit the model parameters by computing the maximum likelihood estimate, and how to evaluate the model goodness-of-fit to the data using the KS plot. We showed that the first model—the Poisson process as a model of spiking with a corresponding exponential distribution of ISIs—did not fit the observed ISI data. A second model—the inverse Gaussian probability model—provided a much more accurate fit to the observed ISIs. The modeling suggests that at least two features of the spiking activity have changed from the low-light to the high-light condition. First, the mean ISI is smaller, and hence the average firing rate is larger, in the high-light condition. Second, the shape of the firing distribution has changed so that the cell is more likely to fire in bursts with short ISIs in the high-light condition.

Problems

8.1. The difference between the autocorrelation functions of the 1 ms increments for the low- and high-light conditions at lags between 20 and 60 ms appears to have some structure but does not rise above the significance lines consistently. Construct a single bootstrap test for a difference in this range of lags as follows:

   a. Pool the ISIs from both conditions.

   b. Construct a bootstrap sample by sampling with replacement 749 ISIs from the pool for the low-light condition and 968 ISIs from the pool for the high-light condition.

   c. Compute the autocorrelation functions for each condition and the difference function from this bootstrap sample, and compute the sum of the difference function over the range of lags from 20 to 60 ms.

   d. Repeat steps (b) and (c) 1,000 times, saving the values of the summed difference function for each sample.

   e. Compare the actual summed difference statistic in the data to the distribution of values from the bootstrap sample. What is the fraction of bootstrap samples that are smaller than the computed statistic? What conclusions can you draw about the data from this procedure?

8.2. Compute the maximum likelihood solution for the inverse Gaussian probability model in two ways:

   a. Use the formula for the log likelihood to solve for the maximum likelihood estimators analytically.

   b. Fit the model empirically by calculating the log likelihood for a range of μ and λ values. Plot the log likelihood surface. Identify the μ and λ values that maximize the log likelihood.

8.3. Load the file `Ch8-spikes-2.mat`, available at

   `http://github.com/Mark-Kramer/Case-Studies-Kramer-Eden`

into MATLAB, and answer the following questions:

a. You will find the variable `spikes`. This variable contains the spike times for a 30 s recording from a single neuron. What is the firing rate?

b. Plot the spike train data (as in figure 8.2a). Examine the entire 30 s of data as well as smaller intervals of data (e.g., 1 s). Summarize briefly the features you observe in the spike trains.

c. Plot the ISI histogram of the data. Describe briefly the features you observe in the histogram.

d. Construct and plot an increment process for the data using an interval size of 50 ms. Explain briefly the features you observe in the plot.

e. Use this increment process to compute the Fano factor of the data. Compare the Fano factor to the value expected for a corresponding Poisson process. What conclusions do you make about the variability of the counts?

f. Compute and plot the autocorrelation function for an increment process created for these data. Be sure to include in your plot the approximate significance lines. Do you observe compelling evidence for autocorrelation at any lag? You may consider an interval size of 50 ms or other choices.

g. Compute and plot the autocorrelation function for the ISIs of the data. Include the approximate significance lines. Do you observe compelling evidence for autocorrelation at any lag?

h. Fit the observed data with a Poisson process model by determining the appropriate value of the model parameter λ. Compare the model to the data by plotting the model and empirical ISI distributions, and by using the KS plot. Does the Poisson process model provide a good fit to the data?

8.4. Load the file `Ch8-spikes-3.mat`, available at

`http://github.com/Mark-Kramer/Case-Studies-Kramer-Eden`

into MATLAB, and answer the following questions.

a. You will find the variable `spikes`. This variable contains the spike times for a 30 s recording from a single neuron. What is the firing rate?

b. Plot the spike train data. Examine the entire 30 s of data as well as smaller intervals of data (e.g., 1 s). Summarize briefly the features you observe in the spike trains.

c. Plot the ISI histogram of the data. Describe briefly the features you observe in the histogram.

d. Construct and plot an increment process for the data using an interval size of 50 ms. Explain briefly the features you observe in the plot.

e. Use this increment process to compute the Fano factor of the data. Compare the Fano factor to the value expected for a corresponding Poisson process. What conclusions do you make about the variability of the counts?

f. Compute and plot the autocorrelation function for an increment process created for these data. Include the approximate significance lines. Do you observe compelling evidence for autocorrelation at any lag? You may consider an interval size of 50 ms, or other choices.

g. Compute and plot the autocorrelation function for the ISIs of the data. Include the approximate significance lines. Do you observe compelling evidence for autocorrelation at any lag?

h. Fit the observed data with a Poisson process model by determining the appropriate value of the model parameter λ. Compare the model to the data by plotting the model and empirical ISI distributions, and by using the KS plot. Does the Poisson process model provide a good fit to the data?

Appendix: Spike Count Mean and Variance for a Poisson Process

In this appendix, we compute the theoretical mean μ and the theoretical variance of the spike count σ^2 for a Poisson process. Let's compute μ using a general formula that makes use of the probability $P(k)$ of observing k spikes,

$$\mu = \sum_{k=0}^{\infty} k P(k). \tag{8.15}$$

Replacing $P(k)$ with (8.6), the expression for a Poisson process, we find

$$\begin{aligned}\mu &= \sum_{k=0}^{\infty} k\left(\frac{\lambda^k e^{-\lambda}}{k!}\right) \\ &= e^{-\lambda}\sum_{k=0}^{\infty} k\frac{\lambda^k}{k!}.\end{aligned}$$

To make progress, let's write out the terms in the summation,

$$\begin{aligned}\mu &= e^{-\lambda}\left(0+\frac{\lambda^1}{1!}+2\frac{\lambda^2}{2!}+3\frac{\lambda^3}{3!}+4\frac{\lambda^4}{4!}+\cdots\right) \\ &= e^{-\lambda}\lambda\left(1+\frac{\lambda^2}{2!}+\frac{\lambda^3}{3!}+\cdots\right) \\ &= e^{-\lambda}\lambda\left(e^{\lambda}\right) \\ &= \lambda,\end{aligned} \tag{8.16}$$

where we have used the fact that $e^x = 1 + \frac{x^2}{2!} + \frac{x^3}{3!} + \frac{x^4}{4!} + \cdots$. Notice that the mean spike count equals the rate parameter of the Poisson process.

To find the spike count variance for a Poisson process, we follow a similar procedure. In general, we compute the variance σ^2 of the spike count k with probability distribution $P(k)$ as

$$\sigma^2 = \sum_{k=0}^{\infty} k^2 P(k) - \Big(\sum_{k=0}^{\infty} k\,P(k)\Big)^2 . \tag{8.17}$$

As in our computation of the mean spike count, we replace $P(k)$ with (8.6), the expression for a Poisson process. Notice that the second term is the square of the expression (8.15), and for a Poisson process we found $\mu = \lambda$. Therefore, let's replace the second term in (8.17) with λ^2 and substitute for $P(k)$ in the first term of (8.17) to find

$$\begin{aligned}\sigma^2 &= \sum_{k=0}^{\infty} k^2 \Big(\frac{\lambda^k e^{-\lambda}}{k!}\Big) - \lambda^2 \\ &= e^{-\lambda} \sum_{k=0}^{\infty} k^2 \frac{\lambda^k}{k!} - \lambda^2 .\end{aligned}$$

To make progress, we follow the same strategy and write out the terms in the summation,

$$\begin{aligned}\sigma^2 &= e^{-\lambda}\Big(0 + \lambda + 2^2\frac{\lambda^2}{2!} + 3^2\frac{\lambda^3}{3!} + 4^2\frac{\lambda^4}{4!} + \cdots\Big) - \lambda^2 \\ &= \lambda e^{-\lambda}\Big(1 + 2\lambda + \frac{3}{2}\lambda^2 + \frac{4}{6}\lambda^3 + \cdots\Big) - \lambda^2 .\end{aligned}$$

Now, we divide this sum of terms into two pieces, a "nice term" (in the first brackets) and "leftovers" (in the second brackets):

$$\sigma^2 = \lambda e^{-\lambda}\Big(\big[1 + \lambda + \frac{\lambda^2}{2!} + \frac{\lambda^3}{3!} + \cdots\big] + \big[\lambda + \frac{2\lambda^2}{2!} + \frac{3\lambda^3}{3!} + \cdots\big]\Big) - \lambda^2 .$$

We can simplify by recognizing that $\big[1 + \lambda + \frac{\lambda^2}{2!} + \frac{\lambda^3}{3!} + \cdots\big] = e^{\lambda}$. Then

$$\begin{aligned}\sigma^2 &= \lambda e^{-\lambda}\Big(e^{\lambda} + \lambda\big[1 + \frac{\lambda^1}{1!} + \frac{\lambda^2}{2!} + \cdots\big]\Big) - \lambda^2 \\ &= \lambda e^{-\lambda}\Big(e^{\lambda} + \lambda e^{\lambda}\Big) - \lambda^2 \\ &= \lambda + \lambda^2 - \lambda^2 \\ &= \lambda ,\end{aligned} \tag{8.18}$$

where again we've used the definition of e^{λ}. We conclude that the spike count variance for a Poisson process equals the firing rate λ.

Combining these results for the mean spike count μ in (8.16) and the spike count variance (8.18), we conclude that for a Poisson process,

$$\mu = \sigma^2 = \lambda,$$

and therefore for a Poisson process, the Fano factor $\sigma^2/\mu = 1$.

9 Modeling Place Fields with Point Process Generalized Linear Models

Synopsis

Data Spike train data recorded in vivo from a place cell in rat hippocampus.

Goal Develop a model that relates the spiking of the neuron to the rat's movement trajectory.

Tools Point process theory, generalized linear models, Kolmogorov-Smirnov plots, likelihood ratio tests.

9.1 Introduction

9.1.1 Background

In chapter 8, we used visualization methods and simple interspike interval models to describe the spiking properties of a retinal neuron that was maintained at constant light and environmental conditions. In other words, we examined a neuron that was firing on its own, without any explicit driving stimuli. In contrast, many neuroscience experiments involve stimulating or perturbing a neural system and recording changes in spiking activity of a set of neurons in response to that stimulus. The stimulation may be a simple signal applied directly to the neural system, such as a current pulse injected into a neuron. Or it may be a more complex or abstract stimulus that is sensed in the peripheral nervous system and influences neural activity elsewhere, such as the presentation of a movie composed of a natural scene to an awake animal, inducing activity patterns in primary visual cortex and downstream areas.

This stimulus-response paradigm relates to the important concept of *neural coding*: that statistical features of spiking activity contain information about the stimuli, behaviors, or other biological signals that influence the activity. From a data analysis perspective, we are interested in modeling the relation between these signals and the observed spiking activity. We can do so through a statistical spike train model. Here we explore a useful class of models based on the statistical theory of point processes. We define the models in terms of a Poisson rate function, which defines the instantaneous likelihood of observing a spike at any point in time as a function of a set of covariates. In particular, we use a class of point process models that can be fitted by maximum likelihood and whose estimators have multiple optimal properties. These are called *generalized linear models* (GLMs). We provide some basic statistical ideas to develop intuition about these types of models, but readers can explore the rich theory underlying this approach via the references mentioned in this chapter.

9.1.2 Case Study Data

A collaborator has contacted us to discuss a new experiment he has performed. As part of this experiment, he has implanted a small bundle of electrodes in a rat's hippocampus and trained the rat to perform a simple spatial task: to run back and forth along a linear maze. During this task, our collaborator believes he has recorded the spiking activity from a place cell, a cell whose activity is position-specific. He has asked us to help characterize these spike train data and support (or refute) the notion that the observed cell is a place cell. He has agreed to provide us with the observed spike train data and the position of the rat as a function of time, recorded during a few minutes of the experiment.

9.1.3 Goal

Our goal is to characterize the properties of the observed cell as the rat runs back and forth in the linear maze. Spiking activity in these cells is known to relate to other variables, such as the speed and head direction of the rat. Here, we focus on modeling the relation between the rat's movement trajectory and the observed spiking activity. In doing so, we select a model through an iterative process of model fitting, evaluation, and refinement.

9.1.4 Tools

In this chapter, we develop a series of generalized linear models. We implement procedures to fit, refine, and compare this series of models. We demonstrate the process of implementing and fitting a Poisson regression model in MATLAB, procedures to evaluate model goodness-of-fit and compare models, and methods to construct confidence intervals for model parameters.

9.2 Data Analysis

9.2.1 Visual Inspection

To access the data for this chapter, visit

```
http://github.com/Mark-Kramer/Case-Studies-Kramer-Eden
```

and load the file `Ch9-spikes-1.mat`. You will find three variables in your workspace. The variable `X` indicates the rat's position (in centimeters) at each moment in time (variable `t`, in seconds). The variable `spiketimes` indicates the times (in seconds) at which spikes occur in the data.

We begin with visual inspection of the data. Let's first plot the rat's movement trajectory as a function of time:

```
load('Ch9-spikes-1.mat')        %Load the place field data
plot(t,X)                        %... and plot the rat's position.
xlabel('Time [s]');              %... with axes labeled.
ylabel('Position [cm]')
```

Figure 9.1 shows that the rat runs back and forth consistently, making about 15 passes during the approximately 3 minute recording. We also observe that the rat moves fairly quickly during each back and forth pass but spends a large amount of time at both ends of the track (near position 0 cm or 100 cm) before turning around and continuing.

Next, we would like to plot the spiking activity in relation to the rat's movement trajectory. However, we cannot simply plot the vector `X` against the vector `spiketimes`; these vectors have different lengths. The length of `X` is the same as the length of `t`, the total number of 1 ms time bins in the recording (177,761 time bins). The length of `spiketimes` is the total number of spikes to occur during the duration of the recording: 220 spikes. Therefore, the first step to visualizing the place-specific spiking activity is to use `spiketimes` to create a new vector, with the same size as `X`, that indicates whether a spike was fired at each time bin. Let's call this vector `spiketrain` and have it contain 1 for each time bin where a spike occurs and 0 for each time bin that lacks a spike.

Q: How would you construct the vector `spiketrain`? Can you think of multiple ways to construct the same vector?

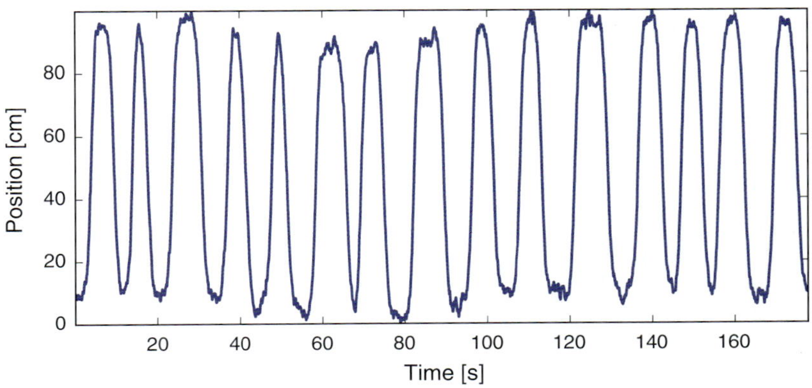

Figure 9.1
Rat position versus time. The rat moves from one end of the linear maze (near position 0 cm) to the other (near position 100 cm).

A: We could use a for-loop to step through each time bin and decide whether a spike occurred. A more efficient approach is to realize that this computation can be performed as a histogram, and use the MATLAB function `hist`:

```
%Histogram spikes into bins centered at times t.
spiketrain = hist(spiketimes,t)';
```

There are many other ways to compute this same vector.

Now we can plot the position and spike train together. In MATLAB,

```
plot(t,X)                       %Plot the position.
hold on                         %Freeze graphics window.
plot(t,10*spiketrain,'g')       %Plot the spikes.
hold off                        %Release graphics window.
xlabel('Time [s]')              %Label the axes.
ylabel('Position [cm]')
```

In the third code line, we have multiplied the values in the vector `spiketrain` by 10 to make them large enough to see in the plot. Inspection of figure 9.2 suggests that spiking tends to occur around the same approximate point during each pass through the track, namely, during the vertical upswing of the position curve in figure 9.2. However, from this visualization alone, it is still difficult to discern the exact animal location when each spike occurs.

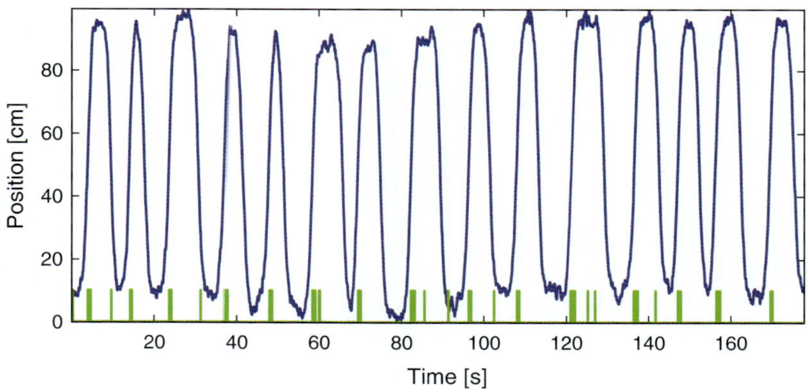

Figure 9.2
Rat position (*blue*) and spikes (*green*) versus time.

Q: Is there a better way to visualize the spiking activity and position?

A: Keeping the variable `spiketrain`, let's compute a new variable, `spikeindex`, using the `find` command:

```
spikeindex=find(spiketrain);     %Deterime index of each spike.
```

The vector `spikeindex` represents the vector indices at which spikes occur. We can use this to index any of the variables whose size matches the number of time bins in the recording. So, another visualization we can now employ is to plot the times and positions of the spikes overlaid on the full movement trajectory. In MATLAB,

```
plot(t,X)                        %Plot the position.
hold on                          %Freeze graphics window.
plot(t(spikeindex),X(spikeindex),'r.')%Plot spikes @ positions.
hold off                         %Release graphics window.
xlabel('Time [sec]')             %Label the axes.
ylabel('Position [cm]')
```

Note that by using the `'r.'` term in the `plot` function, we indicate the times and positions of the spikes as red dots.

From figure 9.3 it is clear that the preponderance of the spiking is occurring whenever the rat is running up the track, in the direction where `X` is increasing, at values of `X` ranging from about 50 cm to about 80 cm. We do not see the same spiking activity in this region when the rat is running back down the track, in the direction where `X` is decreasing. A few

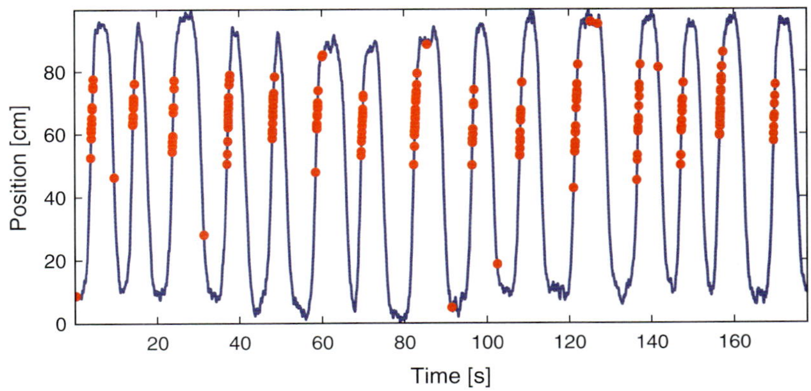

Figure 9.3
Rat position (*blue*) and spikes (*red*) versus time.

spikes occur at other locations, but these appear sparse compared to the place-specific firing in this region.

Another way to visualize this place field structure is to construct an *occupancy normalized histogram* of the spiking activity. To do so, we define a set of position bins spanning the full 100 cm track, count the number of spikes that occur in each location bin, and divide by the occupancy, the total amount of time spent at each location bin. Dividing by the occupancy is important. Otherwise differences in the way the stimulus is presented can bias the characterization of the stimulus response relation. For example, if the rat spent much more time in the 50–80 cm region, we might expect more firing in that region even if the firing does not depend on place at all. Based on our previous visualization, we know that this is not the case for these data, but it is important to keep in mind how the statistics of a stimulus signal might influence the statistics of an output signal.

Let's compute the occupancy normalized histogram in MATLAB:

```
bins=0:10:100;                      %Define spatial bins.
spikehist=hist(X(spikeindex),bins);%Histogram positions @ spikes.
occupancy=hist(X,bins)*0.001;      %Convert occupancy to seconds.
bar(bins,spikehist./occupancy); %Plot results as bars.
xlabel('Position [cm]')             %Label the axes.
ylabel('Occupancy norm. hist. (spikes/s)')
```

In the third line of this code, we multiply the occupancy by 0.001 to put the occupancy in units of seconds and the occupancy normalized histogram in units of spikes per second. From the occupancy normalized histogram in figure 9.4, we see that the firing rate is highest around the 60–70 cm position and falls off rapidly as the rat moves away from that

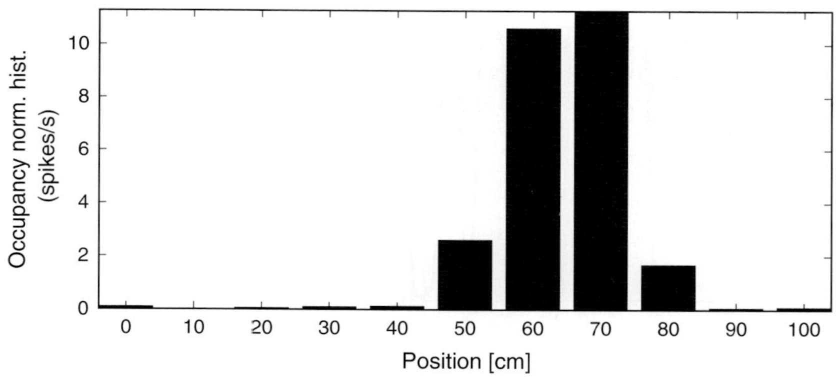

Figure 9.4

Occupancy normalized histogram. Each bar indicates number of spikes occuring in intervals of position, normalized by amount of time spent in each interval of position.

location. This corroborates the results from our previous visualizations. However, there is one feature of the spiking activity that we observed previously for which we are not accounting in this visualization. Here we are relating spiking purely to position without regard for the direction of the rat's movement.

Q: How might we construct a visualization that accounts for spiking as a function of both position and direction?

A: One option is to construct separate occupancy normalized histograms for movement in each direction. To do so, we would need to determine for each time step whether it represents a movement period where the position is increasing or decreasing. How would you do this? (We do so later when constructing place field models.)

9.2.2 Fitting a Point Process Model (Poisson GLM)

Any statistical model that describes data occurring at localized points in time, like spike times, is called a temporal *point process model*. In chapter 8, we constructed a point process model that described the probability distribution of waiting times between spikes for a neuron with no explicit driving stimulus. Here, we would similarly like to construct a statistical model, but in this case the model should characterize how the distribution of the data depends on the covariates of interest: the rat's position and movement direction.

Q: Do the notions of point process model, Poisson model, and rate parameter seem familiar?

A: If not, consider reviewing the case study in chapter 8.

One approach we used to model the spiking data in chapter 8 was a Poisson model, in which we used a rate parameter, λ, to define the expected rate of spiking in any time interval. We then computed the value of λ that maximized the likelihood of observing the recorded spiking activity. For the data of interest here, we extend this concept by defining a rate that varies in time as a function of some set of covariates. These covariates are any variables whose influence on the spiking activity we wish to explore. Our visualizations suggest that useful covariates for our model include the rat's position and its direction of motion.

Let's define some terms. Let $x(t)$ represent the rat's position at time t, and let $d(t)$ represent the direction of motion; we set $d(t) = 0$ when $x(t)$ is decreasing or the rat is stopped, and $d(t) = 1$ when $x(t)$ is increasing. Since these position and direction signals change as a function of time, so does the firing rate. We write $\lambda(t) = g(x(t), d(t))$, where $\lambda(t)$ is called the *Poisson rate function*, and $g(\cdot, \cdot)$ is a function that we need to define the model.

What function should we use for $g(\cdot, \cdot)$? We want something that captures the relation between the covariates and the spiking, and is easy to interpret. The process of finding a model or set of models that are most consistent with the data is called *model identification* or *model selection*. Typically, this is an iterative process where we propose a class of models, find the particular model in that class that best fits the data, assess the quality of that model, and decide whether to refine the model further or to draw conclusions from the model fit. In practice, it is a good idea to begin with descriptive statistics and visualizations of the relation between the covariates and spiking data to select a class of point process models. For the spike train data of interest here, our visualizations suggest a model where the dependence of spiking on position has a mound shape (as in the occupancy normalized histogram, figure 9.4) and which incorporates direction. We start with an overly simple model for pedagogical purposes.

The following is a very basic model inspired by simple linear regression:

$$\text{[Model 1]} \qquad \lambda(t) = \beta_0 + \beta_1 x(t)\,. \tag{9.1}$$

The idea of linear regression is to express a response variable at time t in terms of predictor variables, or *covariates*. Here, β_0 and β_1 are unknown parameters used to characterize a linear dependence between the response variable $\lambda(t)$ and covariate $x(t)$. β_0 represents the expected firing rate at $x(t) = 0$, and β_1 represents the change in firing rate for each unit of increase in position. This initial model does not include any dependence on the rat's movement direction (i.e., there's no $d(t)$ term in (9.1)).

The form of model (9.1) looks like a standard linear regression, which is comforting because methods exist in MATLAB to solve these types of problems (e.g., the function `regress`). However, the observed data are spike events; in discrete time, the data are spike counts. A standard linear regression assumes that the distribution of the data, given the covariates, is normal. Spike counts can take on only non-negative integer values, so their distribution cannot be normal. When the number of spike counts in each time bin is very large, it is possible that the distribution of the data can be approximated by a normal distribution, and in this case, simple regression methods might work. But for the spiking data of interest here, we have very few spikes (0 or 1) in each 1 ms time bin, so a simple regression fit would not be correct.

Instead, we must fit a *Poisson regression model* to the data. If we let $\Delta N(t)$ be the number of spikes that are observed in the interval $(t, t + \Delta t)$, then under the Poisson regression model, $\Delta N(t)$ has a Poisson distribution with a mean parameter equal to the response variable $\lambda(t)$ integrated over the interval $(t, t + \Delta t)$.

How do we fit the Poisson regression model? It turns out that Poisson regression models of a certain form can be fitted efficiently using the theory of *generalized linear models*. In MATLAB, we can fit this model using the `glmfit` function. Before applying this function directly to the data, let's get an overview of the function's inputs and outputs. In MATLAB, we consider the `glmfit` function with four inputs and one output,

```
b = glmfit(X,Y,DISTR,LINK)
```

The first input, X, is a matrix of the covariates on which spiking depends. The size of this matrix is $n \times p$, where p is the number of covariates in the model, and n is the number observations or time steps. In this case, X is an $n \times 1$ matrix, where n is the number of data points $(177,761)$ representing the position of the rat along the track. We already have a variable named X that is exactly this. The second input, Y, is a vector of the spike counts at each time step associated with each row of X. In this case, Y is the vector `spiketrain` that we computed earlier. The third input is a string indicating the distribution of the spike count data in Y. For a Poisson regression model of spike count data, we set the third input to `'poisson'`. In fact, for most neural spike count models fitted using GLM, even those that are not Poisson processes, we use the Poisson distribution. The fourth input is a string indicating a link function between the spiking rate and the covariates. Specifically, if we want to fit a model of the form $h(\lambda(t)) = \beta_0 + \beta_1 x(t)$, then we would say that the function $h(\cdot)$ is the link function. For model 1, this is simply the identity function. In what follows, we show a better way to select this link function.

The output of the `glmfit` function is a vector of numbers representing the maximum likelihood estimates of the model parameters, which for this example we have labeled β_0 and β_1. We use the "hat" notation above a parameter to represent its estimate. The maximum likelihood estimate of β_0 is written $\hat{\beta}_0$, and the maximum likelihood estimate of β_1 is written $\hat{\beta}_1$.

Let's now use this function to fit the parameters of model 1 to the observed location and spiking data. In MATLAB,

```
%Fit Model 1 to the spike train data.
b=glmfit(X,spiketrain,'poisson','identity');
```

Q: Upon executing this command in MATLAB, what do you find? Do the results make sense?

A: Initially, you may notice that MATLAB outputs at the command line a warning that this model—particularly the identity link function—may be inappropriate. Let's ignore this warning and attempt to interpret the resulting parameter estimates. The first of these values is the maximum likelihood estimate for β_0. If we believed this model was accurate, we could interpret this parameter as indicating that the expected firing rate at position $x = 0$ is $\lambda = -0.103 \times 10^{-3}$ spikes per millisecond, or about -0.1 spikes per second, and that as the rat moves in the positive direction, the firing rate increases by $\beta_1 = 0.0272$ spikes per second for every centimeter the rat moves. This result should immediately raise some red flags. The fact that the firing rate is negative indicates that the model becomes uninterpretable for observed values of x. This suggests one major problem with model 1—the firing rate is negative—and motivates changes to the model link function.

To further visualize the quality of this model, we can compare the dependence it defines between position and spike rate to the occupancy normalized histogram we computed earlier. In this case, we use the positions defined by the histogram bins, and compute the modeled spike rate at these positions.

```
bar(bins,spikehist./occupancy);      %Plot occupancy norm. hist.
hold on;                             %Freeze graphics.
plot(bins,(b(1)+b(2)*bins)*1000,'r');%Plot model.
hold off                             %Release graphics.
xlabel('Position [cm]')              %Label the axes.
ylabel('Occupancy norm. hist. (spikes/s)')
```

We see in figure 9.5 that the model spike rate (red line in figure) captures some features of the observed spiking, for example, the fact that the spike rate increases as the rat moves from position $x = 0$ toward position $x = 60$. But the model misses much of the structure, for example, the fact that the spike rate does not change linearly with position and begins to decrease as the rat's position increases beyond $x = 70$. This suggests a second issue with this model: the form of the relation between position and firing rate is wrong.

9.2.3 Refining the Model

We conclude that our initial proposal, model 1 expressed in (9.1), does not represent the data well. Therefore, let's refine the model to account for the identified issues. First, let's choose a link function that is more appropriate for point process modeling. We would like a function that ensures the rate function is non-negative and that is easy to fit. The theory of generalized linear modeling suggests one function in particular: the log link. We set the log of the firing rate to be a linear function of the covariates. If we maintain position as the

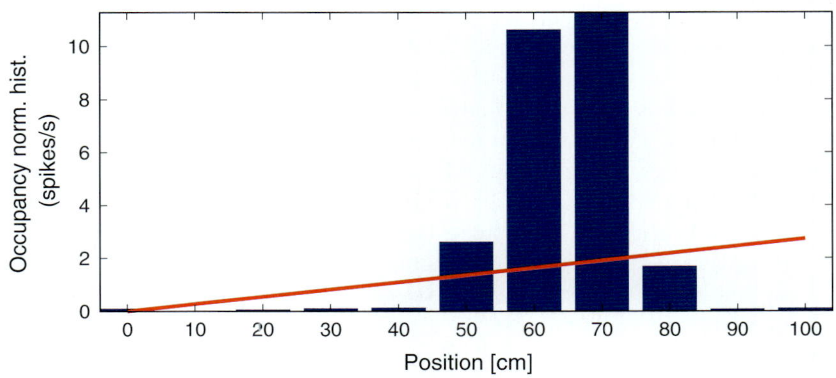

Figure 9.5

Occupancy normalized histogram and results of fitting model 1 (*red*) to these data.

sole covariate, this leads to a model of the form

$$[\text{Model 2}] \qquad \log \lambda(t) = \beta_0 + \beta_1 x(t)\,, \tag{9.2}$$

or equivalently,

$$[\text{Model 2}] \qquad \lambda(t) = e^{\beta_0 + \beta_1 x(t)}\,. \tag{9.3}$$

This link function is called the canonical link for Poisson data. It has a number of appealing properties. As desired, it ensures that the rate function is positive.

Q: Consider the expression for $\lambda(t)$ in (9.3). Why must $\lambda(t)$ always be positive?

The choice of a log link also ensures that the likelihood of the data is concave with respect to the model parameters. This means that the likelihood only has one local maximum value, which is the maximum likelihood (ML) estimate. It can also be shown that in many cases, the parameter estimators will be asymptotically normal, which will allow us to construct confidence intervals and make significance statements about them [6].

To fit model 2 in MATLAB, we use the same `glmfit` command as before but replace the last input (`'identity'`) with `'log'`:

```
%Fit Model 2 to the spike train data.
b2=glmfit(X,spiketrain,'poisson','log');
```

In fact, if we omit the name of the link function in the `glmfit` routine, it will automatically use the canonical link for the selected distribution. Since the log link is canonical for Poisson data, we can simply run the MATLAB command:

```
%Fit Model 2 to the spike train data (omitting last input).
b2=glmfit(X,spiketrain,'poisson');
```

Q: Execute the `glmfit` function with the log link in MATLAB. What do you find?

A: This time, we find that MATLAB does not issue a warning that the link function may be inappropriate. Inspection of the estimated parameter values reveals `b2 = [-7.4389 0.0129]`. These values are markedly different from the parameter values `b` found using model 1. The reason for this difference is that the form of the model has a major impact on the interpretation of the parameter values. In what follows, we discuss the interpretation of these parameter values in detail.

Let's examine the model fit more closely. When $x = 0$, the firing rate under model 2 is

$$\begin{aligned}\lambda(t) &= \exp(\beta_0 + \beta_1 \times 0)\\ &= \exp(\beta_0)\\ &= 0.0006\,\text{spikes/ms}\\ &= 0.6\,\text{spikes/s},\end{aligned}$$

where we have used the value $\beta_0 =$ `b2(1)`. If the rat moves from position $x = 0$ to $x = 1$, the firing rate becomes

$$\begin{aligned}\lambda(t) &= \exp(\beta_0 + \beta_1 \times 1)\\ &= \exp(\beta_0 + \beta_1)\\ &= \exp(\beta_0)\exp(\beta_1)\\ &= 1.013\exp(\beta_0),\end{aligned}$$

where we have used the value $\beta_1 =$ `b2(2)`. That is, a 1 cm increase in position increases the firing rate 1.3%. Because of the link function, position now has a multiplicative rather than an additive effect on the firing rate. Instead of adding to the firing rate, each increase of position leads to a multiplicative modulation of the firing rate at about a 1% increase per cm. Let's see how this model looks by comparing it to the occupancy normalized histogram of the data. In MATLAB,

```
bar(bins,spikehist./occupancy);          %Plot occupancy norm. hist.
hold on;                                 %Freeze graphics.
plot(bins,exp(b2(1)+b2(2)*bins)*1000,'r');  %Plot Model 2.
hold off                                 %Release graphics.
xlabel('Position [cm]')                  %Label the axes.
ylabel('Occupancy norm. hist. (spikes/s)')
```

Q: Consider model 2 and the data (figure 9.6). How do the two compare?

A: Visual inspection suggests that we've solved one problem: the spike rate is no longer negative anywhere. However, the model fit still does not agree with the structure seen in the occupancy normalized histogram. We have improved the link function, but using only the position itself as a covariate leads to a rate that is an exponential function of the rat's position.

There are many variables we might think to add to this model, but what variables could we add to better capture the dependence between firing rate and position, in particular? One thought might be to include nonlinear terms, such as the square of the position value. This

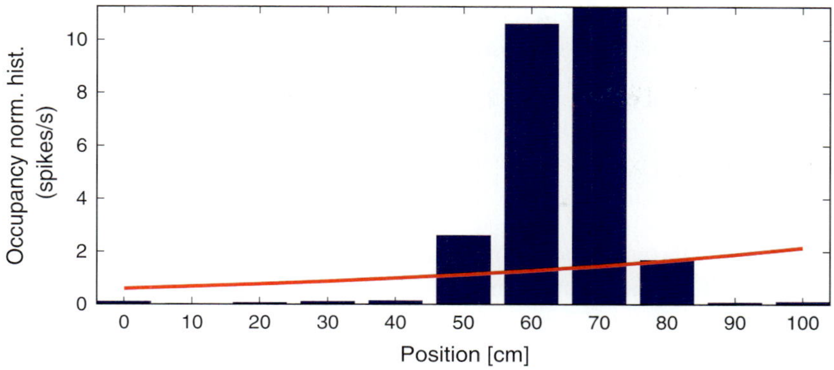

Figure 9.6
Occupancy normalized histogram and results of fitting model 2 (*red*) to these data.

gives us a third candidate model:

$$[\text{Model 3}] \qquad \lambda(t) = e^{\beta_0+\beta_1 x(t)+\beta_2 x(t)^2}. \qquad (9.4)$$

Compared to model 2, we've now included an additional $x(t)^2$ term and unknown coefficient β_2.

Q: We said previously that we would use generalized *linear* models. Does the use of the nonlinear term $x(t)^2$ violate this?

A: It might be better to think of linear in "generalized linear models" as requiring some function of the mean of the response variable to be a linear function of the coefficients (i.e., the β's). The covariates can be linear or nonlinear functions of observed quantities (e.g., the position squared, the sine of the angle of head direction, etc.)

To fit model 3 in MATLAB, we add another column to the matrix of covariates, the first argument of the `glmfit` routine.

```
%Fit Model 3 to the spike train data (omitting last input).
b3=glmfit([X X.^2],spiketrain,'poisson');
```

Q: Compare the first argument to the `glmfit` function for model 3 versus model 2. How are the two inputs similar? How are they different?

Q: Execute the `glmfit` function with the log link in MATLAB. What do you find?

A: As was the case with model 2, we find that MATLAB produces no warnings that the link function may be inappropriate. In this case, there are three estimated parameter values in `b3 = [-26.2791 0.6901 -0.0055]`.

Let's now interpret the parameter estimates for model 3. The estimate of the first parameter is $\hat{\beta}_0 = -26.3$. This means that when the rat is at position $x = 0$, the firing rate is $\lambda(t) = \exp(\beta_0) = \exp(-26.3) \approx 0$. There is almost no chance of observing a spike when the rat is at this location. What happens as the rat moves in the positive direction? This is determined by both $\hat{\beta}_1 = 0.6901$ and $\hat{\beta}_2 = -0.0055$. For every unit increase in position, the firing rate is multiplied by $\exp(\beta_1) = 1.99$, but at the same time, for every unit increase in the square of position, the firing rate is multiplied by $\exp(\beta_2) = 0.99$.

Expressed this way, the values of parameters $\exp(\beta_1)$ and $\exp(\beta_2)$ seem difficult to interpret. Once we visualize this model, we realize that there is another way to express the model so that the parameters are easier to interpret:

```
bar(bins,spikehist./occupancy);         %Plot occupancy norm. hist.
hold on;                                %Freeze graphics.
plot(bins,exp(b3(1)+b3(2)*bins+b3(3)*bins.^2)*1000,'r');%Model 3.
hold off                                %Release graphics.
xlabel('Position [cm]')                 %Label the axes.
ylabel('Occupancy norm. hist. (spikes/s)')
```

We see in figure 9.7 that model 3 aligns much more closely with the occupancy normalized histogram. The firing rate is small at the beginning of the track, increases to a maximum firing rate near 10 Hz about 60 cm along the track, and then decreases as the position increases further. The firing rate model as a function of position looks like the bell-shaped or mound-shaped density that we often associate with the Gaussian (or normal) distribution. The fact that the firing rate is the exponential of a quadratic function of position means that we can rewrite the model in a form that more closely resembles the Gaussian function:

$$\left[\text{Model 3, alternative form}\right] \qquad \lambda(t) = \alpha e^{-\dfrac{(x-\mu)^2}{2\sigma^2}}, \tag{9.5}$$

where $\mu = -\beta_1/(2\beta_2)$ is the point along the track where the firing rate is maximal (the center of the place field), $\sigma^2 = -1/(2\beta_2)$ determines the range over which the firing rate is elevated (the size of the place field), and $\alpha = \exp^{\beta_0 - \beta_1^2/(4\beta_2)}$ is the maximum firing rate at the place field center.

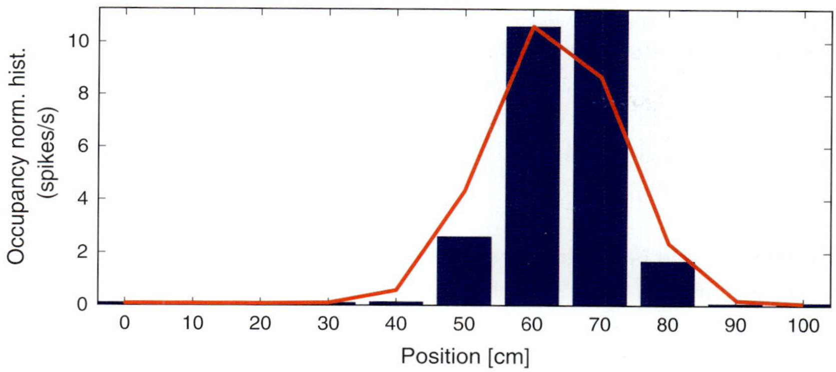

Figure 9.7
Occupancy normalized histogram and results of fitting model 3 (*red*) to these data. Compare with figure 9.5 and fitting 9.6.

In this example, we can use the estimated GLM coefficients to estimate these new model parameters related to the center, size, and maximum firing rate of the place field. The `glmfit` routine has given us the maximum likelihood estimates for β_0, β_1, and β_2. An important result from statistical theory is that the maximum likelihood estimate of any function of model parameters is just that same function of the maximum likelihood estimates of the parameters. This is often called *invariance* or *equivariance* [6]. So $\hat{\mu} = -\hat{\beta}_1/(2\hat{\beta}_2)$ is the maximum likelihood estimate of the place field center, $\hat{\sigma} = \sqrt{-1/(2\hat{\beta}_2)}$ is the maximum likelihood estimate of the place field size, and so on.

Let's now use these expressions to compute the maximum likelihood estimates in MATLAB:

```
%Compute maximum likelihood estimates of
mu=-b3(2)/2/b3(3);                      %...place field center,
sigma=sqrt(-1/(2*b3(3)));               %...place field size,
alpha=exp(b3(1)-b3(2)^2/4/b3(3));       %...max firing rate.
```

Note that the relation between the index of `b3` and subscript of β is shifted; we start the indices at a value of 1 and the subscripts at a value of 0. Executing these lines of code we find `mu = 63.16`, `sigma = 9.56`, and `alpha = 0.011`.

Q: What are the units of each of these parameter estimates?

A: The units for `mu` and `sigma` are centimeters, the same as the units of `x`. The units for `alpha`, the maximum spike rate, are spikes per time step of the vector `spiketrain`. Since each time step is 1 ms, the units are spikes/ms. So a value of $\hat{\alpha} = 0.011$ spikes/ms is equivalent to a maximum firing rate of $\hat{\alpha} = 11$ spikes/s.

We see that the estimated place field center is about 63.2 cm down the track. The estimated place field size, $\hat{\sigma} = 9.6$, suggests that the firing rate decreases about 40% when the rat is about 9.6 cm from the place field center, and decreases about 85% when the rat is about 19 cm from the place field center. The neuron spikes at a rate near $\hat{\alpha} = 11$ spikes/s when the rat is 63 cm along the track, but less than 2 spikes/s when the rat is more than 19 cm away from that position.

9.2.4 Comparing and Evaluating Models

We have fit a number of models for the receptive field of this neuron and compared these models through visual inspection. Ideally, we'd like to go beyond qualitative comparisons and consider quantitative tools to help us evaluate and compare different models. For statistical models, we often use the term *goodness-of-fit* to describe how well a model captures structure in the observed data, and how well the model predicts future data. There is not a single procedure for measuring goodness-of-fit; instead there are many tools that, taken together, can provide a broad perspective on the strengths and weaknesses of a set of models. We explore some approaches for comparing the relative goodness-of-fit between two models and then methods to assess the overall goodness-of-fit of a single model.

Method 1: Comparing AIC Values. Let's say we want to compare the quality of the fit to the data of models 2 and 3. What measure could we use to compare these models? A natural thought is to use the likelihood. We have already used the likelihood to select the parameter values for each of these models; we selected the parameters that maximized the likelihood of the data. We can think of selecting the model parameters as selecting among a set of models with the same model form but different parameter values. In that case, the likelihood was the measure we used to make the selection.

However, there is a major issue with using the likelihood alone to compare goodness-of-fit between different classes of models. It turns out that as we add parameters to the model and make the model more complex, we tend to increase the likelihood of the data, whether or not those additional terms accurately describe the processes that generate the data. If we were to select models that maximize the likelihood of the data without regard for model size, we would tend to get very large models that fit the observed data well but do not predict or describe future data well. This is known as the *overfitting problem*. In order to avoid overfitting, we are compelled to balance the ability to fit complicated datasets with the desire to use simple models with small numbers of parameters. This trade-off is sometimes referred to as the goal of *parsimony* in modeling. We want to be sparing with the number of parameters we allow a model to have. We call a model that describes the data well with the fewest possible parameters a *parsimonious model*.

One common approach for preventing overfitting is *cross-validation*. There are multiple types of cross-validation, but they all share a common idea: split the data into multiple portions, fit the model on one portion of the data (called the training set), and determine how well the resulting model fit describes a separate portion of the data (called the test set). This ensures that the selected model is one that can generalize to additional datasets that were not used to fit the model. One challenge with cross-validation is that it can be computationally expensive. For example, one of the most robust cross-validation approaches, called complete leave-one-out cross-validation, involves sequentially leaving out each data point, fitting a model to the remaining data, and assessing how well the fitted model predicts the excluded data point. This involves fitting n models, where n is the number of data points observed.

Here, instead of fitting a large number of models, we take another approach, which gives results equivalent to cross-validation when the dataset is large. Namely, we use *penalized likelihood* measures to compare model types. These measures make explicit the trade-off between fitting the data well (by increasing the likelihood) and using a small number of parameters (by penalizing large models). Let's consider one such measure, *Akaike's information criterion* (AIC). It is defined as,

$$\text{AIC} = -2\log L(\hat{\theta}_{\text{ML}}) + 2p, \tag{9.6}$$

where $L(\hat{\theta}_{\text{ML}})$ is the likelihood of the data for the selected maximum likelihood parameter estimate $\hat{\theta}_{\text{ML}}$, and p is the number of parameters in the model. We think of the $2p$ in expression (9.6) as a penalty for models with large numbers of parameters.

When comparing models, we compute the AIC for each model separately, and then compute the difference in AICs between models. For models that accurately describe the structure of the data, $L(\hat{\theta}_{\text{ML}})$ will be high, and therefore $-2\log L(\hat{\theta}_{\text{ML}})$ will be small. Parsimonious models will have small numbers of parameters, and therefore $2p$ will be small. Therefore, we are looking for models with AIC values as small as possible.

How do we compute the likelihood or log likelihood of the data for the models in MATLAB? One way is to use the fact that we are modeling the spike train as a Poisson random variable at each time point with rate parameters determined by the model. For model 2, we can compute the AIC as follows:

```
lambda2=exp(b2(1)+b2(2)*X);       %Use Poisson rate function,
LL2=sum(log(poisspdf(spiketrain,lambda2)));%..and log likelihood,
AIC2 = -2*LL2+2*2;                %...to find AIC for Model 2.
```

The first line of this code computes the Poisson rate function for model 2 at each time point. The second line then computes the log likelihood. Recall that the likelihood is the joint distribution of all the data for a specific model. In this case, the number of spikes in each bin is modeled as a Poisson random variable with rate $\lambda(t)$. Therefore, the log likelihood (`LL2`) is just the logarithm of the product of Poisson probability values for the observed spiking under the proposed model (or equivalently, the sum of the log of these Poisson probability values). The third line computes the AIC for this model. Notice that in the last line we use a value of $p = 2$, as there are two parameters in this model (β_0 and β_1).

Similarly, we can compute the AIC for model 3. Let's try to do this in one line of MATLAB code:

```
%Compute the AIC for Model 3.
AIC3 = -2*sum(log(poisspdf(spiketrain, ...
                    exp(b3(1)+b3(2)*X+b3(3)*X.^2))))+2*3;
```

Q: Consider the definition of `AIC3`. Can you explain in words the different terms? What does the term `2*3` represent? How many parameters are there in model 3?

Finally, we can compute the difference between the AIC values for these two models:

```
dAIC=AIC2-AIC3;       %Difference in AICs between Models 2 and 3.
```

We find a value of `dAIC = 636.0145`. This difference indicates that the AIC of model 3 is smaller than that of model 2, suggesting that model 3 is superior. How should we interpret the value of the difference? The answer depends on the probability model we are using, and generally we are just interested in which model has the lowest AIC without worrying about the magnitude of the difference. However, one rough way of thinking about this value is in terms of the penalty. The fact that model 3 has an AIC of about 636 less than the AIC of model 2 suggests that model 3 would still be preferable to model 2 even if model 3 had $636/2 = 318$ more parameters than it actually does.

It turns out that there is a simpler way to compute the difference in AICs between two GLMs. Whenever MATLAB (and most other computational software packages) computes the maximum likelihood solution for a GLM, it also computes the model deviance. The model deviance is a measure of lack of fit between the model and data, which is defined by

$$\text{Deviance} = -2\log L(\hat{\theta}_{\text{ML}}) + C,$$

where C is a constant.[1] Therefore the difference in AICs between two models can be computed as

$$\Delta\text{AIC} = \text{AIC}_1 - \text{AIC}_2 = \left(\text{Dev}_1 + 2p_1\right) - \left(\text{Dev}_2 + 2p_2\right), \tag{9.7}$$

where AIC_1, Dev_1, and p_1 are the AIC, deviance, and number of parameters for the first model, and AIC_2, Dev_2, and p_2 are the AIC, deviance, and number of parameters for the second model. The constant C cancels out when computing the difference in AIC values.

1. The constant C is equal to -2 times the log likelihood of a saturated model that has as many parameters as points in the data. Such a model is guaranteed to overfit the data, namely, it will fit the observed data as well as possible but not generalize well to new data.

To compute the deviance, we add a second output to the `glmfit` routine. While we're at it, let's also add a third output to the `glmfit` commands for our models, `stats`, which will provide a set of useful statistics related to the model fits.

```
%Fit Model 2 and Model 3, and compute difference in AICs.
[b2 dev2 stats2]=glmfit(X,spiketrain,'poisson');
[b3 dev3 stats3]=glmfit([X X.^2],spiketrain,'poisson');
dAIC=(dev2+2*2)-(dev3+2*3);
```

The resulting difference in AICs (variable `dAIC`) matches the value we computed earlier.

Method 2: Chi-Square Test for Nested Models. The AIC provides a method for identifying parsimonious models and comparing between models but does not, on its own, indicate whether a particular model provides a statistically significant improvement in its description of a dataset. For example, we might add a predictor to a model that has no real connection to the observed data and yet decreases the AIC by chance. In order to assess whether a model provides a significant improvement over another, we can use hypothesis tests based on the model likelihoods.

In particular, there is a general class of hypothesis tests called *maximum likelihood ratio tests* (MLRTs) that often provide the most statistically powerful comparison between models. In general, it can be challenging to compute the test statistic and its sampling distribution for MLRTs. However, it becomes easy to perform this test in cases where we are comparing two nested GLMs, that is, when one of the models can be made equivalent to the other by setting some parameters to specific values. For example, it is possible to make model 3 equivalent to model 2 by setting $\beta_2 = 0$. We say that model 2 is nested in model 3. However, there is no way to set any parameters to make model 2 equivalent to model 1 or vice versa, so these models are not nested. It can be shown that when we compare two nested Poisson GLMs for spike train data, the MLRT will asymptotically be a simple chi-square (χ^2) test. The proof for this result can be found in many textbooks on GLMs, such as [29].

Let's specify the components of this hypothesis test. Assume that the nested model has n_1 parameters, $\{\beta_1, \ldots, \beta_{n_1}\}$ and that the larger model has n_2 parameters, $\{\tilde{\beta}_1, \ldots, \tilde{\beta}_{n_2}\}$. The null hypothesis for this test is H_0: $\tilde{\beta}_{n_1+1} = \cdots = \tilde{\beta}_{n_2} = 0$, that all the additional parameters not contained in the nested model are equal to zero. The alternative hypothesis is that at least one of these additional parameters are different from zero. The test statistic for this MLRT is equivalent to the difference in the deviances between the nested model (here, Dev_1) and the larger model (here, Dev_2),

$$\Lambda = \text{Dev}_1 - \text{Dev}_2 .$$

Under the null hypothesis, this statistic should asymptotically have a chi-square distribution with $n_2 - n_1$ degrees of freedom. We can compute the p-value for a test comparing two nested GLMs for spiking data using the `chi2cdf` command in MATLAB.

Let's compute the p-value for a MLRT comparing models 2 and 3:

```
p=1-chi2cdf(dev2-dev3,1);  %Compare Models 2 and 3, nested GLMs.
```

In this case, the difference in parameters between model 2 and model 3 is 1; model 3 has one additional parameter. We therefore set the degrees of freedom of the chi-square distribution to 1, the second input to the function `chi2cdf`. We find the computed p-value is zero, to the precision that MATLAB is able to compute the chi-square distribution. In practice, this means that the p-value for this test is not exactly zero but is smaller than 10^{-10}. We have a great deal of evidence that the additional, quadratic parameter in model 3, β_2, is nonzero.

Method 3: Confidence Intervals for Individual Model Parameters. If we want to test directly for whether a single parameter contributes significantly to the model, we can examine its interval estimate. The GLM fitting procedure not only computes the maximum likelihood estimator for each model parameter but also computes the *Fisher information*, a quantity related to the curvature of the likelihood, which can be used to compute confidence intervals about any individual parameters or any combination of parameters. We do not discuss the Fisher information in detail here (for more, see [6]), but the basic idea is intuitive. If the likelihood is very flat at its maximum, then changing the parameter values slightly will not decrease the likelihood substantially. Therefore, there is a potentially large range of parameter values that could make the data likely. If the likelihood is very peaked at its maximum, then a slight change in the parameter values would cause a large change in the likelihood, and therefore a much narrower range of parameter values would be consistent with the data.

In the MATLAB `glmfit` routine, all the results used to express parameter uncertainty are contained in the third output, which we usually label `stats`. The output is a structure, that contains a variety of useful components. Two components that are useful for examining the significance of individual model parameters are `stats.se` and `stats.p`. The first, `stats.se`, provides the standard error of each parameter estimate. Since maximum likelihood estimators have approximately normal distributions with enough data, an approximate 95% confidence interval for any parameter β_i would be $\hat{\beta}_i \pm 2\hat{\sigma}_{\beta_i}$, where $\hat{\beta}_i$ is the parameter estimate and $\hat{\sigma}_{\beta_i}$ is the estimated standard error.

In estimating model 2, we computed the output variables `b2` and `stats2`. Let's now use these outputs to compute confidence intervals for the parameters of model 2:

```
%Compute 95% CI for parameters of Model 2.
CI2=[b2-2*stats2.se b2+2*stats2.se];
```

The top row of the variable `CI2` is the confidence interval for β_0, and the bottom row is the confidence interval for β_1. How should we interpret these confidence intervals? Just as before, they will be more interpretable if we exponentiate first.

```
eCI2 = exp(CI2);     %Exponentiate Model 2 CIs.
```

The confidence interval for β_0 describes the uncertainty in the firing rate at position $x = 0$. At that position, we are 95% certain that the rate is between 0.0004 and 0.0008 (the first row of `eCI2`) spikes per millisecond, or between 0.4 and 0.8 spikes per second. The confidence interval for β_1 describes the uncertainty in the effect of a unit change in position on the firing rate. Every time we increase x by 1, the rate gets modulated by a value between 1.009 and 1.0171 (the second row of `eCI2`). In other words, each centimeter increase in position increases the firing rate between about 0.9% and 1.7%.

Another use for the confidence intervals is to express the statistical significance of individual parameters within the model. If the true value of a parameter is zero, then the covariate corresponding to that parameter does not contribute to the prediction of the data in the GLM. If we compute a confidence interval for a parameter and it does not contain zero, we have sufficient evidence (at the confidence level used to construct the interval) that the true parameter value differs from zero, and that the covariate for that parameter has a significant contribution within the GLM. We can use this once again to determine whether the addition of the quadratic term in model 3 provides a significant improvement over model 2. To do so, let's use the computed output variables `b3` and `stats3` for model 3 to determine the confidence intervals for each parameter in model 3.

```
%Compute 95% CI for parameters of Model 3.
CI3 = [b3-2*stats3.se b3+2*stats3.se];
```

The resulting variable `CI3` consists of three rows. The bottom row is the confidence interval for β_2. We see that this interval (`CI3[3,:]=[-0.0063 -0.0046]`) does not contain zero, so the quadratic term is significant at the 95% confidence level.

How significant is this term? To answer this, we can conduct a hypothesis test for whether β_2 is different from zero. This test, based on the maximum likelihood estimate of a model parameter and its standard error, is called a *Wald test*. The significance level of this test is also output in the `glmfit` routine under the output variable `stats.p`. For model 3, the significance level for parameter β_2 is

```
p_beta2 = stats3.p(3); %Significance level of Model 3 parameter.
```

We find `p_beta2 = 4.12e-38`, which is very close to zero. This result is consistent with our previous finding, via the MLRT, that the improvement of model 3 over model 2 suggested that the quadratic model component was significant at a level of $p < 10^{-10}$.

Method 4: KS Test for Model Goodness-of-Fit. The goodness-of-fit methods we have developed so far are useful for comparing models and for determining whether individual variables contribute significantly to a model. However, these methods do not tell us whether the model captures the structure of the spiking data well overall. There are a number of methods to evaluate the overall goodness-of-fit between a model and data, but here we focus on an approach specific to point process (spike) data. The method is based on an important result known as the *time-rescaling theorem*.

In many statistical analyses, we assume that the data are identically distributed, that is, that all the data come from the same probability model and have the same mean. A common goodness-of-fit method is then to compare the empirical distribution of the data to that single probability model. However, for most spike trains, the data are not identically distributed. For example, the rate of spiking may change as a function of time or with other covariates, so the expected interspike interval changes for each data point. The time-rescaling theorem provides a transformation of the data that is identically distributed if the rate model is correct.

We first state the time-rescaling theorem for Poisson processes.[2]

Time-Rescaling Theorem: Consider a collection of spike times $(S_1, S_2, \ldots, S_n)$ from a Poisson process with rate function $\lambda(t)$. Then let

$$Z_1 = \int_0^{S_1} \lambda(t)dt \qquad \text{and} \qquad Z_i = \int_{S_{i-1}}^{S_i} \lambda(t)dt\,,$$

for $i = 2, \ldots, n$. By the time-rescaling theorem, the rescaled variables, $Z_1, Z_2, \cdots, Z_n$, are independent and identically distributed random variables from the exponential distribution with parameter 1.

We do not prove the time-rescaling theorem here but note that it comes from the change-of-variables formula from basic probability theory; see [6] for a detailed proof.

How do we use the time-rescaling theorem to analyze spike train data? If we have a collection of spike times, $S_1, S_2, \ldots, S_n$, and any Poisson rate model, say, from a fit GLM, then we can compute the rescaled waiting times, $Z_1, Z_2, \cdots, Z_n$. We then perform any standard goodness-of-fit method to compare $Z_1, Z_2, \cdots, Z_n$ to the exponential probability model. If the Poisson rate model describes the data well, then the exponential model should describe the rescaled times well.

The actual goodness-of-fit technique we use for the rescaled data is the *Kolmogorov-Smirnov (KS) plot*. Recall from chapter 8 that the KS plot compares the empirical

2. This result is extended to general point process data in chapter 10.

cumulative distribution function (CDF) of the data to a model CDF. In that chapter, we compared the empirical CDF for observed interspike intervals against various model CDFs. In this case, we compare the empirical CDF of the rescaled waiting times to an exponential CDF.

Let's now apply the time-rescaling theorem to evaluate model 3. First we must compute the rescaled waiting times. In MATLAB,

```
N = length(spiketimes);                   %Define no. of of spikes.
lambda3=exp(b3(1)+b3(2)*X+b3(3)*X.^2);   %Evaluate Model 3.
Z(1)=sum(lambda3(1:spikeindex(1)));       %1st rescaled waiting time,
for i=2:N                                 %... and the rest.
 Z(i)=sum(lambda3(spikeindex(i-1):spikeindex(i)));
end
```

Q: Do you understand how the variable `Z` approximates the integral of the model rate function?

Next, we compute the empirical CDF of these rescaled waiting times using the MATLAB function `ecdf`:

```
%Compute empirical CDF from rescaled waiting times.
[eCDF, zvals] = ecdf(Z);
```

The first output, `eCDF`, is the empirical CDF, evaluated over an interval of possible Z values `zvals`. Finally, we compute the model CDF at `zvals`, and construct the KS plot:

```
mCDF = 1-exp(-zvals);                     %Model CDF at z values.
plot(mCDF,eCDF)                           %Create KS plot.
hold on                                   %Freeze graphics window.
plot([0 1], [0 1]+1.36/sqrt(N),'k')       %Upper confidence bound.
plot([0 1], [0 1]-1.36/sqrt(N),'k')       %Lower confidence bound.
hold off                                  %Release graphics window.
xlabel('Model CDF')                       %Label the axes.
ylabel('Empirical CDF')
```

Q: Does the definition of the model CDF (variable `mCDF`) make sense? Remember that by the time-rescaling theorem, we expect that the rescaled variables $(Z_1, Z_2, \cdots Z_n)$ are from the exponential distribution with parameter 1.

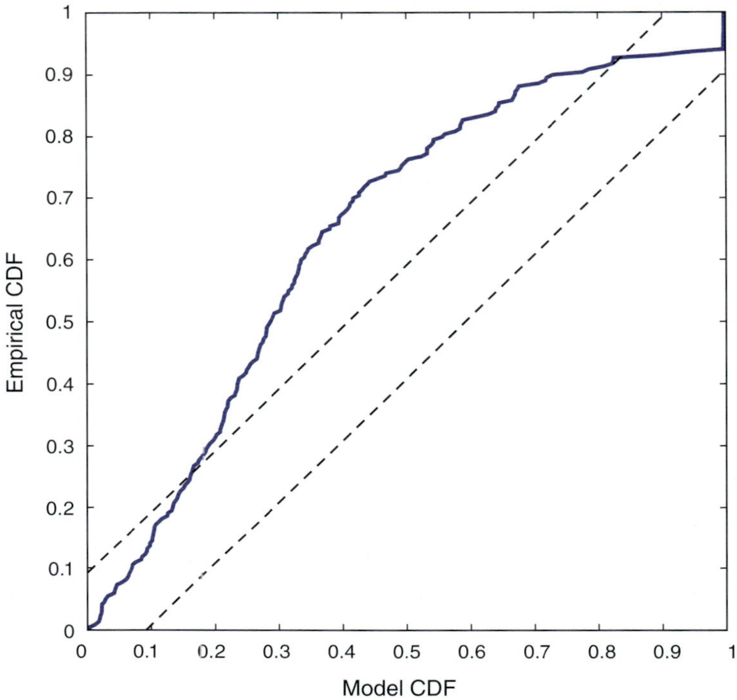

Figure 9.8
KS plot of rescaled data for model 3, with upper and lower confidence bounds (*dotted lines*).

Q: Consider the KS plot in figure 9.8. By the time-rescaling theorem, is model 3 an adequate model of the data?

A: If the spiking data came from a Poisson process with a rate given by model 3, then we would expect the KS plot to remain within the 95% confidence bounds, or any departures from these bounds to be small. We find in figure 9.8 that the KS plot takes on values far outside of the 95% confidence bounds. Therefore, we conclude that our model is not completely capturing the structure of the spiking data.

Method 5: Residual Analysis. Residuals represent the difference between the data and the model prediction at the level of individual data points. While quantities such as the deviance or KS plot are useful for getting an overall picture of how well the model fits the data as a whole, residual analysis is essential for understanding which components of a dataset are well or ill fit by the model. It is therefore one of the best tools for determining what is missing in a model.

There are many types of residuals that can be computed for point process data (including raw residuals, Pearson residuals, and deviance residuals [29]). We do not go into detail about the advantages of each type of residual. Instead, let's focus on one type of residual that is particularly useful for spiking data: the cumulative raw residual process. In continuous time, we would compute this residual process, $R(t)$, as

$$\begin{aligned} R(t) &= \text{total observed no. of spikes at time } t - \text{total expected no. of spikes at time } t \\ &= N(t) - \int_0^t \lambda(u)du\,, \end{aligned}$$

where $N(t)$ is the counting process that gives the total number of spikes fired up to time t, and $\lambda(t)$ is the Poisson rate model at time t. The residual process $R(t)$ compares what's observed (i.e., the spikes recorded in the data) to what the model produces. Since we are working in discrete time, at any time point t_k we compute this as

$$R(t_k) = \sum_{i=1}^{k} \Delta N_i - \lambda(t_i)\Delta t\,, \tag{9.8}$$

where ΔN_i is the number of spikes that are observed in the discrete-time interval t_i.

In MATLAB, we compute $R(t_k)$ by taking the cumulative sum of the raw residuals, which are returned as an output of the `glmfit` function. For model 3, we labeled this output variable `stats3`. Then, to compute the cumulative sum of these residuals, we use the MATLAB function `cumsum` as follows:

```
R = cumsum(stats3.resid);   %Cumulative sum of Model 3 residuals.
```

To inspect how the cumulative residual process evolves in time, let's plot it:

```
plot(t,R)
```

If model 3 were correct, then the residuals should be zero mean and uncorrelated with any covariates. We see in figure 9.9a that the cumulative residual process ends at zero, suggesting that the residuals sum to zero over all time steps. However, we also identify a pattern in the residual process, which suggests that there is still some structure in the data that is not captured by model 3. More specifically, visual inspection suggests a relation between the residuals and a model covariate: the position (plotted in figure 9.9b). We observe that the cumulative residual process seems to increase whenever the rat in moving in the positive direction and to decrease whenever the rat is moving in the negative direction. This analysis suggests exactly what has been missing from our models: direction.

9.2.5 Refining the Model (Continued)

Our goodness-of-fit analysis suggested that model 3 provided a significant improvement over model 2. However, analysis of the cumulative residual process (figure 9.9) revealed that model 3 misses an essential feature of the data: model 3 does not account for the dependence of spiking on direction. So, let's refine the model by including a new covariate that captures the direction of the rat's movement.

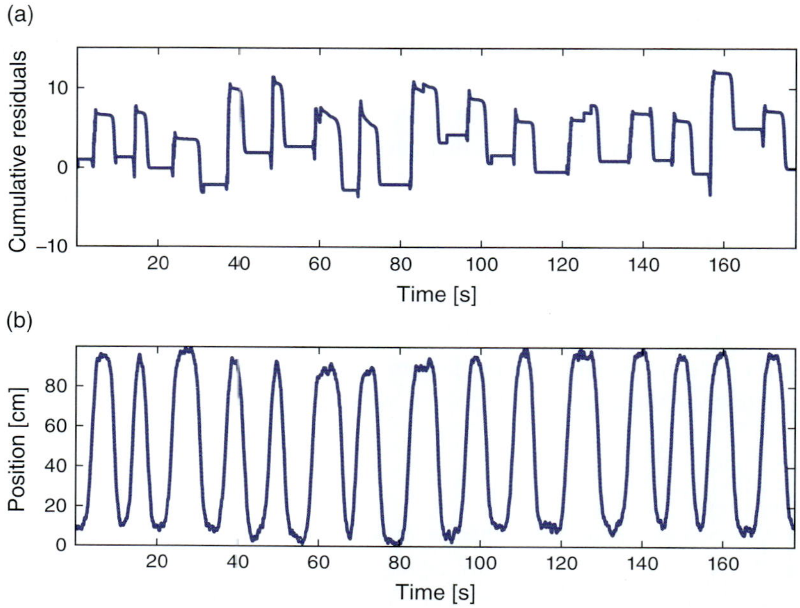

Figure 9.9
(*a*) Cumulative residual process for model 3. (*b*) Rat position versus time.

We must first define a simple indicator function for the direction of movement. If the value of $X(t)$ increased since the last time step, we set this variable to 1; otherwise we set it to 0. For the very first time step, we don't know from which direction the rat came, so we set it to 0 arbitrarily. In MATLAB,

```
dir = [0; diff(X)>0];
```

> **Q:** Does the variable `dir` indicate the direction of movement with a 0 or 1, as we desire? *Hint:* Consider the MATLAB function `diff` and how this function is used to find times when $X(t)$ increases.

With the indicator function for the direction of movement now defined, we must incorporate this new signal into the model. The simplest solution is to add the `dir` variable directly as a new predictor. This would lead to a new model,

$$[\text{Model 4}] \qquad \lambda(t) = e^{\beta_0+\beta_1 x(t)+\beta_2 x(t)^2+\beta_3 \text{dir}}. \tag{9.9}$$

We then fit this model and interpret the parameter estimates. With our previous experience in this chapter, fitting the model in MATLAB is now relatively straightforward:

```
%Fit Model 4, and return estimates and useful statistics.
[b4,dev4,stats4]=glmfit([X X.^2 dir],spiketrain,'poisson');
```

Q: Do you see how, in this code, the three covariates (position, position squared, and direction) are included as inputs to the `glmfit` function?

We are particularly interested in the final parameter estimate. When the direction indicator variable is equal to zero—when the rat is moving in the negative direction or standing still—this component does not affect the rate. When the direction indicator is 1, this component of the model modulates the rate by e^{β_3}. For our estimated parameter we find `b4(4) = 3.2753` and therefore that

```
exp(b4(4)) = 26.4521
```

Q: What is the impact of direction on the firing rate?

A: Under this model, when the rat is moving in the positive direction, the firing rate is more than 26 times higher than the firing rate when the animal is moving in the negative direction. Since we only observe the rat long enough for the place cell to fire a few hundred spikes, we expect the majority of these spikes to occur when the animal is moving in the positive direction.

We could perform the same change of parameters for this model as for the alternative form of model 3. In (9.5) we expressed model 3 using a mean, width, and maximum firing rate for the place field. If we performed this change of parameters for model 4, we would still have just a single mean, width, and maximum rate parameter, and one additional parameter related to the modulation of direction, whose estimated value would not change.

9.2.6 Comparing and Evaluating Models (Continued)

Now that we have a new model that attempts to capture the dependence of spiking on movement direction, let's compare the resulting model fit to the previous model fits and evaluate the overall goodness-of-fit of this new model. First, we compare the AIC values between models 3 and 4:

```
dAIC=(dev3+2*3)-(dev4+2*4);%Difference in AIC between Models 3&4.
```

Q: Why do we use the terms `2*3` for model 3 and `2*4` for model 4?

We find `dAIC = 233.8806`, a difference that is positive and large, suggesting a significant improvement in the fit of model 4 compared to model 3. To evaluate the significance, we can once again perform an MLRT. Model 3 is clearly nested in model 4, since under the null hypothesis that $\beta_3 = 0$, we obtain model 3. Because there is only one parameter to fix, the MLRT statistic would have a chi-square distribution with 1 degree of freedom if the null hypothesis were true. So we compute the p-value for this test as

```
p=1-chi2cdf(dev3-dev4,1); %Compare Models 3 and 4, nested GLMs.
```

and find $p = 0$. Again, we see that the addition of a parameter leads to a highly significant improvement in the fit to the data.

Next, let's compute the confidence interval and significance for the β_3 parameter:

```
%For Model 4, compute 95% CI for last parameter,
CI_beta3 = [b4(4)-2*stats4.se(4) b4(4)+2*stats4.se(4)];
p_beta3 = stats4.p(4);  %... and significance level.
```

We find that the confidence interval (`CI_beta3 = [2.5550, 3.9957]`) does not contain the value zero, and is highly significantly different (`p_beta3 = 9.5421e-20`) from zero. These goodness-of-fit analyses corroborate each other in suggesting that including the direction term provides a significant improvement in capturing the observed spiking structure.

Next, we investigate the overall goodness-of-fit of this model using the time-rescaling theorem and by constructing a KS plot. In MATLAB,

```
lambda4=exp(b4(1)+b4(2)*X+b4(3)*X.^2+b4(4)*dir);%Eval. Model 4.
Z(1)=sum(lambda4(1:spikeindex(1))); %1st rescaled waiting time.
for i=2:N                           %... and the rest.
  Z(i)=sum(lambda4(spikeindex(i-1):spikeindex(i)));
end
[eCDF, zvals] = ecdf(Z);            %Define empirical CDF,
mCDF = 1-exp(-zvals);               %...and model CDF,
plot(mCDF,eCDF)                     %...to create KS plot.
hold on                             %Freeze graphics window.
plot([0 1], [0 1]+1.36/sqrt(N),'k') %Upper confidence bound.
plot([0 1], [0 1]-1.36/sqrt(N),'k') %Lower confidence bound.
hold off                            %Release graphics window.
xlabel('Model CDF')                 %Label the axes.
ylabel('Empirical CDF')
```

The KS plot for model 4 (figure 9.10) looks much better than the one we constructed for model 3 (figure 9.8). We notice that the KS plot for model 4 stays within the 95% bounds

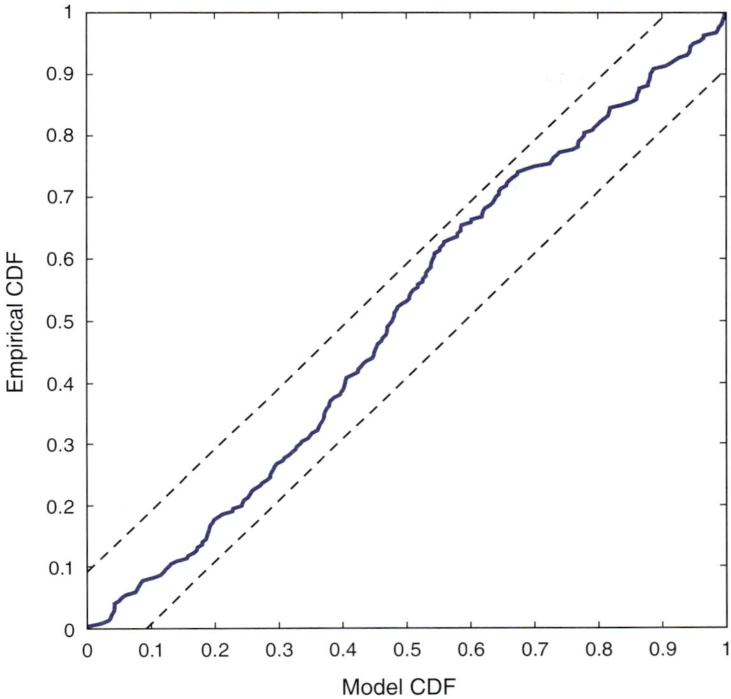

Figure 9.10
KS plot of the rescaled data for model 4, with upper and lower confidence bounds (*dotted lines*).

everywhere, unlike the KS plot for model 3. This does not mean that the model is correct, but there is not significant evidence of lack-of-fit from the KS plot.

Finally let's compute and examine the cumulative residual process:

```
R = cumsum(stats4.resid);  %Cumulative sum of Model 4 residuals.
plot(t,R)                  %Plot it.
```

Figure 9.11 plots the cumulative residual process for model 4 and the movement trajectory. We find that the residual process is now centered at zero. There may still be some structure as a function of time, but it is no longer as closely associated with the movement trajectory as for model 3.

9.2.7 Drawing Conclusions from the Model

It is likely that we could refine this model further, perhaps by adding additional covariates or including different dependence structure on the covariates we are currently using. The

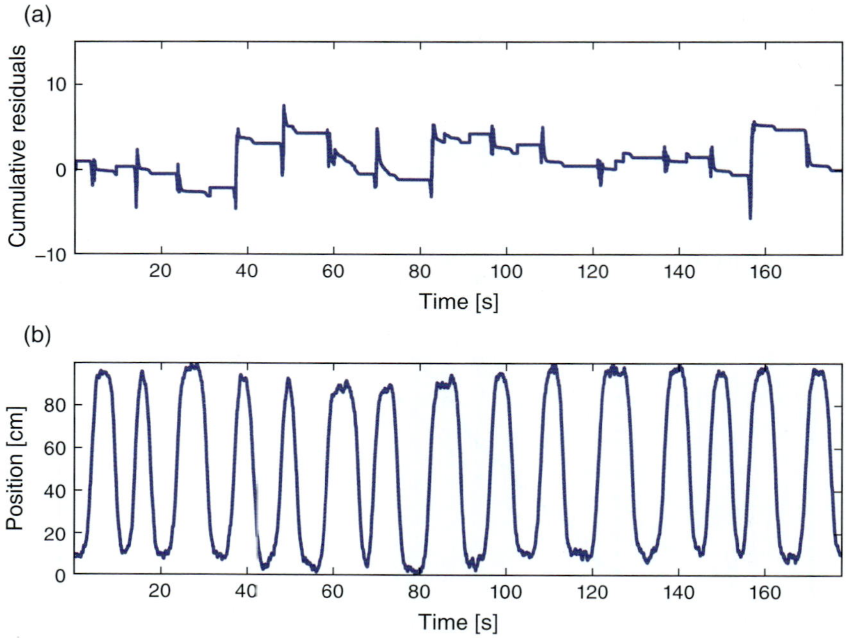

Figure 9.11

(*a*) Cumulative residual process for model 4. (*b*) Rat position versus time.

process of model refinement is not about identifying a single correct model. Instead, it is about building a model that sufficiently captures features of the data in which we are interested. For this analysis, let's decide that based on our multiple goodness-of-fit tools, model 4 is good enough. We now use this model to better understand the structure of the place cell that generated these data.

The process of model refinement and comparison has helped us identify important features of this neuron's receptive field. It was clear from the initial visualizations that this neuron's firing activity is position sensitive; it is more likely to fire when the animal is at certain positions than others. Our modeling analysis further showed that this position dependence could be well described by an exponentiated quadratic function of position, that is, a Gaussian-shaped function.

Let's now estimate parameters for the center, width, and maximum firing rate of this place field. To do so, we make use of the expressions derived for the alternative form of model 3 in (9.5). Computing these values for the estimates of model 4 (variable `b4`),

```
%For Model 4, compute maximum likelihood estimates of
mu=-b4(2)/2/b4(3);                      %...place field center,
sigma=sqrt(-1/(2*b4(3)));               %...place field size,
alpha=exp(b4(1)-b4(2)^2/4/b4(3));       %...max firing rate.
```

Q: Interpret the values you computed. What do you find?

A: The place field is centered approximately 63 cm down the track (`mu = 63.18`), and most of the firing activity occurs in a region between about 63 ± 19 cm (`sigma = 9.58`, so `2 * sigma = 19.16`). At first, the maximum firing rate value of about `alpha = 8.2e-4` spikes/ms, or 0.8 spikes/s seems too small, until we realize that this is the maximum firing rate only when `dir=0`. When `dir=1`, this is multiplied by $e^{\hat{\beta}_3}$ = `exp(b4(4)) = 26.45`. The maximum firing rate when the rat moves in the positive direction is about 22 spikes/s.

Let's also visualize the firing rate as a function of both position and direction. We use the MATLAB function `glmval` to do this. This function uses both the parameter estimates and the `stats` structure that is output by `glmfit`, and computes the estimated Poisson rate function and its confidence intervals for any covariate values. In MATLAB,

```
xs=(0:100)';                       %Define interval of positions,
Ns=size(xs);                       %...and number of positions.
%Evaluate Model 4 in direction 0 (X decreases).
[lambda4_0 up0 low0]=glmval(b4,[xs xs.^2 zeros(Ns)],'log',stats4);
%Evaluate Model 4 in direction 1 (X increases).
[lambda4_1 up1 low1]=glmval(b4,[xs xs.^2 ones(Ns)],'log',stats4);
plot(xs,lambda4_0,'b')             %Plot Model 4, X decreasing,
hold on                            %...freeze graphics,
plot(xs,lambda4_0+up0,'b--')       %...add upper CI,
plot(xs,lambda4_0-low0,'b--')      %...and lower CI.
plot(xs,lambda4_1,'r')             %Plot Model 4, X increasing,
plot(xs,lambda4_1+up1,'r--')       %...add upper CI,
plot(xs,lambda4_1-low1,'r--');     %...add lower CI.
hold off                           %Release graphics.
```

Q: Consider the inputs to the function `glmval` in this code. How are the covariates of position, position squared, and direction defined? What distinguishes model 4 evaluated for increasing position versus decreasing position?

Q: Consider the estimated Poisson rate function versus the covariate of position for model 4 plotted in figure 9.12. How does the firing rate vary with position and direction?

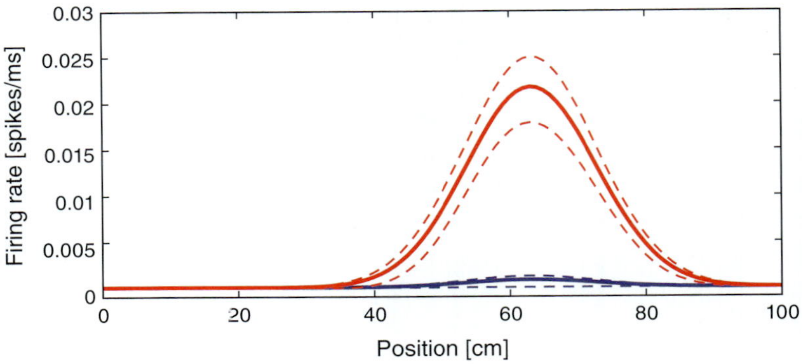

Figure 9.12

Visualization of model 4 firing rate versus position. Firing rate varies with position and direction. Red curve indicates position increasing; blue curve, position decreasing. Dashed curves indicate upper and lower confidence intervals.

> **A:** Inspection of figure 9.12 reveals that the firing rate depends on both position and direction. The rate is largest near position 63 cm and when the position is increasing. When the position is decreasing, the increase in rate near position 63 cm is much smaller. These model 4 results, evaluated using parameters fitted to the observed data (i.e., variables `b4` and `stats4`) are consistent with our initial visualizations of the data (e.g., figure 9.3).

We can also draw some conclusions about the data from elements not in the fitted model. As we discuss in chapter 10, the defining feature of Poisson models is that they have no history-dependent structure; the probability of a spike at any moment can depend on a variety of factors, but it does not depend on past spiking. The fact that we were able to achieve a good fit (based on the KS plot analysis) from a Poisson model suggests that past spiking dependence is not required to capture much of the essential statistical structure in the data. Similarly, other covariates, such as movement speed, were not required to capture the place field structure of this neuron. The neuron may still code for these variables; however we can describe the spiking structure in terms of other variables.

Summary

In this case study, we used an iterative process to identify a statistical model for a hippocampal place cell. We started by visualizing the data, and then proposed point process models, compared and evaluated those models, refined the models, and finally drew inferences from a model that sufficiently captured the structure in the data that we were trying to explain. Each step of this procedure should inform the next. The goal of the visualization analysis is to identify covariates and receptive field structure to include in the first proposed models. The goodness-of-fit analysis should provide insight into features of the statistical model that are lacking and suggest new models. In this case study, we settled on a good model after just two iterations of model refinement. In practice, it may take many more iterations—or even iterations between model identification and new experiments—to identify good models.

It is worth noting that more than half of this case study is devoted to model interpretation and goodness-of-fit methods. This is common for advanced statistical analyses; fitting models to data is often the easy part. The challenge often comes in being able to interpret and evaluate the results of the fitting procedure. Since there is not a single correct way to evaluate a model, we instead use a range of tools to evaluate multiple aspects of model quality. Here we focused on a few that are generally useful for point process data and for Poisson modeling in particular. In the next chapter, we look at additional goodness-of-fit methods, in particular, ones to get at history dependence and rhythmicity in spiking data.

Problems

9.1. Load the file `Ch9-spikes-2.mat` available at

`http://github.com/Mark-Kramer/Case-Studies-Kramer-Eden`

into MATLAB. You will find three variables. The variable `spiketimes` corresponds to the times of recorded spikes (in seconds). The variables `t` and `X` correspond to time (in units of seconds) and position (in units of centimeters), respectively. Use these data to answer the following questions.

a. Visualize the spike train data as a function of position and time. What features do you observe? Do you detect evidence for a place field in your visualizations?
b. Plot the occupancy normalized histogram of the data. Do you observe any evidence for increased spiking at particular positions?
c. Fit the GLM proposed in model 1 (9.1) to these spike train data. What values do you find for the model parameters? How well does the fit perform?
d. Fit the GLM proposed in model 2 (9.2) to these spike train data. What values do you find for the model parameters? Does model 2 fit the data better than model 1?
e. Fit the GLM proposed in model 3 (9.4) to these spike train data. What values do you find for the model parameters? Does model 3 fit the data better than models 1 or 2?

f. Fit the GLM proposed in model 4 (9.9) to these spike train data. What values do you find for the model parameters? Does model 4 fit the data better than models 1, 2, or 3?

g. Describe your visualization and model results, as you would to a colleague or collaborator.

9.2. Using the original (9.4) and alternative (9.5) forms of model 3, show that the expressions for μ, σ^2, and α following (9.5) are correct.

9.3. Compute the differences in the AIC between model 1 and models 2 and 3. Does model 2 fit the data better than model 1 in the AIC sense? Does model 3 fit better than model 1? Interpret these results.

9.4. Let's update model 3 to include a new covariate (y). Define the new covariate in MATLAB as

```
Y = randn(size(X));
```

so that the variable `Y` has the same dimensions as the position variable `X` but is random. Then implement the new model:

$$[\text{Model } Y] \qquad \lambda(t) = e^{\beta_0+\beta_1 x(t)+\beta_2 x(t)^2+\beta_3 y(t)}. \tag{9.10}$$

Conduct an MLRT comparing your new model to model 3. Compute a confidence interval for the parameter corresponding to your new signal. Does this interval contain zero? Compute the p-value for this parameter under a Wald test. Would you be able confidently to exclude the new random signal from your model?

9.5. Expand model 3 to express distinct place field structure for the two directions. Write down a mathematical expression for your model, and work out a MATLAB expression to fit your model.

10 Analysis of Rhythmic Spiking in the Subthalamic Nucleus during a Movement Task

Synopsis

Data Spiking activity from one neuron over 50 trials each of 2 s duration.

Goal Characterize and interpret features of rhythmic spiking activity.

Tools ISI histograms, autocorrelation plots, point process spectral estimators, history-dependent generalized linear models.

10.1 Introduction

10.1.1 Background

The subthalamic nucleus (STN) is a component of the basal ganglia that is thought to play an important role in the regulation of motor function. Neurons in the STN often have rhythmic spiking dynamics, and aberrations in the spike rhythms in STN have been associated with tremor and other motor symptoms in Parkinson's disease. In fact, the STN is one of the primary targets for deep brain stimulation (DBS) to treat Parkinsonian disorders. While the mechanisms by which DBS works are largely unknown, many researchers have hypothesized that altering rhythmic activity in the STN is an important component. Therefore, it is important to be able to characterize spike rhythms in STN neurons and to know with statistical confidence when these rhythms have changed.

In previous chapters, we developed and used methods to uncover rhythmic dynamics in field recordings. Many of the methods used to characterize rhythms in spiking data will be familiar, like autocorrelation functions and spectral estimators, but will require additional care to interpret. In some cases, rhythmic spiking dynamics can be much more subtle than the rhythms we found previously, and statistical modeling methods will be needed to make powerful inferences from spiking data.

10.1.2 Case Study Data

A clinical neurosurgeon contacts you to help her analyze spiking data she collects from patients with Parkinson's disease during surgery to implant DBS electrodes. She is interested in making sure that the electrode is localized properly in the STN, and in studying the role that STN activity plays in movement planning and execution for these patients. She asks each patient to perform a simple hand movement task during the electrode implantation procedure, and records the resulting neural activity.

The data consist of single unit spiking activity from one STN neuron recorded over 50 trials of a hand movement task. In each trial, a patient held a joystick and watched cues appear on a screen. The first cue indicated whether the patient should move the joystick to the left or right. The second cue, the GO cue, indicated the time to start moving the joystick. The dataset contains 2 s of activity for each trial, 1 s before the GO cue and 1 s after. We label the period before the GO cue comes on the *planning period*, and the period after the GO cue the *movement period*.

10.1.3 Goal

It is well known that the spiking activity of some STN neurons in patients with Parkinson's disease shows rhythmic properties in the beta frequency range, 11–30 Hz [30, 31]. The neurosurgeon hypothesizes that information related to the movement task, such as the planning versus the movement period and whether the movement is to the left or right, will influence this rhythmic spiking activity. Our goal is to characterize the spiking properties, identify whether rhythmic activity is present and statistically significant, and how such activity is influenced by the task variables.

10.1.4 Tools

We extend some of the tools already developed for studying rhythms and modeling spike trains. Many of the basic visualization tools, such as raster plots, ISI histograms, and autocorrelation functions are useful without modification. We compute spectral estimators for spiking data using the same methods as before, but we have to think carefully about how to interpret the spectrum of a point process. Finally, we extend the GLM methods we previously developed for spike train data to allow us to model history-dependent, non-Poisson spiking processes.

10.2 Data Analysis

10.2.1 Visual Inspection

To access the data for this chapter, visit

`http://github.com/Mark-Kramer/Case-Studies-Kramer-Eden`

and download the file `Ch10-spikes-1.mat`. Let's start by loading the data into MATLAB:

```
load('Ch10-spikes-1.mat')   %Load the spike data.
```

We find three variables:

- `direction`. A `[50 x 1]` vector of indicators for the movement direction for each of 50 trials. A value of 0 indicates movement to the left, and a value of 1 indicates movement to the right.
- `t`. A `[1 x 2000]` vector of time stamps in milliseconds indicating the time into the trial. Time 0 indicates the GO cue.
- `train`. A `[50 x 2000]` matrix of spike counts in each trial and time bin. The rows of the matrix correspond to trials, while the columns correspond to time.

Q: How many trials involve movment to the left? to the right?

A: To answer this, we must count the number of zeros and ones in `direction`. The `sum` function provides an efficient method to do so:

```
sum(direction)
```

which returns a value of `25`. We conclude that half of the 50 trials involve right movement, and half involve left movement.

Our first goal is to construct a raster plot to visualize the spikes through time for every trial. There are many ways to create this plot in MATLAB. One simple way is to realize that the matrix consists of a collection of zeros and ones as a function of time and trial, corresponding exactly to the information for the raster plot. We can treat this matrix as an image and visualize it using the `imagesc` function in MATLAB:

```
imagesc(t,1:50,train)                  %Construct a raster plot.
colormap(1-gray);                      %...and change the colormap.
```

By default, the resulting image would place a red tick at each time and trial that a spike appears. Here, we have adjusted the MATLAB `colormap` to draw black ticks on a white background for clarity and ease of visual inspection.

Visually, we do not observe much obvious structure in the raster plot (figure 10.1). There does not seem to be substantially more spiking either before or after the GO cue. It is also difficult to see any effect of the movement direction for each trial, since left and right trials are interspersed. We can remedy this by grouping all 25 of the left trials together, and all 25 of the right trials together, and generating a new set of raster plots for each trial type.

```
Ltrials = find(direction==0);          %Find left trials,
Rtrials = find(direction==1);          %...and right trials.
subplot(2,1,1)
imagesc(t,1:25, train(Ltrials,:))      %Image left trials,
subplot(2,1,2)
imagesc(t,1:25, train(Rtrials,:))      %...and right trials.
```

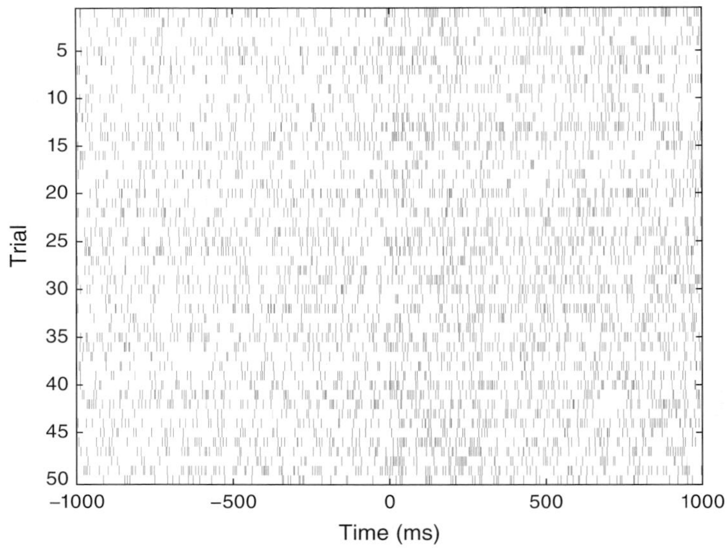

Figure 10.1
Raster plot of spiking as a function of time and trial. Time 0 ms indicates the GO cue.

Q: Consider the raster plots for the left and right trials in figure 10.2. Do you notice any difference in spiking between the two trial types? Is rhythmic spiking apparent?

A: In the separate raster plots we observe slightly more spiking during left trials than during right trials, both during the planning (before time 0 s) and movement (after time 0 s) periods. However, we do not observe in this raster plot clear evidence of rhythmic spiking. There are places where you might convince yourself that rhythmic spiking is present, but it is not clear whether this structure is real or could occur by chance.

Often when people think about rhythms in spiking data, they think about spiking occurring almost perfectly regularly at some fixed frequency, like a metronome. However, it is rare to find neural spiking systems with this kind of regularity. Instead, if we think about spiking activity as arising from a random point process, we can consider rhythmic spiking as the influence of past spiking on the probability of observing a spike at some time in the future. In other words, another way to think about rhythmic spiking is as a history-dependent point process. Later, we build specific probability models to capture this history dependence. But first we visualize the other influences in the data, those related to planning and movement, and to direction.

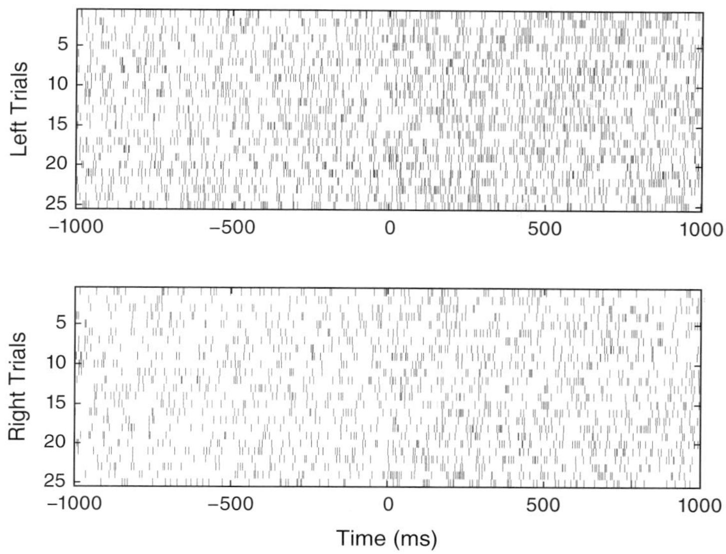

Figure 10.2
Raster plots for left and right trials.

We initially construct a *peristimulus time histogram*, (PSTH). These PSTHs are useful for visualizing the relation between spike rates and the time relative to a specific time point in repeated trial data. In this case, we look at the spiking rate relative to the time of the GO cue. To compute the PSTH, we partition the time interval into bins, add up the number of spikes that occur within each bin over all trials, and then divide by the number of trials and by the length of the bin. Most often, we select a fixed time width for all the bins.

The matrix `train` already contains the binned spiking data, where the bins are each 1 ms in duration. If we want to use 1 ms bins for the PSTH, all that needs to be done is to sum over all the trials and scale by the number of trials and bin length. In MATLAB,

```
PSTH = sum(train)/50/1e-3;                %Compute PSTH, 1 ms bins.
```

Here we divide by `50` because we sum the spiking over 50 trials, and we divide by `1e-3` = 0.001 because the bin width is 1 ms = 0.001 s, and we would like to express the rate in units of spikes/s, or Hz. However, this method will only work for 1 ms bins. A more general approach is to find the times of all the spikes and use the `hist` command to compute the PSTH:

```
[spiketimes spiketrials]=find(train');   %Find when spikes occur,
PSTH=hist(spiketimes, 1:2000)/50/1e-3;   %... compute histogram.
```

Q: Why must we transpose the matrix `train` before applying the find function?

Q: Consider the inputs to the `hist` function. Why does the second input (`1:2000`) correspond to 1 ms bins?

Either method produces the same PSTH result. Now let's use the `bar` function to produce a bar graph of the PSTH:

```
bar(t,PSTH);                                    %Display the PSTH.
```

Although it was not immediately clear from the raster plot, it is evident in the PSTH (figure 10.3) that there is a difference in the firing rate between planning and movement periods. The firing rate during planning (time < 0 ms) appears to average around 60 spikes/s, and the firing rate during movement (time > 0 ms) appears to be slightly elevated, averaging around 80 spikes/s. However, this PSTH may be a bit misleading because of the very small bin widths used. In each bin, the PSTH only appears to take a few discrete values (e.g., 40, 60, 80 spikes/s) with nothing in between. Also, there are many bins where the PSTH value is equal to zero; these bins are not obvious in figure 10.3 because they are

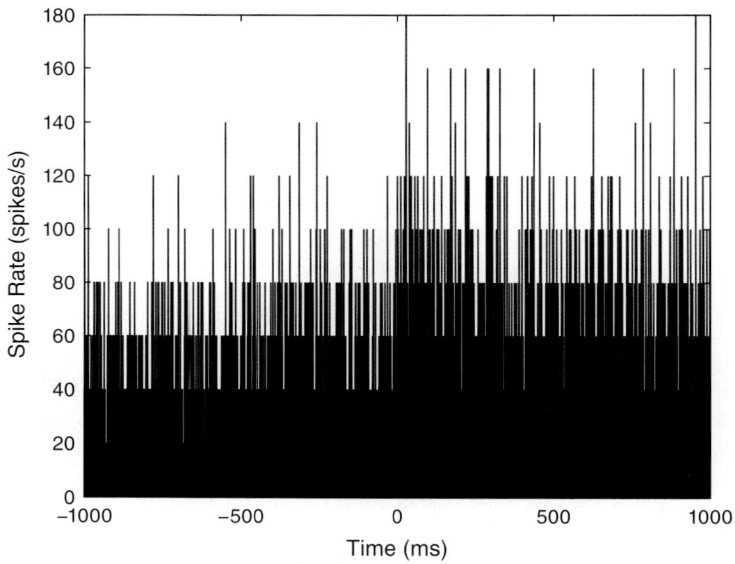

Figure 10.3
PSTH of spiking activity relative to GO cue, using 1 ms bins and all trials.

masked by nearby bins. The initial PSTH therefore gives a false sense of the average spike rate. To investigate further, let's increase the PSTH bin width:

```
PSTH10 = hist(spiketimes, 1:10:2000);      %Compute histogram.
bar(t(1:10:2000),PSTH10/50/10*1000);       %...and display PSTH.
```

Q: How does the second input to the `hist` function specify 10 ms bins?

With 10 ms bin widths, the difference in firing rate between planning and movement periods is still evident (figure 10.4), but we see that we had overestimated the firing rate because bins with small PSTH values were masked by neighboring bins with higher values. Now we find that the average firing rate during planning is closer to 35 spikes/s, while the rate during movement is closer to 55 spikes/s.

Q: Construct a PSTH for these data with 100 ms bins. How does the result compare to the 10 ms bin PSTH in figure 10.4? What might be some issues that impact PSTHs with large bin widths? What do you think might be a good way to select bin widths for a PSTH?

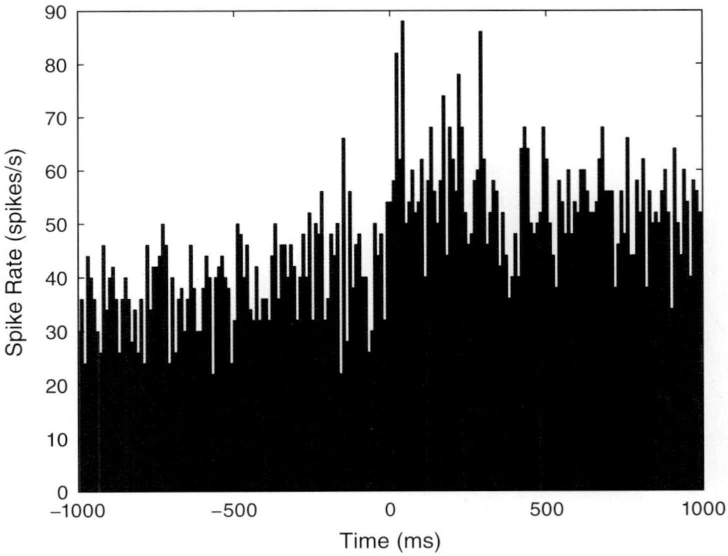

Figure 10.4
PSTH of spiking activity relative to GO cue, using 10 ms bins and all trials.

We can also compute the average spike rate in the movement and planning periods directly. To do so, we need to average the spiking data over both trials and time in each period. In MATLAB,

```
i_plan = find(t < 0);                      %Indices for planning.
i_move = find(t >= 0);                     %Indices for movement.
%Compute the average spike rate,
PlanRate=mean(mean(train(:,i_plan)))/1e-3;%...during planning,
MoveRate=mean(mean(train(:,i_move)))/1e-3;%...during movement.
```

Q: You might notice that we call the function `mean` twice in this code. What is the effect of these two (nested) function calls?

A: The variable `train` is a matrix. The first application of `mean` computes the mean of each column of `train`, corresponding to the mean over trials. The result is a (row) vector. The second application of `mean` operates on this row vector and computes the mean over time to produce a scalar.

Executing these commands, we find `PlanRate` = `38.9600`, and `MoveRate` = `54.9600`. These results are consistent with our estimation of the average firing rates through visual inspection of figure 10.4.

In addition to the planning and movement periods, the task can be broken down based on the direction of movement that is cued. The variable `direction` contains an indicator variable for each trial, which is 0 for left trials and 1 for right trials. Since the left and right trials are interspersed, we need to first find which trials correspond to each direction, and then compute the firing rates for each type of trial. In MATLAB,

```
Ltrials = find(direction==0);              %Find left trials,
Rtrials = find(direction==1);              %... and right trials,
LRate=mean(mean(train(Ltrials,:)))/1e-3;%... and compute rates.
RRate=mean(mean(train(Rtrials,:)))/1e-3;
```

Q: What do you conclude about the firing rates for the two types of trials?

A: Upon executing these commands, we find `LRate` = `58.6600` and `RRate` = `35.2600`. These results suggest that the spiking rates for planning and movement in left trials are higher than the rates for right trials.

We have now shown that the average firing rate changes between planning and movement periods, and between left and right trials. One question that arises is how the changes pertaining to trial period and to trial direction are related. We have four categories to explore: planning for left trials, planning for right trials, movement on left trials, and movement on right trials. We could compute the mean firing rates in each of these categories. Instead, let's examine the trial-to-trial distribution of firing rates in each category. Let's sum the spiking over each 1 s planning or movement period for each trial type, and generate a box plot that visualizes these distributions:

```
PlanL = sum(train(Ltrials,i_plan),2);  %Firing rate L, planning.
PlanR = sum(train(Rtrials,i_plan),2);  %Firing rate R, planning.
MoveL = sum(train(Ltrials,i_move),2);  %Firing rate L, movement.
MoveR = sum(train(Rtrials,i_move),2);  %Firing rate R, movement.
boxplot([PlanL PlanR MoveL MoveR], ... %Display results.
  'labels',{'Plan Left','Plan Right','Move Left','Move Right'});
```

Q: Have you used the function `boxplot` before? If not, look it up in MATLAB Help.

Q: How do the variables in this code correspond to firing rates?

A: In this case, the duration of the interval is 1 s for both planning and movement periods. Therefore, we may think of dividing each sum by 1 s, which does not affect the numerical values in the variables.

The box plot in figure 10.5 corroborates our previous findings that the firing rate is increased in left trials and in the movement period of each trial. Further, it suggests that these effects may occur independently and additively; the difference between left and right trials seems approximately the same for both planning and movement periods. At this stage, it would be natural to ask whether these differences between mean firing rates are statistically significant. We explore issues of significance later, and discuss why some traditional approaches for comparing means across groups (such as *t*-tests between pairs or analysis of variance between all groups) might not capture the structure in the data very well.

While these visualizations have helped elucidate the long-term spiking structure over periods of about 1 s, they have not provided much information about short-term rhythmic structure. In chapter 8, we used interspike interval (ISI) histograms to help visualize fine time scale history dependence in spiking activity. Let's do so again here. To compute the ISI histogram, we first need to compute the ISIs. One approach is to write a for-loop that iterates through each trial in the `train` matrix, finds the spike times, and computes the ISIs. However, in constructing the PSTH, we computed the variable `spiketimes`, which

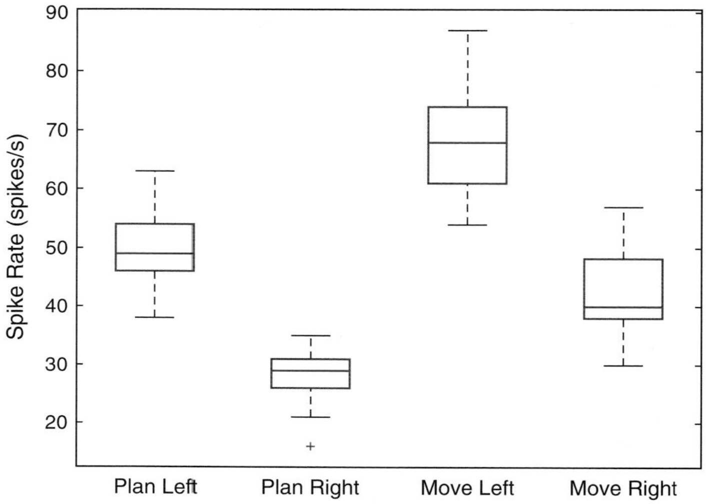

Figure 10.5
Trial-to trail variability of spike rates for planning and movement periods for left and right trials. Median firing rates (*red lines*), 25th and 75th percentiles of firing rates across trials (*blue boxes*), minimum and maximum firing rates (*dotted lines*) excluding possible outliers (+).

already contains the spike times in trial 1, followed by the spike times in trial 2, and so on, all in one array. If we compute the differences in these spike times, we will have the ISIs for each trial mixed with an occasional negative value when an element of `spiketimes` is the first spike time for any trial. Let's eliminate these spurious values and plot a histogram of the ISIs:

```
ISIs=diff(spiketimes);  %Determine ISIs for all time & trials.
ISIs=ISIs(find(ISIs>0));%Remove spurious values between trials,
hist(ISIs,0:250)        %...and display results.
```

Q: Can you think of a way this code might lead to incorrect computation of the ISIs? Did this happen for this dataset?

A: If the last spike in any given trial occurred at an earlier time than the first spike in the subsequent trial, then the ISI computed between trials would not be negative and would not be removed. This never happens in these data, but it suggests that we should be careful when we code various procedures to think about the possible outcomes.

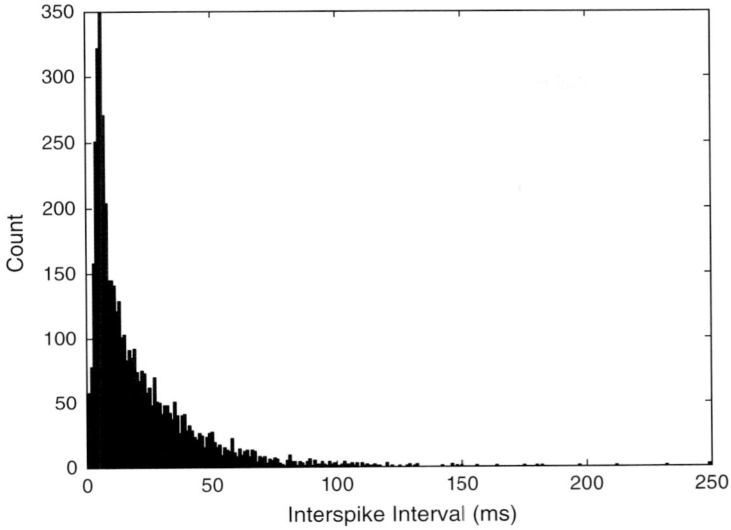

Figure 10.6
ISI histogram for full dataset, including planning and movement periods, and left and right trials.

Q: Find another way to compute the set of ISIs for these data. Confirm that your method produces the same result as the preceding code does.

The ISI histogram in figure 10.6 reveals a few interesting features, including a short relative refractory period (fewer ISIs near 0 ms) and a large number of ISIs near about 6 ms, suggesting bursting activity (i.e., times of rapid spiking with small intervals between spikes). However, it is difficult to see any structure that looks like clear rhythmic spiking, such as a large second peak at longer ISIs. Let's see if there is a difference in the ISI distributions for planning and movement periods:

```
%In the planning period,                         %...find spikes,
[spiketimesPlan spiketrials]=find(train(:,i_plan)');
PlanISIs = diff(spiketimesPlan);                 %...compute ISIs,
PlanISIs = PlanISIs(find(PlanISIs>0));           %...drop spurious ones,
subplot(211); hist(PlanISIs,0:250)               %...and plot it,
ylabel('Count')                                  %...with axes labeled.
title('Planning')
```

```
%In the movement period,                     %...find spikes,
[spiketimesMove spiketrials] = find(train(:,i_move)');
MoveISIs = diff(spiketimesMove);             %...compute ISIs,
MoveISIs = MoveISIs(find(MoveISIs>0));       %...drop spurious ones,
subplot(212); hist(MoveISIs,0:250)           %...and plot it,
xlabel('Interspike Interval [ms]')           %...with axes labeled.
ylabel('Count')
title('Movement')
```

Q: What features differ between the ISI histograms for the planning and movement periods, shown in figure 10.7?

A: There are a few notable differences between the ISI histograms in the two periods. The total number of ISIs is larger during the movement period, corroborating our earlier finding that the firing rate is higher during this period. The peak in the ISI distribution around 6 ms seems more pronounced during movement, suggesting a

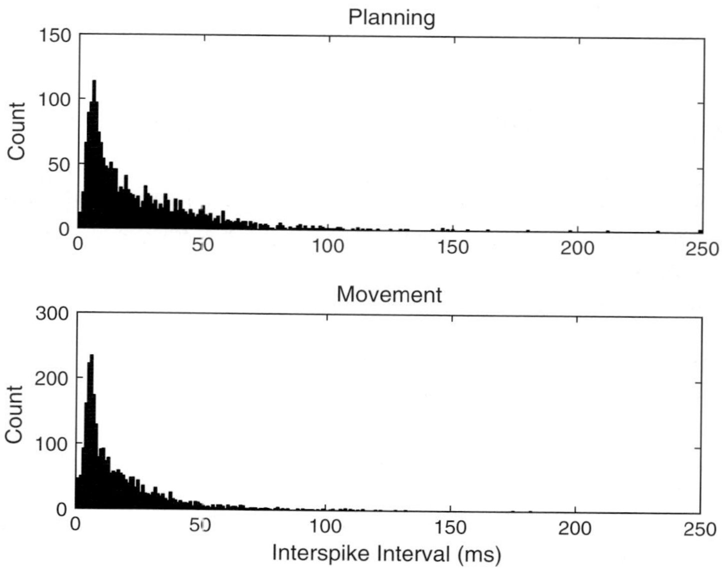

Figure 10.7
ISI histograms for planning (*top*) and movement (*bottom*) periods, including left and right trials.

higher tendency to fire in bursts. There may be some differences in the tail structure of the ISI distribution at ISI values above 20 ms, but it is difficult to relate this to differences in rhythmic spiking structure.

In chapter 8, we also used the sample autocorrelation function (ACF) to visualize history-dependent spiking properties. Recall that the ACF shows the correlation between a signal at two points in time separated by a fixed lag, evaluated over different values of the lag. For these spike train data, let's use the increments (i.e., the number of spikes in each time bin) as the signal, compute the ACF for each trial, and average the results over all the trials. In MATLAB,

```
for k=1:50                                        %For each trial,
 plan = train(k,i_plan);                          %...get planning data,
 move = train(k,i_move);                          %...get movement data,
 acf1(k,:)=xcorr(plan-mean(plan),'coeff');%...compute ACFs,
 acf2(k,:)=xcorr(move-mean(move),'coeff');
end                                               %...and plot results,
subplot(211); stem(1:100,mean(acf1(:,1001:1100)));
ylabel('Autocorrelation')                         %... with axes labeled.
subplot(212); stem(1:100,mean(acf2(:,1001:1100)));
ylabel('Autocorrelation');  xlabel('Lag [ms]')
```

Q: Why do we examine the variables `acf1` and `acf2` at columns `1001:1100`?

A: The columns of both matrices specify the lags, which extend from -1 s to $+1$ s. We choose indices `1001:1100` to investigate positive lags beginning at 1 ms and ending at 100 ms. Notice that we compute the mean of both matrices across trials (rows) before displaying the results.

The autocorrelation plots for both the planning and movement periods show very clear structure (figure 10.8). In both periods we see negative correlation for lags of 1–3 ms, followed by positive correlation at lags of 1–8 ms, with a peak at a lag of 6 ms. During the planning period, we see clear structure continuing at higher lags, with an extended period of negative correlations at lags of about 15–35 ms and positive correlations at lags of about 50–70 ms. For the movement period, this structure at lags longer than about 10 ms seems reduced or even absent. These results suggest a change in the rhythmic firing properties between the planning and movement periods. The second peak in the autocorrelation plot for the planning period around 60 ms suggests a $1/60$ ms $\approx$ 17 Hz rhythm that disappears during movement.

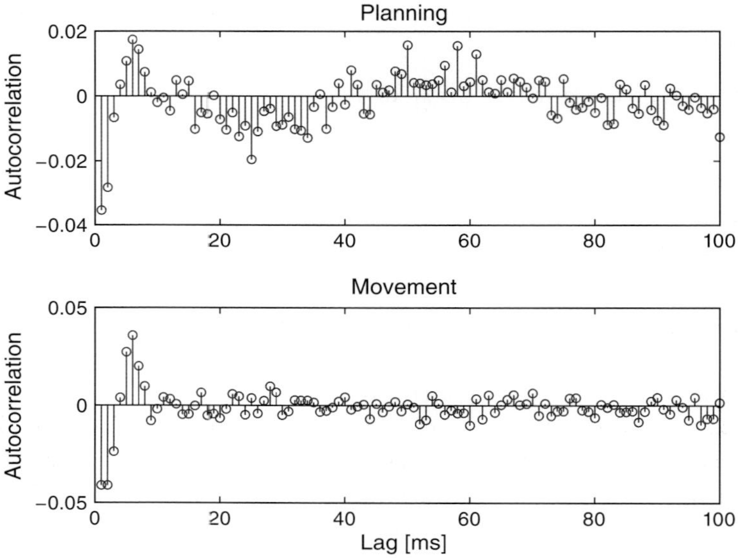

Figure 10.8
Autocorrelation of spike increments for lags up to 100 ms between planning (*top*) and movement (*bottom*) periods.

10.2.2 Spectral Analysis of Spiking Data

So far, we have focused on statistical characterizations of spike trains in the time domain. Another approach to visualizing and describing structure in point processes focuses on estimating rhythmic (or harmonic) features of the data in the frequency domain. In chapters 3 and 4, we discussed methods to decompose field data (or continuous valued functions and random processes) into harmonic components. The same type of approach can be used to decompose point processes and to determine the relative contributions to their variability attributable to each frequency. Point process spectral estimation provides a useful tool for spike train analysis but should be performed with care. On the one hand, computing spectral estimates for discrete spike train data is similar to the approaches used for continuous valued signals. On the other hand, the interpretation of point process spectra can differ from continuous process spectra, and blindly applying intuition from continuous spectral estimation to point process data can lead to incorrect interpretations.

Point Process Spectral Theory. In this section, we develop some of the theory needed to understand and interpret spectral estimators for spiking data. As with the spectrum for continuous valued processes (chapter 3), the spectrum for a point process can be expressed as the Fourier transform of its autocovariance function. In discrete time, assume we have binned the spiking activity into intervals of length Δt, and that ΔN_i represents the number of spikes fired in the ith interval, which is called the ith increment. The autocovariance sequence is then

$$\begin{aligned}\gamma(h) &= E\Big[(\Delta N_i - \lambda_0 \Delta t)(\Delta N_{i+h} - \lambda_0 \Delta t)\Big] \\ &= E\Big[\Delta N_i \, \Delta N_{i+h}\Big] - \big(\lambda_0 \Delta t\big)^2 ,\end{aligned}$$

where $\lambda_0 \Delta t$ is the expected number of spikes in a single time bin. The discrete spectrum is the discrete Fourier transform of this sequence. Using this definition, we can compute the theoretical autocovariance and spectrum for any stationary point process model.

Example 1: Spectral density of a homogeneous Poisson process. For a homogeneous Poisson process, the increments ΔN_i are independent, and each has a Poisson distribution with the same mean. Assume the mean number of spikes in any bin is $\lambda_0 \Delta t$. Since the mean and variance of a Poisson random variable are the same (see chapter 8 appendix), this is also the variance of any increment. The autocovariance at a lag $h = 0$ is just the variance of ΔN_i, which is $\lambda_0 \Delta t$. At any other lag, the increments are independent and therefore the autocovariance is zero. The autocovariance function is given by

$$\gamma(h) = \begin{cases} \lambda_0 \Delta t & \text{if } h = 0 , \\ 0 & \text{otherwise} . \end{cases} \tag{10.1}$$

The spectral density of this homogeneous Poisson process is the Fourier transform of this autocovariance sequence in the limit as $\Delta t \to 0$, which is

$$F(\omega) = \lambda_0 . \tag{10.2}$$

The Fourier transform $F(\omega)$ does not vary as a function of ω, that is, the spectral density is flat. This suggests that constant power is required at all frequencies to describe a homogeneous Poisson process. This is the point process analogue of white noise.

Example 2: Spectral density of a refractory point process. Assume now that we modify the spiking process to include a relative refractory period. We do this by adding a negative autocorrelation at small lags that diminishes in magnitude as the lag increases. Thus, a spike 1 ms in the past would substantially decrease the probability of observing another spike in the current time interval, but a spike many milliseconds in the past would have little to no effect on the current spike probability. Mathematically, we assume that the autocovariance function has the form

$$\gamma(h) = \begin{cases} \lambda_0 \Delta t - \alpha f_\sigma(h) & \text{if } h = 0 , \\ -\alpha f_\sigma(h) & \text{otherwise} , \end{cases}$$

where α is a constant representing the degree of refractoriness, and $f_\sigma(h)$ is the Gaussian function (or equivalently, the normal probability density function) with a mean of zero, and

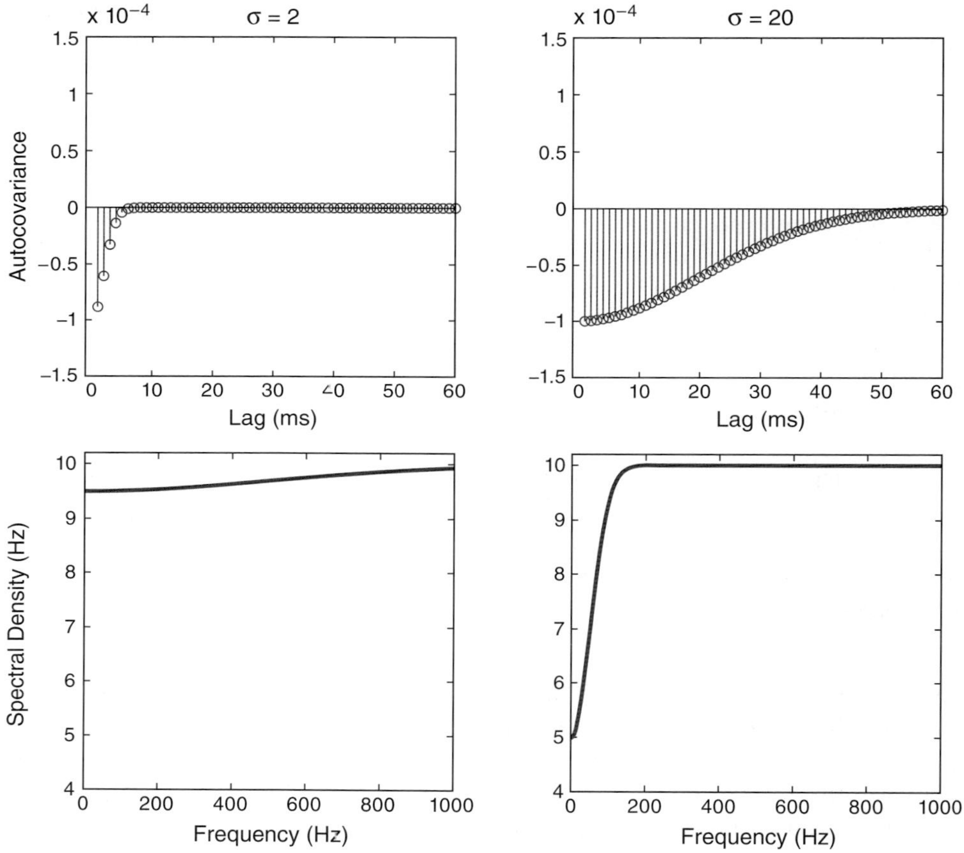

Figure 10.9

Autocovariance functions (*top*) and spectral density functions (*bottom*) for refractory spiking processes with Gaussian-shaped refractory structure with $\sigma = 2$ (*left*) or $\sigma = 20$ (*right*).

a standard deviation of σ. Examples of $\gamma(h)$ are shown in the top panels of figure 10.9 for the values $\sigma = 2$ ms and $\sigma = 20$ ms.

The reason we chose the Gaussian function is that it captures a smooth decay of the refractory effect and its Fourier transform is easy to compute. The spectral density $F(\omega)$ of this refractory point process is

$$F(\omega) = \lambda_0 - \alpha\sqrt{2\pi}\sigma \cdot f_{1/\sigma}(\omega)\,.$$

This spectral density does depend on ω; as ω becomes large, $f_{1/\sigma}(\omega)$ goes to zero and the spectrum approaches the constant level of a Poisson process; however, as ω becomes small, the spectrum decreases below this level. Therefore, the effect of the refractory period on the spectrum is a dip at low frequencies. The bottom panels of figure 10.9 show the spectral density for the refractory point processes with the corresponding autocovariance functions in the top panels. We observe that the extent of this decrease in power in the frequency domain is inversely proportional to the extent of the refractory period in the time domain. When $\sigma = 2$ ms, the dip in the spectrum is small but extends all the way out beyond 500 Hz. When $\sigma = 20$ ms, the dip in the spectrum is larger but only extends out to about 100 Hz.

It may seem strange that considerable power remains at high frequencies even though the refractory period decreases the probability that the spike train fires at frequencies higher than $1/\sigma$. The explanation for this apparent incongruity is that the high-frequency power is not only related to the probability of firing at high frequencies but is also necessary to generate the impulse profile of each spike event. As frequency becomes large, the power spectral density of any point process will asymptote to λ_0, the expected spike rate, not to zero.

This example illustrates a counterintuitive result. To understand the mean firing rate of a spiking process, we look to the high end of the spectrum, and to understand whether refractoriness prevents high-frequency spiking, we look to the low end of the spectrum. This is one reason it is important to be careful when interpreting spectral estimators for point processes.

Spectral Estimates for STN Data. With a new appreciation for interpreting point process spectra, let's compute spectral estimators for the STN data for the planning and movement periods. As in chapters 4 and 5, we use multitaper spectral estimators computed in the Chronux software package [2] (see section 1.24 of chapter 1). Since the data follow a trial structure, we compute estimators for each trial and average the results across trials. The planning and movement periods for each trial are 1 s in duration, so we can achieve a 4 Hz frequency resolution by setting the time-bandwidth product to 4. This allows us to use seven tapers, each providing independent and informative estimates. In MATLAB,

```
%Set the parameters of the MTM.
TW = 4;                        %Choose time-bandwidth product of 4,
ntapers=2*TW-1;                %...which sets no. of tapers.
params.Fs=1000;                %Define sampling frequency,
params.tapers=[TW,ntapers];    %...time-band product, no. of tapers.
params.fpass=[0 500];          %Define frequency range to examine.
params.trialave=1;             %Perform trial averaging.

%Compute the coherence during planning & movement.
[SPlan,f]=mtspectrumpb(train(:,i_plan)',params);
[SMove,f]=mtspectrumpb(train(:,i_move)',params);
```

```
%Plot the spectra ... with axes labeled.
subplot(211); plot(f,SPlan); ylabel('Power [Hz]')
subplot(212); plot(f,SMove); ylabel('Power [Hz]')
xlabel('Frequency [Hz]')
```

We set the parameters used by the multitaper method with the variable `params`. This variable is a *structure array*, with data containers called fields; see MATLAB Help for more details about structure arrays. In this case, we specify in `params` the time-bandwidth product (`TW`) and number of tapers (`ntapers`), the sampling frequency (`1000` Hz), the interval of frequencies to return (`0` to `500` Hz), and to average the spectra across trials (`params.trialave=1`). We use the Chronux function `mtspectrumpb` to compute the power spectral densities of the spike trains using the multitaper method. This function returns outputs corresponding to the spectra (`SPlan` and `SMove`) and the frequency axis (`f`).

A number of distinct features are evident in these spectral estimates for the planning and movement periods (figure 10.10). At higher frequencies, the spectral density for the planning period asymptotes to about 35 Hz, while for the movement period it asymptotes closer to 55 Hz. This corroborates our previous results based on the observed mean firing rates. Both spectral estimates show a decrease at frequencies below 200 Hz, suggesting

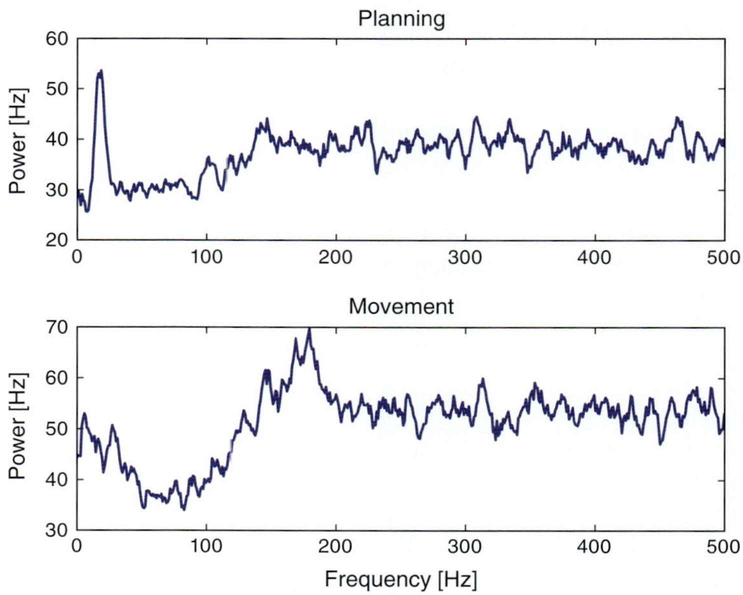

Figure 10.10

Trial-averaged spectra of spiking data in planning (*top*) and movement (*bottom*) periods.

an approximately 5 ms refractory period. However, in the planning period, we see a large peak at about 18 Hz that does not appear during the movement period. This suggests that the STN neuron spikes rhythmically during the planning period and that this rhythm is attenuated during movement.

Here we have separated each trial into planning and movement periods and computed separate spectral estimates for each period. This assumes that the data are stationary—that the mean and autocovariance structure do not change in time—during the planning period, that there is a sudden change between the planning and movement period, and that the data are stationary again during the movement period. Is this a reasonable assumption? One way to explore this is to compute a spectrogram, as we did in chapter 3. Let's use a moving window of 500 ms duration and a step size of 50 ms.

Q: Since the duration of the moving window is 500 ms, if we want to maintain a frequency resolution of 4 Hz, what is the time-bandwidth product and number of tapers?

A: The time-bandwidth product is 0.5 s × 4 Hz = 2. In this case, we therefore have three tapers that provide informative, independent estimates of the spectrum.

We also focus on the frequency range from 0 to 50 Hz, which includes the large peak near 18 Hz that we previously observed during the planning period (figure 10.10). Then, to compute the spectrogram in MATLAB,

```
%Set the parameters of the MTM.
movingwin = [.5 .05];          %Define window duration & step size,
params.fpass = [0 50];         %...frequency range to examine,
params.tapers = [2 3];         %...time-band product, no. of tapers.
[S,T,F]=mtspecgrampb(train',movingwin,params); %Get spectrogram.
T = T+t(1)/1000;               %Set time axis,
imagesc(T,F,S')                %...and display it,
xlabel('Time [s]')             %...with axes labeled.
ylabel('Frequency [Hz]')
```

We compute the spectrogram using the Chronux function `mtspecgrampb`. As inputs to this function, we specify the duration and step size of the moving window (`movingwin=[.5 .05]`) as well as the frequency band to examine (`params.fpass`) and the time-bandwidth product and number of tapers (`params.tapers`). The function returns the spectrogram (`S`), a time axis (`T`), and a frequency axis (`F`). The time axis indicates the center time of each window, and Chronux begins counting time at 0 s. To scale the time axis appropriately for our measurements, we add to `T` the starting time for the data (−1 s).

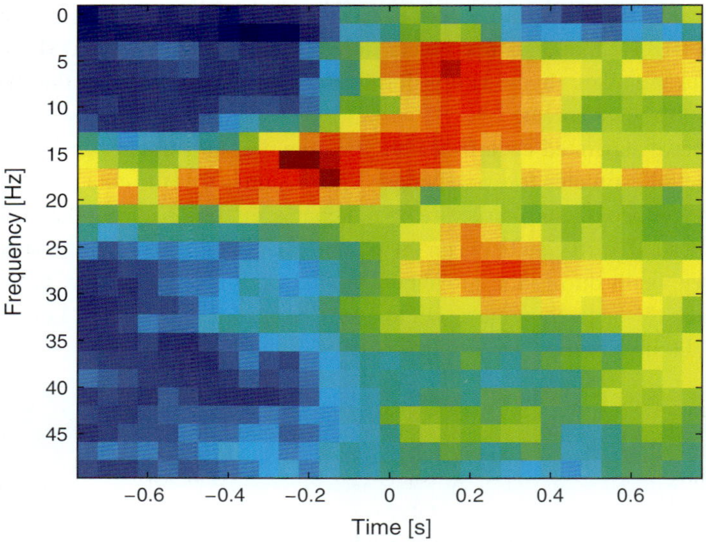

Figure 10.11
Trial-averaged spectrogram of spiking data across planning and movement periods. Warm (cool) colors indicate high (low) power.

Q: What features do you observe in the spectrogram of the spike train data (figure 10.11): How do these features change between the planning and movement periods?

A: The period before time 0 s, corresponding to the planning period, shows a clear peak at 15–20 Hz. As we approach time 0 s, there is a rapid change in the spectral structure. The peak starts to vanish, not only because the spectral density at 15–20 Hz decreases but also because the spectral density across all frequencies increases. This reflects a combination of a change in rhythmic spiking and a change in the overall firing rate. The fact that the spectral content seems to undergo a sudden shift from one regime (planning) to another regime (movement) suggests that it is reasonable to construct models with different spike dynamics in these two periods.

10.2.3 Point Process Models

So far we have focused on descriptive statistics and visualization methods to characterize the properties of the spike train data. In chapter 9, we developed point process models to understand the factors influencing a spiking process. We can use the same types of models to describe how past spiking activity influences future activity to produce rhythmic spiking and other forms of history dependence. In doing so, we utilize all the tools that we previously developed for point process modeling: tools to fit parameters and construct confidence bounds, tools to measure the goodness-of-fit between the model and the data, and tools for testing whether individual model components or combinations of components provide significant improvements in fitting the data.

In this case study, we continue to use a specific class of models, GLMs for point processes. As discussed in chapter 9, GLMs have a number of attractive properties, including the fact that we can guarantee that the likelihood of the data as a function of the model parameters is convex and therefore has one peak that is easy to find. In MATLAB we use the `glmfit` routine to find the maximum likelihood estimators for each of the models.

Let's revisit how to use the `glmfit` function. The first input to this function is the design matrix or predictors. This is an $n \times p$ matrix, where n is the total number of time steps in the dataset, and p is the number of predictors in the model. For these data, we have 50 trials, each comprising 2,000 time steps, so $n = 50 \times 2000 = 100,000$ data points. The second input to `glmfit` is the response vector—the spike train—which is an $n \times 1$ vector. We have to reshape the spike train to be one long vector containing all the trials sequentially, and also construct the design matrix accordingly. The third input is the GLM distribution to use. Here we assume that the spike count in each bin is approximately Poisson, as we did in chapter 9. Recall that by selecting the Poisson GLM, we implicitly use a log link function. In other words, we set the log of the firing rate to be a linear function of the predictors and model parameters. Equivalently, we construct a model of the form $\lambda(t) = e^{\beta X(t)}$, where $\lambda(t)$ is the spiking intensity at time t, $X(t)$ is a $p \times 1$ column vector of the predictors at time t, and β is a $1 \times p$ column vector of the model parameters. Since each bin contains either 0 or 1 spike, we could also use the binomial distribution with a maximum of one count in each bin. However, choosing the Poisson count distribution will make it easier to interpret the resulting models.

One important note: Although we are approximating the number of spikes in each bin as Poisson, this does not mean we are modeling the data as a Poisson process. A Poisson process is one that has Poisson spike counts in each interval but also requires that the counts in two separate intervals be independent. Here, we explicitly model the spike count in one bin as a function of the number of spikes fired in previous bins. Therefore, the spike counts will not be independent between bins.

Let's build a model to compare the firing rate between the planning and movement periods; subsequent models will also include the effects of history dependence. In this model, the design matrix consists of one column that is zero whenever $t < 0$, corresponding to the planning period, and that is 1 whenever $t \geq 0$, corresponding to the movement period. The resulting model has the form

$$\text{[Model 1]} \qquad \lambda(t) = e^{\beta_0 + \beta_1 I_{\text{move}}(t)}, \tag{10.3}$$

where $I_{\text{move}}(t)$ is the movement period indicator function. We must also remember to reshape the spike train matrix into a single response vector `y`. In MATLAB,

```
K = 50;                                       % no. of trials.
T0 = length(t);                               % no. of time points.
Imove=ones(K,1)*[zeros(1,T0/2) ones(1,T0/2)]; %Move indicator.
Imove=reshape((Imove)',K*T0,1);               %Reshape indicator.
y = reshape(train',K*T0,1);                   %Reshape spikes.
[b1 dev1 stats1] = glmfit(Imove,y,'poisson');%Fit Model 1.
```

Q: Consider the variable `Imove`, initially defined in the third line of this code. How does this variable indicate the movement period?

A: The time variable `t` is negative for the first 1,000 indices (i.e., first half of `t`) and non-negative for the next 1,000 indices (i.e., the second half of `t`). The term `[zeros(1,T0/2) ones(1,T0/2)]` indicates these two time intervals for each trial. We then perform matrix multiplication (by the term `ones(K,1)`) to repeat the interval indicator for each of the 50 trials. Note that in the fourth line of this code, we reshape the indicator matrix to become an indicator vector.

The `glmfit` function returns three outputs: `b1` is the fitted parameter vector, `dev1` is the model deviance, and `stats1` contains a set of statistical results related to the model fit (see chapter 9). Let's examine the model parameters, remembering to exponentiate them to improve interpretability.

The first parameter estimate indicates that the firing rate during the planning period is `exp(b1(1))=0.0390` spikes/ms or 39 spikes/s. The second parameter estimate indicates that during the movement period the firing rate is modulated by `exp(b1(2))=1.4107`; it increases by 41%.

Q: Is this 41% increase significant, or is it likely to occur by chance under this model?

A: To answer this, let's use the output `stats1` to construct a 95% confidence interval for the second parameter.

```
%Compute CI for firing rate modulation during movement period,
CI_lower = exp(b1(2)-2*stats1.se(2));    %...lower CI,
CI_upper = exp(b1(2)+2*stats1.se(2));    %...upper CI.
```

We find the 95% confidence interval `[1.3295 1.4968]`, which suggests that the firing rate during the movement period is between 33% and 50% higher than the planning period. The fact that this interval does not contain 1 suggests that this effect is significant at the 0.05 level. More directly, we can examine the *p*-value for this parameter (based on a Wald test) having a nonzero modulation. In MATLAB,

```
pval = stats1.p(2);       %p-value from Wald test.
```

We find `pval = 3.3866e-31` and are therefore quite certain that the difference in firing rate between the planning and movement periods is real.

In other case studies (see chapters 8 and 9), we explored a number of tools to assess goodness-of-fit of spiking models. Here, we examine just a few of these tools to investigate how well the initial simple model fits the spiking data. First, let's construct a KS plot of the rescaled interspike intervals. We compute the estimated firing rate based on the model:

```
lambda1=exp(b1(1)+b1(2)*Imove);      %Firing rate of Model 1.
```

Next, we compute the rescaled intervals by integrating, or in this case summing, the estimated rate between spike times:

```
spikeindex=find(y);                   %Find spikes,
N=length(spikeindex);                 %...and total no. of spikes.
Z(1)=sum(lambda1(1:spikeindex(1))); %1st rescaled waiting time,
for i=2:N,                            %...and the rest.
     Z(i)=sum(lambda1(spikeindex(i-1)+1:spikeindex(i)));
end
```

Finally, we compute the empirical CDF of the rescaled spike times (`Z`) and compare it to the model CDF, which is an exponential distribution with parameter 1 if the model is correct (see chapter 9):

```
[eCDF, zvals] = ecdf(Z);              %Empirical CDF at z values.
mCDF = 1-exp(-zvals);                 %Model CDF at z values.
plot(mCDF,eCDF)                       %Create KS plot.
hold on                               %Freeze graphics window.
plot([0 1], [0 1]+1.36/sqrt(N))      %Upper confidence bound.
plot([0 1], [0 1]-1.36/sqrt(N))      %Lower confidence bound.
hold off                              %Release graphics window.
xlabel('Model CDF')                   %Label the axes.
ylabel('Empirical CDF')
```

The KS plot (figure 10.12) reveals that the initial model tends to fit the distribution of the larger rescaled ISIs well, but not the shorter ISIs. Therefore, we conclude that model 1 does not completely capture the structure of the spiking data.

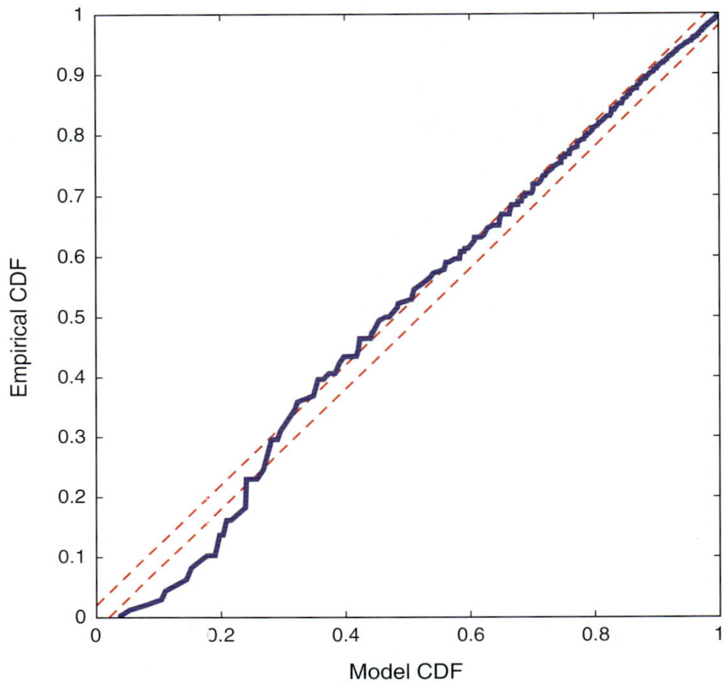

Figure 10.12
KS plot for model 1 based on planning versus movement periods, with 95% confidence bounds (*dotted lines*).

Q: What features of the spiking structure could be missing from this model?

A: We have a few options from prior visualizations, including movement direction (left or right) as well as past spiking history. We may improve the model by incorporating either (or both) features.

Refining the Model: Movement Direction. With an eye toward improving the initial model, we first visualize the impact of movement direction on model 1. We create a column vector for the design matrix that indicates the trial movement direction for each time step:

```
%Create indicator for trial movement direction.
xdir = reshape((direction*ones(1,T0))',K*T0,1);
```

Q: That's a busy line of MATLAB code, What's happening here? Hint: Remember that `direction` is a `[50,1]` vector of zeros and 1 s, and note that we perform matrix multiplication (`*`) and apply the `reshape` function to convert a matrix to a vector.

We could add the variable `xdir` directly to the initial model (10.3), and proceed immediately with a model fit. But first let's examine the point process residuals for model 1, and whether these residuals depend on the trial movement direction. To compute and display the integrated residuals for left versus right trials in MATLAB,

```
i0 = find(xdir==0);        %Left movement trial.
i1 = find(xdir==1);        %Right movement trial.
R = cumsum(stats1.resid);%Cumulative sum of Model 1 residuals.
plot(i0,R(i0),'.')         %Plot residuals for L trials,
hold on                    %...freeze graphics,
plot(i1,R(i1),'g.')        %...and plot residuals for R trials,
hold off                   %...release graphics,
xlabel('Trial')            %...label axes.
ylabel('Integrate Point Process Residual')
```

We find in figure 10.13 that the integrated residuals tend to increase for left trials, suggesting that model 1 consistently underestimates the firing rate for these trials, and tend to decrease for right trials, suggesting that model 1 overestimates the rate for those trials. Including trial direction may therefore help improve the model. Let's now add the predictor `xdir` to the initial model (10.3) to create the second model,

$$\text{[Model 2]} \qquad \lambda(t) = e^{\beta_0+\beta_1 I_{\text{move}}(t)+\beta_2 I_{\text{dir}}(t)}, \tag{10.4}$$

and fit this model in MATLAB:

```
%Fit Model 2, and return estimates and useful statistics.
[b2 dev2 stats2] = glmfit([Imove xdir],y,'poisson');
```

When adding the direction for each trial to the model, we find that the baseline firing rate parameter changes to `exp(b2(1))=0.0487` = 48.7 spikes/s. The baseline condition now corresponds to the planning period only (`Imove=0`) of trials to the left (`xdir=0`). The modulation due to changing from planning to movement periods (`exp(b2(2))=1.41`) stays about the same, suggesting a 41% increase in firing during the movement period. The final parameter relates to the modulation when changing from a left trial to a right trial. The modulation of `exp(b2(3))=0.6011` indicates that on right trials the firing rate is reduced by about 40%. Let's check the significance of this last parameter (via a Wald test):

```
pval=stats2.p(3);  %Significance of Model 2 direction parameter.
```

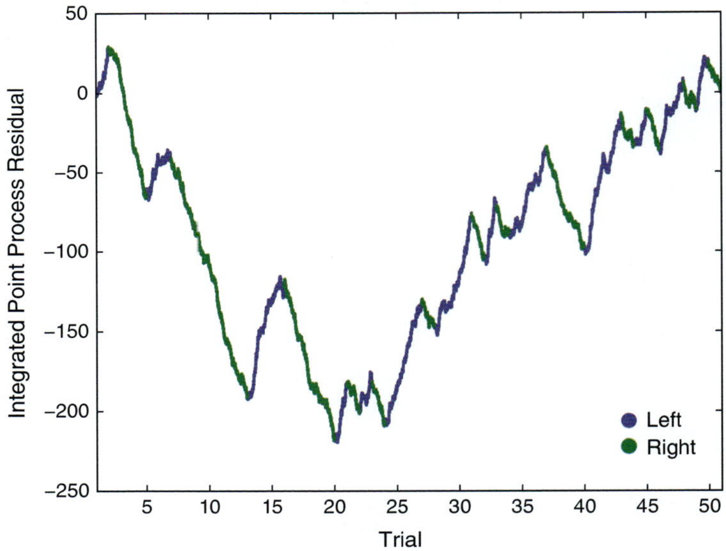

Figure 10.13
Point process residuals for model 1 for left (*blue*) and right (*green*) trials.

We find `pval = 5.2818e-64` and conclude that the decrease in firing rate is highly significant. Finally, let's construct a KS plot for model 2:

```
lambda2=exp(b2(1)+b2(2)*Imove+b2(3)*xdir);  %Evaluate Model 2.
Z(1)=sum(lambda2(1:spikeindex(1)));  %1st rescaled waiting time,
for i=2:N                              %... and the rest.
  Z(i)=sum(lambda2(spikeindex(i-1)+1:spikeindex(i)));
end
[eCDF, zvals]=ecdf(Z);                 %Empirical CDF.
mCDF = 1-exp(-zvals);                  %Model CDF.
plot(mCDF,eCDF)                        %Create KS plot.
hold on                                %Freeze graphics.
plot([0 1], [0 1]+1.36/sqrt(N))        %Upper confidence bound.
plot([0 1], [0 1]-1.36/sqrt(N))        %Lower confidence bound.
hold off                               %Release graphics window.
xlabel('Model CDF')                    %Label the axes.
ylabel('Empirical CDF')
```

Q: Compare the KS plot for model 2 (figure 10.14) to the KS plot for model 1 (figure 10.12). Does the updated model provide an improvement over the original model?

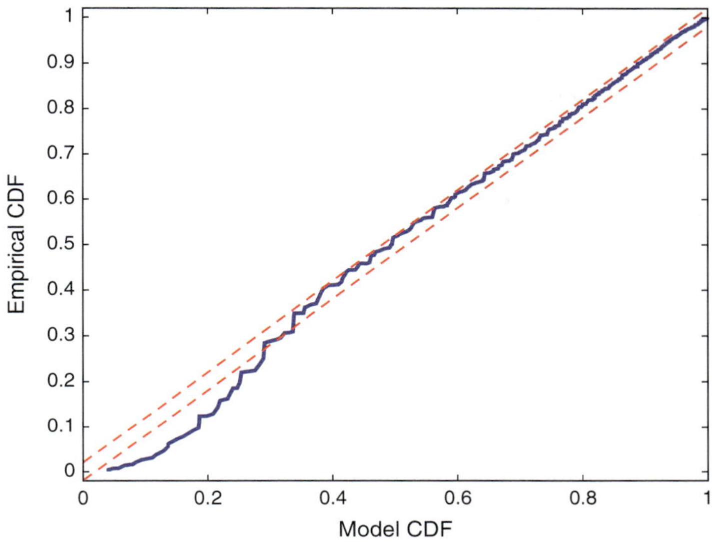

Figure 10.14
KS plot for model 2, including trial period and movement direction.

A: This KS plot of model 2 looks similar to that of model 1. We observe some improvements at intermediate rescaled ISIs, but we are still seeing too few small rescaled ISIs. The updated model represents a slight improvement over the first model but is still well outside the confidence bounds of the KS plot.

Q: What other covariates could we add to improve the model?

A: The previous autocovariance and spectral analysis suggests that another feature of these data relates to history-dependent spiking. Perhaps we should add columns to the design matrix that depend on past values of the spiking activity.

Refining the Model: History Dependence. Up to this point, we have used Poisson regression models for the spiking data that did not incorporate any history-dependent terms. For those models, the number of spikes in each of the small discrete-time bins is assumed to have a Poisson distribution, and the number of spikes in any two nonoverlapping bins is assumed to be independent. A simple result from probability theory is that the sum of independent Poisson random variables also has a Poisson distribution. Therefore, for those previous models, the number of spikes over multiple time bins, which is the sum of the number of spikes in each bin, still has a Poisson distribution, and all those previous models would be considered Poisson process models.

Now, we would like to add predictors to the model related to the past history of spiking. Thus, we do not expect the number of spikes in nonoverlapping bins to be independent; a spike at the current time will change the probability of seeing a spike in the near future. Even if the number of spikes in a small time bin is assumed to be Poisson, it is no longer the case that the number of spikes in any larger interval will be Poisson. Thus, we would no longer call the resulting spike train a Poisson process. We still use the notation $\lambda(t)$ to represent the expected number of spikes in the tth time bin, but this is no longer called a Poisson rate function. Instead, we call it the *conditional intensity function* and write $\lambda(t|H_t)$, where H_t represents the past history of spiking.

There are many different functions of past spiking that we could use in the model. The simplest option is to use the binned spike train itself, lagged by different amounts of time. Mathematically, we are building a probability model for ΔN_i as a function of ΔN_{i-j} for $j = 1, \ldots, k$ for some model order k.

How should we select the model order? We return to this question later and propose principled approaches to create *parsimonious* models, models that fit the data well with as few parameters as possible. But we start by selecting a model order based on spectral analysis. We found a peak in the spectral density during the planning period at 15–20 Hz. A history-dependent model that extends to lags of 70 ms should be able to capture rhythms above 15 Hz. Let's therefore define a variable `ord` to represent the model order:

```
ord = 70;                    %Set the model order.
```

To predict the spiking at the current time, we need to observe the spiking history at least `ord` lags in the past. Therefore, we redefine the variables `y`, `xdir`, and `Imove` starting from index `ord+1`, and then build up the past spiking history into a matrix called `xHist`. In MATLAB,

```
T1 = T0-ord;                 %Update no. of time points.
%Redefine observable & predictors to support history dependence.
y=reshape(train(:,ord+1:end)', K*T1,1);                %Data.
xdir=reshape((direction*ones(1,T1))',K*T1,1);          %Direction.
Imove=ones(K,1)*[zeros(1,T0/2-ord) ones(1,T0/2)];     %Period,
Imove=reshape((Imove) ,K*T1,1);                        %...reshaped.
xHist = [];                  %Create the history predictor,
for i = 1:ord                %...for each step in past.
    xHist=[xHist reshape(train(:,ord+1-i:end-i)',K*T1,1)];
end
```

Q: Consider the variable `xHist`. Can you explain how this variable represents the past spiking activity?

We could then include the three predictors, `xdir`, `Imove`, and `xHist`, in the GLM separately as follows:

```
%Fit Model 3, with history dependence.
[b3 dev3 stats3] = glmfit([Imove xdir xHist],y,'poisson');
```

This model would have a total of 73 parameters—one intercept parameter, one movement parameter, one direction parameter, and 70 parameters representing the columns of the history matrix. Mathematically,

$$[\text{Model 3}] \qquad \lambda(t|H_t) = e^{\beta_0+\beta_1 I_{\text{move}}(t)+\beta_2 I_{\text{dir}}(t)+\sum_{j=1}^{70}\beta_{2+j}\Delta N_{t-j}}. \tag{10.5}$$

However, model 3 implies that the influences of the trial period ($I_{\text{dir}}(t)$) and of the past history (ΔN_{t-j}) are multiplicatively separable; that is, that the only difference between the planning and movement period is a change in the average firing rate and the spike rhythms are the same in both periods. The autocovariance and spectral analyses suggest that the influence of past spiking differs between these trial periods. Therefore, we should instead construct a model that has an interaction between the `Imove` and `xHist` variables. We do this by including columns that represent the product of these two variables. Mathematically, we propose a model of the form

[Model 4]

$$\lambda(t|H_t) = e^{\beta_0+\beta_1 I_{\text{move}}(t)+\beta_2 I_{\text{dir}}(t)+\sum_{j=1}^{70}\left(1-I_{\text{move}}(t)\right)\beta_{2+j}\Delta N_{t-j}+\sum_{j=1}^{70}\beta_{72+j}I_{\text{move}}(t)\Delta N_{t-j}}. \tag{10.6}$$

Model 4 has 143 parameters—one intercept parameter, one parameter for each of the movement and direction indicator variables, 70 parameters describing the history dependence in the planning period, and 70 more parameters describing the history dependence in the movement period. Whenever we construct a point process model that includes history dependence, we call it a *conditional intensity model* rather than a rate model. To fit this model in MATLAB,

```
%Fit Model 4, with history dependence in each period.
[b4 dev4 stats4] = glmfit([Imove xdir ...     %Period & direction.
  ((1-Imove)*ones(1,ord)).*xHist ...          %History in planning.
  (Imove*ones(1,ord)).*xHist], y,'poisson');%History in movement.
```

Let's examine the exponentiated values directly. For the first three:

```
%Examine first three exponentiated parameters of Model 4.
exp(b4(1:3)) = [0.0480    1.3819    0.6063]
```

Since model 4 has 140 additional parameters related to history dependence, inspecting these printed values directly would be too overwhelming. Instead, let's plot their exponentiated values as a function of lag for each period:

```
subplot(211); plot(1:ord,exp(b4(4:ord+3)));         %Planning,
ylabel('Modulation')                                %..axes labeled.
subplot(212); plot(1:ord,exp(b4(ord+4:end)));       %Movement,
ylabel('Modulation'); xlabel('Lag [ms]')            %..axes labeled.
```

Each of these values represents a modulation to the firing intensity when a previous spike has occurred at the given lag. For both periods (figure 10.15), the value at lag 1 ms is near zero, suggesting that the probability of spiking again immediately after a previous spike is greatly reduced—the neuron is refractory. At lags greater than 1 ms, the probability of spiking now is increased when there is a spike at the given lag. For example, if there is a spike 6 ms in the past during the movement period, the probability of spiking is modulated by about 1.8, or it increases by about 80%. A modulation value near 1 suggests no effect

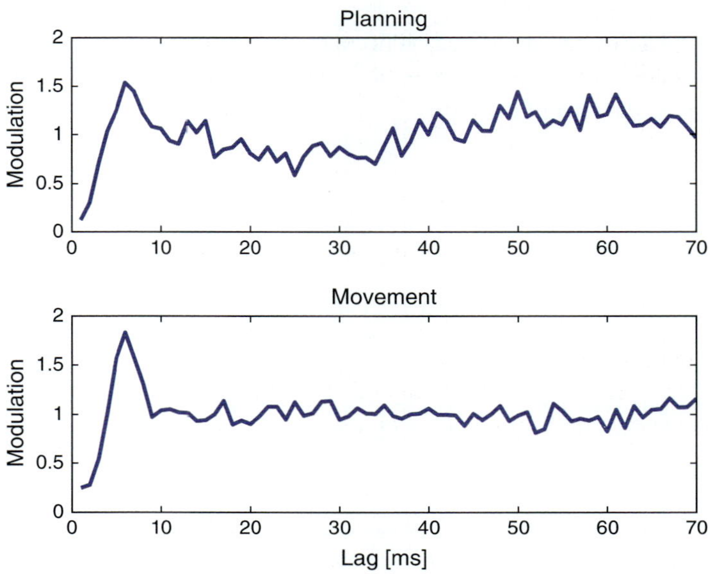

Figure 10.15

Exponentiated GLM parameters related to history dependence in model 4 for planning (*top*) and movement (*bottom*) periods.

on the spike intensity related to a spike at that lag. For the movement period, we see a refractory period followed by a period of increased intensity 6 ms after a spike. However, any spiking 10 ms in the past or longer appears to have no effect on the current spiking intensity. During the planning period, we see a similar pattern of refractoriness followed by increased intensity up to lag 10 ms. At longer lags, however, we see a pattern of decreased spiking probability about 15–35 ms after a spike, followed by increased spiking probability about 45–65 ms after a spike. The parameter values for these periods suggest that the modulation for these lags is relatively small, about a 10%–15% decrease or increase in the firing intensity.

Q: Are these history effects statistically significant?

A: One approach we could take is to visualize the *p*-values (based on the Wald *t*-test) for each parameter. Since it can be hard to tell visually whether *p*-values are sufficiently close to zero, let's plot the negative of the natural logarithm of the *p*-values. Any value of the negative log above 3 indicates a *p*-value below 0.05. In MATLAB,

```
subplot(211)
plot(1:ord,-log(stats4.p(4:3+ord)),'.')  %p-values for planning,
hold on                                  %...freeze graphics,
plot([1 ord],[-log(.05) -log(.05)]);     %...draw threshold,
hold off                                 %...release graphics.
subplot(212)
plot(1:ord,-log(stats4.p(4+ord:end)),'.')%p-values for movement
hold on                                  %...freeze graphics,
plot([1 ord],[-log(.05) -log(.05)]);     %...draw threshold,
hold off                                 %...release graphics.
```

We see in figure 10.16 that for both the planning and movement periods, the *p*-values corresponding to lags up to about 8 ms tend to be significant. For the movement period, most of the parameters for lags above 8 ms do not suggest significant modulation. For the planning period, we find a few parameters for higher lags that are significant on their own. However, in this case we have to be careful about multiple comparisons. As discussed in chapter 8, if we control the probability of a single test or comparison being significant by chance at 0.05, then after performing multiple tests or comparisons, the probability that any of them are deemed significant by chance becomes much higher. For 140 independent comparisons, even if no effects are actually present, the probability of finding at least one to be significant by chance is $1 - 0.95^{140} = 0.99$.

A common approach for dealing with multiple comparisons is the *Bonferroni correction* method, where instead of setting the significance level to control for the probability of a single significant result by chance, we divide by the total number of comparisons to

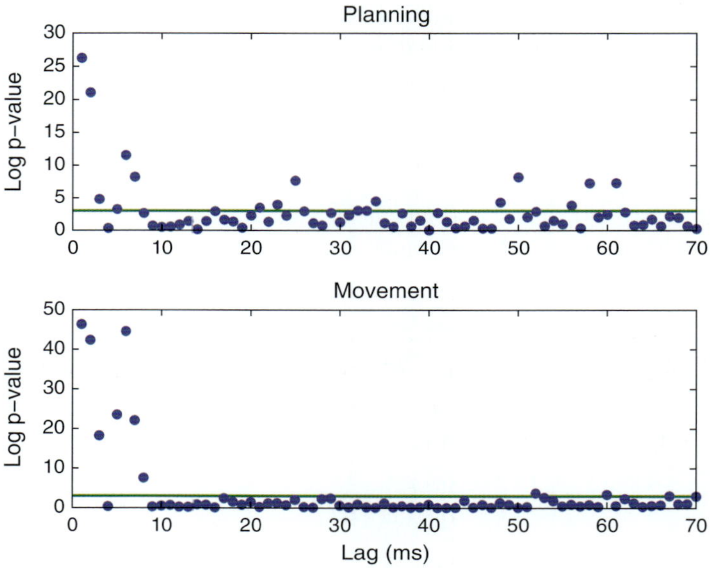

Figure 10.16

Negative log p-values for model 4 parameters related to history dependence. Values above green line indicate p-values below 0.05 in planning (*top*) and movement (*bottom*) periods.

control for the probability of having any comparison significant by chance. In this case, the corrected significance level would be $0.05/140 = 0.00036$. The negative log of this corrected significance level is about 8. So, in figure 10.16, we would declare significant only points above 8. Of the parameters corresponding to lags above 10 ms, only the one at lag 50 for the planning period would remain significant.

Bonferroni's method is often much more conservative than necessary. In this case, we can try another option. Instead of 140 separate tests for each lag in each period, we can construct a single test to answer a question of direct interest: Is there a significant difference between the complete sets of history parameters in the planning and movement periods? In chapter 9, we discussed maximum likelihood ratio tests (MLRTs) for point process models. We showed that for nested models (where one model can be made equivalent to the other by setting some parameters to specific values), under the null hypothesis that the data arise from the smaller model, the difference in the deviance between the two models should have a chi-square distribution with the number of degrees of freedom equal to the number of extra parameters in the larger model. In this case, let's compare model 3, for which the history dependence is the same for the planning and movement periods, with model 4. To compute the p-value for this test in MATLAB,

```
pval = 1-chi2cdf(dev3-dev4,ord);          %Compare two nested GLMs.
```

Q: Why do we set the second input to the function `chi2cdf` equal to `ord`?

We find `pval = 2.3190e-08`, and we would be very unlikely to see this result if model 3—the model with identical history dependence structure—were correct.

Finally, let's construct a KS plot to see how well model 4 fits the data. When we introduced the time-rescaling theorem in chapter 9, we assumed that the data came from a Poisson process. Now that we are modeling history-dependent point processes, we need to make a slight update to the time-rescaling theorem:

Time-Rescaling Theorem for History-Dependent Point Processes. Let $S_1, S_2, \ldots, S_n$ be a collection of spike times from any point process with conditional intensity function $\lambda(t|H_t)$. Let

$$Z_1 = \int_0^{S_1} \lambda(t|H_t)dt \qquad \text{and} \qquad Z_i = \int_{S_{i-1}}^{S_i} \lambda(t|H_t)dt,$$

for $i = 2, \ldots, n$. Then the rescaled variables, $Z_1, Z_2, \ldots, Z_n$, are independent and identically distributed random variables from the exponential distribution with parameter 1.

This looks nearly identical to the time-rescaling theorem for Poisson processes, with the Poisson rate functions replaced by history-dependent conditional intensity functions. However, the fact that the rescaled times, Z_i, are still independent is perhaps more surprising in this case, since the original point process (i.e., S_1, S_2 ... S_3) does have history dependence. This result suggests that if we have the right model, we can rescale the spiking process to account for history dependence as well as changes in the firing intensity related to time and other covariates.

Let's now return to constructing a KS plot to see how well model 4 fits the data. We use the `glmval` routine in MATLAB to compute the model intensity at each time step:

```
lambda4 = glmval(b4,[Imove xdir ... %Evaluate Model 4.
    ((1-Imove)*ones(1,ord)).*xHist ...
    (Imove*ones(1,ord)).*xHist],'log');
spikeindex=find(y);                  %Find spikes,
N=length(spikeindex);                %...and total no. of spikes.
Z(1)=sum(lambda4(1:spikeindex(1))); %1st rescaled waiting time,
for i=2:N                            %... and the rest.
  Z(i)=sum(lambda4(spikeindex(i-1)+1:spikeindex(i)));
end
[eCDF, zvals]=ecdf(Z);               %Empirical CDF.
```

```
mCDF = 1-exp(-zvals);                    %Model CDF.
plot(mCDF,eCDF)                          %Create KS plot.
hold on                                  %Freeze graphics.
plot([0 1], [0 1]+1.36/sqrt(N))          %Upper confidence bound.
plot([0 1], [0 1]-1.36/sqrt(N))          %Lower confidence bound.
hold off                                 %Release graphics window.
xlabel('Model CDF')                      %Label the axes.
ylabel('Empirical CDF')
```

Q: Compare the KS plot for model 4 (figure 10.17) to the KS plot for model 1 (figure 10.12). Does the updated model provide an improvement over the original model?

A: We see a substantial improvement in the fit according to the KS plot at smaller rescaled ISIs. The model still does not quite pass the KS test, as can be seen when the KS plot veers outside of the confidence band around the model CDF value of 0.2. However, the inclusion of history dependence has clearly improved the fit of the short rescaled ISIs.

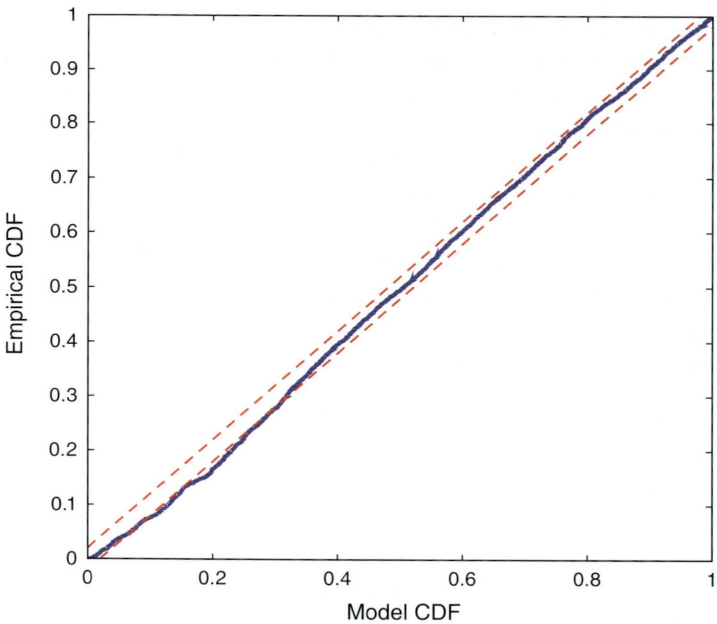

Figure 10.17
KS plot for model 4 with separate history dependence structure for planning and movement periods.

Q: Based on prior visualization analyses, are there any other variables that we could include to further improve the goodness of fit of the model?

Choice of Model Order. Before we settle on model 4, let's return to the question of how to select the order of a model such as (10.6). We chose somewhat arbitrarily a model order of 70 for the number of history-dependent terms for both the planning and movement periods, based on the visualization analyses. In subsequent modeling, we found that many of these parameters did not contribute significantly to the model fit on their own. One approach to selecting model order is to start by fitting a higher-order model than you suspect is necessary, and then paring down the parameters that are not significant on their own. In this example, this process would leave a few parameters at small lags for both the planning and movement periods, and one parameter corresponding to a lag of 50 ms for the planning period. However, the MLRT example showed that although many of the history-dependent parameters in this model are not significant on their own, together they substantially improve the model.

Instead of starting with a large model and paring it down, another approach to model selection is to start with a small model and step up the model order while examining the change in goodness-of-fit for each subsequent model. Of course, we have to be careful in comparing goodness-of-fit, since larger models tend to fit the data better even if they do not better predict future data. In chapter 9, we used Akaike's information criterion (AIC) to compare the goodness-of-fit of models of different sizes. Let's examine the change in AIC as we change the model order.

Before comparing models of different sizes, let's consider the classes of models to explore. For the most recent model, we have separate history terms related to the planning and movement periods. The visualization analyses and previous modeling results suggest that the extent to which past spiking influences current spiking likely differs between these two periods. Therefore, let's fit models with different orders for the history dependence in these separate periods. Mathematically, we would search for the value of the pair, (p, q), that minimizes the AIC of the model

$$\lambda(t) = e^{\beta_0+\beta_1 I_{\text{move}}(t)+\beta_2 I_{\text{dir}}(t)+\sum_{j=1}^{p}\left(1-I_{\text{move}}(t)\right)\beta_{2+j}\Delta N_{t-j}+\sum_{j=1}^{q}\beta_{p+2+j} I_{\text{move}}(t)\Delta N_{t-j}}.$$

This might be the best choice if the goal were to find the most parsimonious model that captures the full dataset.

Alternatively, we could decide to fix a single value for the model order and force the history-dependent component to be the same size in both periods. In that case, we would find the single value p that minimizes the AIC of the model

$$\lambda(t) = e^{\beta_0+\beta_1 I_{\text{move}}(t)+\beta_2 I_{\text{dir}}(t)+\sum_{j=1}^{p}\left(1-I_{\text{move}}(t)\right)\beta_{2+j}\Delta N_{t-j}+\sum_{j=1}^{p}\beta_{p+2+j} I_{\text{move}}(t)\Delta N_{t-j}}.$$

This might be the best choice if the goal were to compare the rhythmic spiking features between the planning and movement periods with a parsimonious model, since it allows us to test for differences, as with the preceding MLRT.

In this case, let's consider a simpler problem: for the data in the planning period alone, let's find the model order that minimizes the AIC of

$$\lambda(t) = e^{\beta_0+\beta_2 I_{\mathrm{dir}}(t)+\sum_{j=1}^{p}\beta_{1+j}\Delta N_{t-j}}. \tag{10.7}$$

Recall that the AIC is related to the deviance of the fitted model (modulo a fixed constant) plus 2 times the number of parameters in the model. We will build models of order 1 up to order 100 and compute the AIC for each model. Because we are fitting many models, each with a large dataset, the following MATLAB code might take a few minutes to run:

```
maxord = 100;              %Maximum model order.
%Redefine observable & predictors to support history dependence.
yplan=reshape(train(:,maxord+1:T0/2)',K*(T0/2-maxord),1);
plandir=reshape((direction*ones(1,T0/2-maxord))',...
    K*(T0/2-maxord),1);
planHist=[];               %Create the history predictor,
for i=1:maxord             %...for each step in past,
 planHist=[planHist ...    %...define history,
   reshape(train(:,maxord+1-i:T0/2-i)',K*(T0/2-maxord),1)];
                           %..fit the model,
 [b0 dev0 stats0]=glmfit([plandir planHist],yplan,'poisson');
                           %...and compute the AIC.
 aic(i) = dev0+2*length(b0);
end
plot(1:maxord,aic);        %Plot the AIC,
xlabel('Model Order')      %...with axes labeled,
ylabel('AIC')
```

We see in figure 10.18 that the AIC drops sharply (with one little blip) from order 1 through order 6, after which the AIC starts to increase. However, as the model order continues to increase, the AIC plot undergoes a number of additional drops around lags of 25–35 ms and around 60 ms, reaching a minimum at a lag of 62 ms. The AIC continues to increase after that point up to lag 100 ms. Of course, it is possible that the AIC undergoes additional drops beyond a lag of 100 ms.

Q: Given the AIC results in figure 10.18, what model order would you choose?

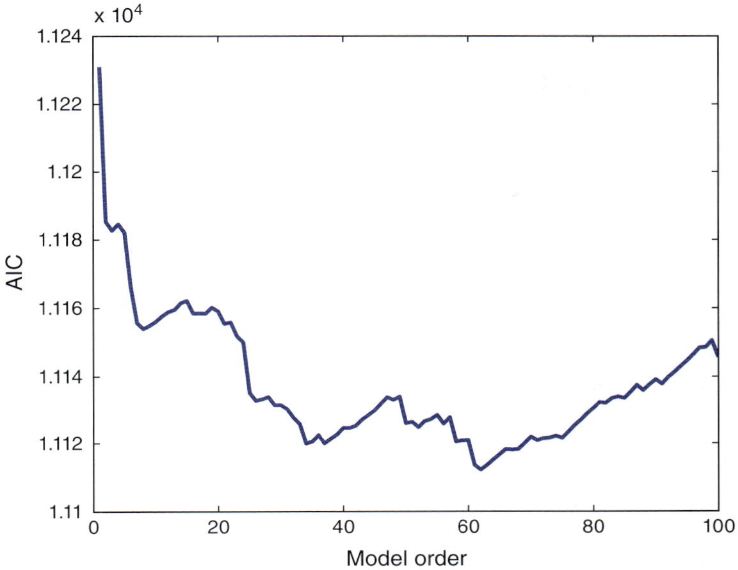

Figure 10.18
AIC plot of models with increasing number of history-dependent components for planning period.

Refining the Model: Smooth History Dependence. Let's examine one more approach for constructing a parsimonious model that still captures history dependence in these data. Looking back at the parameter estimates in figure 10.15, we see that nearby parameters have quite variable estimates. Do we really believe that the influence of a spike 25 ms in the past can be very different from the influence of a spike 26 ms in the past? If not, it might be preferable to define a model with fewer free parameters that guarantees a smooth modulation as a function of lag.

Motivated by this observation, let's replace the separate terms for each single millisecond lag. We do so with a small set of basis functions on the spiking history that can flexibly capture smooth history dependence structure. Mathematically, we fit a model of the form

[Model 5]

$$\lambda(t|H_t) = e^{\beta_0+\beta_1 I_{\text{move}}(t)+\beta_2 I_{\text{dir}}(t)+\sum_{j=1}^{p}\left(1-I_{\text{move}}(t)\right)\beta_{2+j}\, g_j(H_t)+\sum_{j=1}^{p} I_{\text{move}}(t)\beta_{p+2+j}\, g_j(H_t)}, \tag{10.8}$$

where H_t is the full past firing history, and $g_j(H_t)$ is a small collection of basis functions that together provide a smooth influence of the history on the current spiking intensity.

There are many sensible choices for these basis functions. A common choice is a smoothing spline basis. Here we use a set of Gaussian kernel functions with means at different lags as the basis functions. Mathematically, we set $g_j(H_t) = \sum_{\tau} \Delta N_{t-\tau} e^{-(\tau-\mu_j)^2/(2\sigma^2)}$, where we let $\mu_j = 10j - 15$ for $j = 1$ to 8, and $\sigma = 5$. This choice of basis function is rather arbitrary; Gaussian kernels lack many of the nice properties of other basis functions. In this case, it was primarily chosen for ease of implementation in MATLAB but also because of some simple intuition. The Gaussian kernel provides a way of smoothing the history dependence over a time scale set by the σ term. Setting $\sigma = 5$ and spacing the means at 10 ms lags apart ensures that distinct features in the history modulation function will not be closer than about 10 ms in lag.

We construct this set of basis functions in MATLAB by building a matrix of these Gaussian functions for each lag:

```
for i=1:ord        %Construct the Gaussian kernels.
    C(i,:) = normpdf(-5:10:ord,i,5);
end
```

Q: Plot the columns of matrix `C`. Do you see the Gaussian functions?

The result in `C` is a `[70 x 8]` matrix that converts the previous 70-column `xHist` matrix into a new set of 8 columns to include in the design matrix, `xHist*C`.

```
%Fit Model 5, with Gaussian kernel basis.
nparams = size(C,2);
[b5 dev5 stats5] = glmfit([Imove xdir ...         %Period & direction.
  ((1-Imove)*ones(1,nparams)).*(xHist*C)  ...%History in planning.
  (Imove*ones(1,nparams)).*(xHist*C)], ...    %History in movement.
  y,'poisson');
```

Now let's plot for this model the modulation in the spiking intensity due to a previous spike at any lag by multiplying the corresponding set of parameters by `C`.

```
subplot(211);plot(1:ord,exp(C*b5(4:nparams+3)));%Planning,
ylabel('Modulation')                             %..axes labeled.
subplot(212);plot(1:ord,exp(C*b5(nparams+4:end)));%Movement,
ylabel('Modulation'); xlabel('Lag [ms]')         %..axes labeled.
```

The model fit in figure 10.19 captures much of the same structure that we saw for model 4 (figure 10.15). However the fit in model 5 is much smoother and requires only 8 parameters per curve instead of 70.

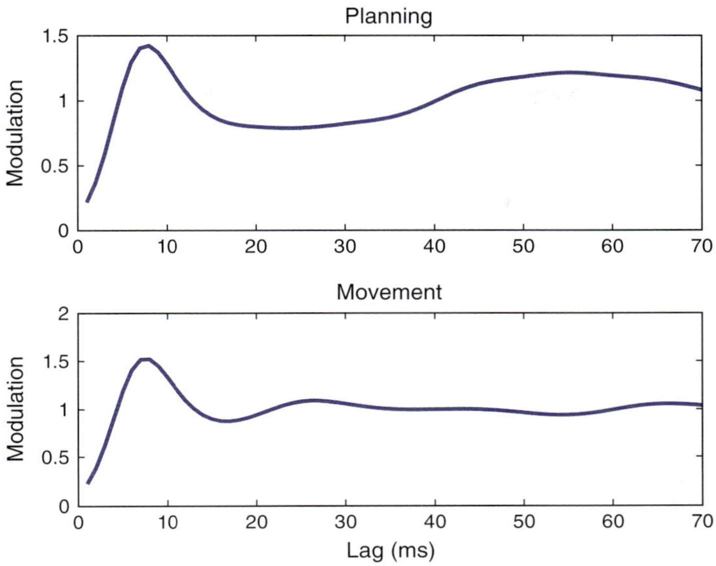

Figure 10.19

Exponentiated GLM parameters related to history dependence for planning (*top*) and movement (*bottom*) periods using Gaussian kernel basis functions of model 5.

Q: How well does model 5 fit the data?

To answer this, let's construct one more KS plot. In MATLAB,

```
lambda5 = glmval(b5,[Imove xdir ... %Evaluate Model 5.
  ((1-Imove)*ones(1,nparams)).*(xHist*C) ...
  (Imove*ones(1,nparams)).*(xHist*C)],'log');
spikeindex=find(y);                  %Find spikes,
N=length(spikeindex);                %...and total no. of spikes.
Z(1)=sum(lambda5(1:spikeindex(1)));  %1st rescaled waiting time,
for i=2:N                            %... and the rest.
  Z(i)=sum(lambda5(spikeindex(i-1)+1:spikeindex(i)));
end;
[eCDF, zvals]=ecdf(Z);               %Empirical CDF.
mCDF = 1-exp(-zvals);                %Model CDF.
plot(mCDF,eCDF)                      %Create KS plot.
hold on                              %Freeze graphics.
```

```
plot([0 1], [0 1]+1.36/sqrt(N))      %Upper confidence bound.
plot([0 1], [0 1]-1.36/sqrt(N))      %Lower confidence bound.
hold off                             %Release graphics window.
xlabel('Model CDF')                  %Label the axes.
ylabel('Empirical CDF')
```

Q: Compare the KS plot for model 5 (figure 10.20) to the KS plot for model 4 (figure 10.17). How does the model with Gaussian kernel basis functions compare to the previous model?

A: Model 5 appears to provide an improvement compared to the previous model, even though we are using far fewer parameters. However, model 5 still does not pass the KS test completely, as can be seen by the KS plot exiting the 95% confidence region at a model CDF value around 0.1.

Although model 5 does not completely pass the KS test, it does capture many of the features that we sought to identify through the analyses. We might therefore accept this as the final

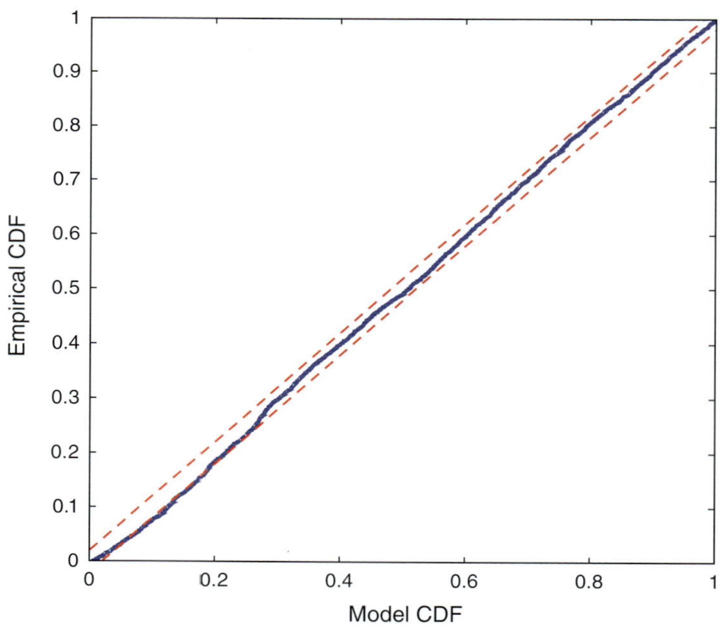

Figure 10.20
KS plot for model 5 with history modulation terms using Gaussian kernel basis functions.

model, for the time being, and try to interpret its components. We can always return to the model selection procedure and continue to refine the model if we find that it fails to capture some important features of the data.

10.2.4 Drawing Conclusions from the Model

At last, let's examine and interpret the model 5 parameters. The best guess for the baseline firing intensity—for the planning period during left trials with no previous spiking in the past 70 ms—is `exp(b5(1))=0.048` spikes/ms = 48 spikes/s. The firing intensity, excluding the effect of history dependence, increases about `exp(b5(2))=1.3913`, or 39%, when we move from the planning period to the movement period. This increase is highly significant (`stats5.p(2)=1.2e-07`) according to the model. The relative firing intensity between left and right trials is `exp(b5(3))=0.6048`; we find there is an approximate 40% decrease in the firing intensity for right trials as compared to left trials. This difference is again highly significant (`stats5.p(3)=1.1424e-50`) according to the model.

The estimated modulation due to past spiking in model 5 for both the planning and movement periods was shown in figure 10.19. We observe a refractory period followed by a period of bursting (i.e., increased spiking intensity) around 6 ms in both the planning and movement periods. During the planning period, we also observe a notable pattern of decreased firing intensity at 20–30 ms after a previous spike, followed by increased firing intensity at 50–60 ms after a previous spike. This pattern seems absent during the movement period. Are the individual parameters related to this history dependence significant? To address this, let's plot the *p*-values:

```
subplot(211)                                %p-values for planning,
plot(1:nparams,-log(stats5.p(4:3+nparams)),'.')
hold on                                     %...freeze graphics,
plot([1 nparams],[-log(.05) -log(.05)])     %...draw threshold,
hold off                                    %...release graphics.
subplot(212)                                %p-values for movement,
plot(1:nparams,-log(stats5.p(4+nparams:end)),'.')
hold on                                     %...freeze graphics,
plot([1 nparams],[-log(.05) -log(.05)])     %...draw threshold,
hold off                                    %...release graphics.
```

For model 5, all the parameters related to history modulation during the planning period are significant at the 0.05 level (figure 10.21). However, this leads to the same multiple comparisons problem discussed earlier. For model 4 we performed 140 tests, while here we have only 16 tests, far fewer, so the multiplicity problem is slightly reduced but still present. Again, we can avoid this problem by instead focusing on a single test to address the question of whether the history dependence structure differs between the planning and movement periods. Let's fit a reduced model that uses the same basis functions for history

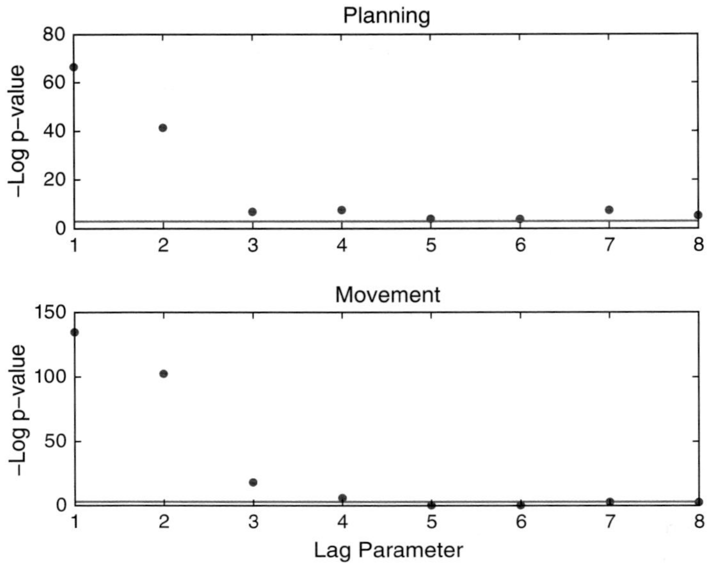

Figure 10.21

Negative log *p*-values for model 5 parameters related to history dependence. Values above green line indicate *p*-values below 0.05 in planning (*top*) and movement (*bottom*) periods.

dependence as model 5 but includes only a single set of parameters for both the planning and movement periods.

```
%Fit a reduced version of Model 5,
[b6 dev6 stats6] = glmfit([Imove xdir xHist*C],y,'poisson');
```

Next, let's compute the MLRT statistic as the difference in the deviance between model 5 and its reduction, and compute its significance based on the chi-square distribution,

```
p=1-chi2cdf(dev6-dev5,nparams); %Compare Model 5 and reduction.
```

Once again, we end up with a very significant result that provides confidence that the spiking dynamics really do change between periods. The rhythmic structure that we see during planning is attenuated during the movement period.

Summary

In this case study, we explored a variety of approaches for identifying and quantifying rhythmic spiking dynamics. One key point is that we can think about spike rhythms in terms of how the likelihood of spiking depends on its own history. Many of the most useful visualization tools that we explored, such as the autocorrelation plot and the spectral estimates, implicitly define the covariability of the spiking process with its past. The GLM methods allowed us to model this history dependence explicitly and to explore multiple features of the data simultaneously.

In the STN data, we also saw that rhythmic spiking structure can be subtle and hard to detect if we do not use the proper tools. The dynamic rhythms in these data were nearly impossible to see in raster plots or ISI histograms but became quite apparent in the autocorrelation plot, spectral density estimates, and model fits. Many people think of rhythmic spiking data as being like the beats of a metronome: perfect beats at a specific frequency. But we saw that very subtle changes in the firing probability, perhaps just a 10% increase or decrease at a particular lag, can lead to robust rhythms, as identified by visualization and modeling tools. Real spike rhythms often look more like the examples in this chapter than like the rhythms of a beating metronome.

Problems

10.1. In this problem, we perform a spectral analysis of the spiking activity recorded from a pair of neurons in rat barrel cortex. The vibrissae (whiskers) of a rat are arranged in an ordered pattern of rows and columns along its face. The barrel cortex contains a topographic representation of this pattern of whiskers. Stimulation of a single whisker causes neurons in the corresponding cortical column, known as a barrel, to generate spikes. Load the file `Ch10-spikes-2.mat`, available at

`http://github.com/Mark-Kramer/Case-Studies-Kramer-Eden`

into MATLAB. This dataset contains simulated spiking activity of two neurons in separate barrels while a bar was swept across the rat's whiskers in a rostral to caudal direction (from the nose toward its tail). The stimulation was regular and periodic, but the experimenter forgot to write down the frequency of stimulation. Each spike train was recorded at a sampling frequency of 1000 Hz.

a. Plot the spiking activity of both neurons to show the relative timing of their spikes. Describe the structure of the spiking activity for each neuron and the relation of the spiking activity between them.

b. Compute and plot the autocovariance function for the spiking activity of each neuron as a function of lag. What do these functions suggest about the stimulus driving these neurons and the relation between the two point processes?

c. Compute and plot the power spectral density of the spiking activity for each neuron. What is the stimulation frequency? Over what range of frequencies is bursting behavior observed?

d. Using the results of this spectral analysis as a starting point, write down an appropriate conditional intensity model for the firing of each of these two neurons.

What variables should be part of this model? Find maximum likelihood estimates of the parameters of your model, and measure the goodness of fit between your model and the data. How do the estimated model parameters relate to the structure observed in the frequency domain analysis?

10.2. Figure 10.6 displayed the ISI histogram for the full dataset. Compute four separate ISI histograms for the planning and movement periods during the left and right trials separately. What do you expect to observe? Do you find any notable differences?

10.3. Simulate a Poisson spike train that lasts 10 s with an average spike rate of 100 Hz. Compute an estimate of the spectrum and compare it to the theoretical spectrum. Then from this simulated Poisson spike train remove any spikes that occur within 10 ms of another spike, and compute the spectrum of the resulting spike train. Interpret the resulting spectral estimate.

10.4. Figure 10.18 showed an AIC plot for the data from the planning period. Repeat the analysis to generate an AIC plot for the spike train data from the movement period. Which order model minimizes the AIC for this period? Which model would you use to fit the combined data from both periods?

10.5. The Mexican hat basis functions are computed as the difference of a Gaussian kernel with a smaller variance and another Gaussian kernel with a larger variance. Implement a modification of model 5 that uses a set of Mexican hat basis functions for the history dependence. How does this change in the basis set affect the model fit?

11 Analysis of Spike-Field Coherence during Navigation

Synopsis

Data 100 trials of 1 s of local field potential and spike train data sampled at 1000 Hz.

Goal Characterize the coupling between the spike and field activity.

Tools Fourier transform, spectrum, coherence, phase, generalized linear models.

11.1 Introduction

11.1.1 Background

In the previous chapters, we focused on two types of data: field data (e.g., EEG, ECoG, LFP) and spiking data (i.e., action potentials), and we developed techniques to analyze these data. In this chapter, we consider the simultaneous observation of both data types. We analyze these multiscale data using the techniques developed in previous chapters and focus specifically on computing the coherence between the spike and field recordings. Understanding the relations between activity recorded at different spatial scales (i.e., a macroscopic field and microscopic spikes) remains an active research area.

11.1.2 Case Study Data

Our experimental collaborator has implanted an electrode in rat hippocampus as the animal performs a task requiring navigation and decision making. From these data, he is able to extract the local field potential (LFP) as well as the spiking activity of a single neuron. He would like to characterize how these multiscale data—the population field activity and the single neuron spiking activity—relate. Based on existing evidence in the literature and experimental intuition, he expects that rhythmic activity in the LFP impacts the probability that a spike will occur. As his collaborator, we will help him to develop tools to examine this hypothesis. He provides us with 100 trials of simultaneous LFP and spike train data with a sampling frequency of 1000 Hz. The duration of each trial is 1 s, corresponding to a fixed temporal interval following a particular decision of the rat.

11.1.3 Goal

Our goal is to understand the coupling between the spiking activity and the LFP following the stimulus. To do so, we analyze the multiscale data recorded simultaneously. To assess this coupling, we will start with two visualizations of the data: the spike-triggered average and the field-triggered average. We then compute the *spike-field coherence*, a coupling measure that builds upon previous development of the Fourier transform and spectrum. We also examine how the firing rate impacts measures of coupling and how to mitigate this impact.

11.1.4 Tools

In this chapter, we focus primarily on computing the spike-field coherence. Development of this measure makes use of skills developed in previous chapters. In computing the spike-field coherence, we continue to utilize the Fourier transform. We also consider how generalized linear models (GLMs) can be used to construct a measure of spike-field association with an important advantage over the spike-field coherence.

11.2 Data Analysis

11.2.1 Visual Inspection: Spike-Triggered Average and Field-Triggered Average

To access the data for this chapter, visit

http://github.com/Mark-Kramer/Case-Studies-Kramer-Eden

and download the file `Ch11-spikes-LFP-1.mat`. We begin the analysis by visualizing examples of the simultaneously recorded spike train and LFP data. Let's load these multiscale data into MATLAB and plot the activity of the first trial:

```
load('Ch11-spikes-LFP-1.mat')  %Load the multiscale data,
plot(t, y(1,:))                %...and plot LFP in 1st trial,
hold on                        %...freeze graphics,
plot(t, n(1,:),'k')            %...and plot spikes in 1st trial,
hold off                       %...release graphics,
xlabel('Time [s]')             %...and label axis.
```

The data file consists of three variables, which correspond to the LFP data (`y`, in units of millivolts), the simultaneously recorded spiking activity (`n`), and a time axis (`t`, in units of seconds). Notice that the data are matrices, in which each row indicates a separate trial, and each column indicates a point in time. In this case, the variable `n` is binary; `n(k,i)=1` indicates a spike in trial `k` at time index `i`.

Q: What is the sampling frequency for these data?

A: We are given the time axis `t`. To compute the sampling frequency, we compute the sampling interval: `dt = t(2)-t(1)` and find `dt = 0.001`. The sampling interval is therefore 1 ms, so the sampling frequency (f) is $f = 1/dt = 1000$ Hz.

The data for this trial are plotted in figure 11.1. Visual inspection immediately suggests that the LFP data exhibit a dominant rhythm. By counting the number of peaks (or troughs) in 1 s of data, we estimate the dominant rhythm to be ≈ 10 Hz. However, careful inspection suggests that other features appear in the LFP from this first trial of data (i.e., additional lower-amplitude wiggles in the signal). Let's keep this in mind as we continue the analysis. To visualize the spikes from the neuron, we plot the activity in the first row of the data matrix (black curve in figure 11.1); this is a crude representation of the activity but sufficient for the initial inspection.

Q: Continue your visual inspection for other trials of the data. What do you observe?

Visual inspection suggests that the neuron is active (i.e., it spikes) during the trial. Of course, we may visualize and analyze features of the spike train and the LFP using the methods described in earlier chapters (e.g., see problem 1). However, our goal here is to characterize the relation (if any) between the LFP and spikes. Let's consider a relatively simple characterization of this relation, the spike-triggered average.

Spike-Triggered Average. The *spike-triggered average* (STA) is a relatively simple procedure to visualize the relation between the LFP and spiking data. To compute the STA, we implement the following procedure.

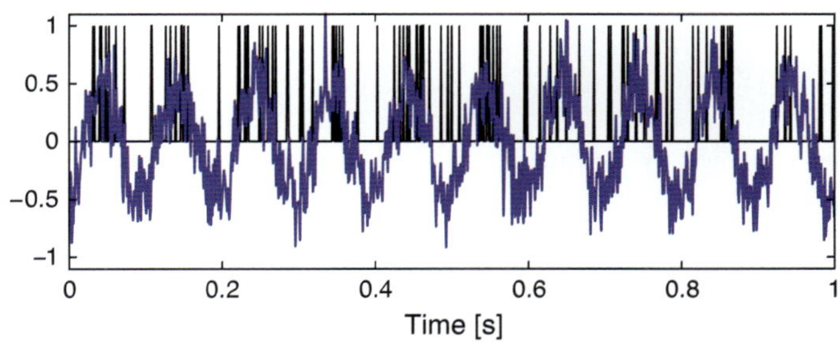

Figure 11.1
Traces of the LFP data (*blue*) and spike train (*black*) recorded in the first trial.

For each trial $k = \{1, \ldots, K\}$, do the following:

1. Identify the time of each spike occurrence $t_{k,i}$, where $i \in \{1, , N_k\}$, and N_k is the number of spikes in the kth trial.
2. For each spike time $t_{k,i}$, determine the LFP within a small temporal interval near the spike time $LFP_{k,i}$.
3. Average $LFP_{k,i}$ across all spikes.

Despite this seemingly complicated procedure, the STA is a relatively intuitive measure. The intuition is to find each spike and determine how the LFP changes nearby. The procedure to compute the STA for each trial is relatively straightforward to perform in MATLAB:

```
win = 100;                  %Choose a window size of +/- 100 ms.
K = size(n,1);              %Define the no. of trials.
N = size(n,2);              %Define no. of data points per trial.
STA=zeros(K,2*win+1);       %Create a variable to hold the STA.
for k=1:K                   %For each trial,
  spks=find(n(k,:)==1);     %...find the spikes,
  for i=1:length(spks)      %...and for each spike,
    if (spks(i) > win & spks(i)+win<N)   %...average the LFP.
     STA(k,:)=STA(k,:)+y(k,spks(i)-win:spks(i)+win)/length(spks);
    end
  end
end
```

In this MATLAB code, we must be careful to include only appropriate time intervals when computing the STA.

> **Q:** What is the purpose of the *if-statement*:
>
> ```
> if (spks(i) > win & spks(i)+win<N)
> ```
>
> in the code?

Notice that the variable `STA` is a matrix, with each row corresponding to a separate trial. Figure 11.2 shows the results for the STA in four trials and in all trials. The individual trial results suggest an approximate rhythmicity in the STA; visual inspection reveals that the STA fluctuates with a period of approximately 100 ms. However, these fluctuations are not phase-locked across trials. For some trials, the LFP tends to be positive when the cell spikes (i.e., at $t = 0$ in figure 11.2), while in other trials the LFP tends to be negative when the cell spikes. The initial results do not suggest a consistent relation exists between the spikes and the LFP across trials.

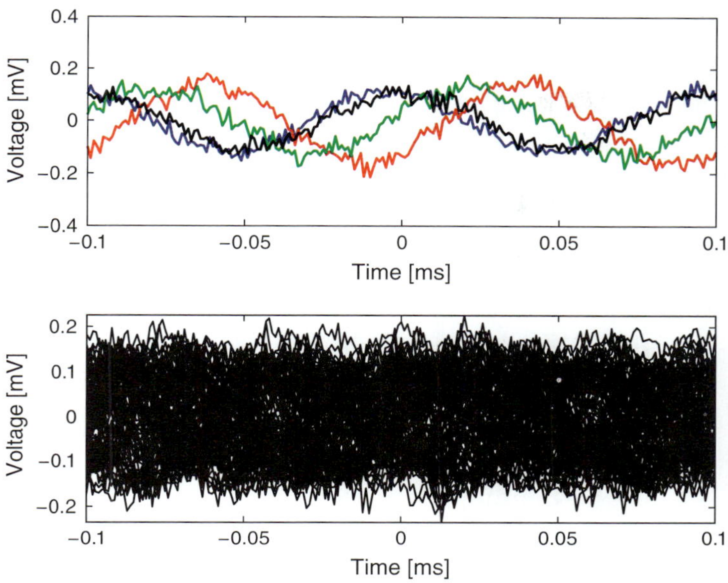

Figure 11.2

Spike-triggered average (*top*) for individual trials (trial 1 in *blue*, 6 in *red*, 10 in *green*, 16 in *black*) and (*bottom*) all 100 trials.

However, let's not abandon all hope yet. We might be concerned that the rhythmicity in the STA (figure 11.2) is consistent with the dominant rhythm of the LFP (figure 11.1). Because the STA is an average of the LFP, we might expect the largest-amplitude features of the LFP to make the biggest impact on the STA. Perhaps this large-amplitude rhythm in the LFP is hiding more subtle features embedded in lower-amplitude activity in the LFP. Let's continue the search.

> **Q:** How would you update the preceding MATLAB code to compute both the average LFP (i.e., the STA) and the standard deviation of the LFP across spikes for each trial?

Field-Triggered Average. Let's now implement another visualization, the *field-triggered average* (FTA). The FTA is similar in principle to the STA. However, for the FTA, we use the field to organize the activity of the spikes (i.e., we use the field to trigger the spikes). Here we choose a particular feature of the field: the phase. We have already investigated the important role of phase in organizing neural rhythms (see chapter 5). Now we examine the role of the LFP phase in organizing the spiking activity.

For each trial $k = \{1, \ldots, K\}$, do the following:

1. Filter the LFP data in trial k into a narrow band, and apply the Hilbert transform to estimate the instantaneous phase.
2. Sort the spiking data in trial k according to the phase of the LFP.

We discussed the Hilbert transform and instantaneous phase in the analysis of cross-frequency coupling in chapter 7. We apply the same procedures here, but to a different end. To implement computation of the FTA in MATLAB,

```
dt = t(2)-t(1);          %Define the sampling interval.
fNQ = 1/dt/2;            %Define Nyquist frequency.
Wn = [9,11]/fNQ;         %Set the passband,
ord  = 100;              %...and filter order,
b   = fir1(ord,Wn);      %...build bandpass filter.

FTA=zeros(K,N);          %Create a variable to hold the FTA.
for k=1:K                %For each trial,
 Vlo = filtfilt(b,1,y(k,:));%... apply the filter,
 phi = angle(hilbert(Vlo)); %... and compute the phase,
 [~,indices] = sort(phi);  %... then sort the phase,
 FTA(k,:) = n(k,indices);   %... and store the sorted spikes.
end
%Plot the average FTA versus phase.
plot(linspace(-pi,pi,N), mean(FTA,1))
```

Notice the steps to set up the filter. We choose a bandpass filter from 9–11 Hz. We choose this interval to focus on the LFP rhythm of largest amplitude ($\approx$ 10 Hz), which we identified through visual inspection (figure 11.1). For each trial, we apply the filter to the LFP and then use the Hilbert transform (`hilbert`) to estimate the phase. Finally, we sort this phase and use the sorted indices to arrange the spikes and store the results. We show the FTA averaged across all trials in figure 11.3. In this case, no modulation in the number of spikes is apparent across trials. Instead, the number of spikes at each phase appears equally likely.

We may apply the FTA analysis to different frequency intervals of the LFP. Choosing a frequency interval may be motivated by our knowledge of the neural system generating the activity or by inspection of the field and spiking data.

> **Q:** Investigate different frequency bands in the FTA analysis. Do you observe any interesting features? *Hint*: Consider frequencies near 45 Hz.

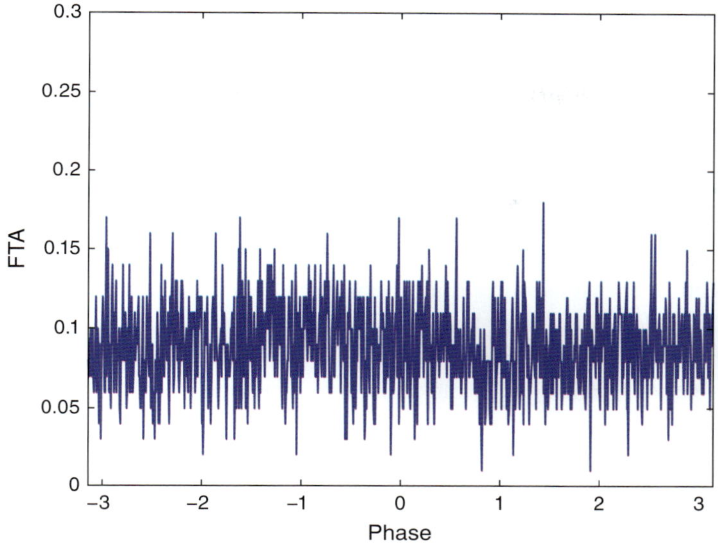

Figure 11.3
Field-triggered average for LFP data in the frequency range 9–11 Hz, averaged across all trials.

One final note about the FTA. The purpose of this measure is visualization, not statistical testing. Hopefully, this visual inspection will provide some insight into the data and guide continuing studies in promising directions. In what follows, we consider approaches to test for significant effects when we build a GLM to assess spike-field relations.

11.2.2 Spike-Field Coherence

To characterize the relation between the LFP and spikes, we have so far visualized the data and computed relatively simple and intuitive aids to visualization. Now we examine a more sophisticated and powerful method: the spike-field coherence. We investigated the coherence when applied to field activity (namely, ECoG data in chapter 5); we may refer to this type of coherence as field-field coherence to distinguish it from spike-field coherence of interest here. In practice, this distinction is usually unnecessary, as in most cases the context is clear. However, in this chapter, we are careful to distinguish field-field coherence from spike-field coherence.

The *field-field coherence* is a frequency domain measure of linear association between two continuous time series.[1] We showed in chapter 5 that two fields are coherent across

1. In practice, we observe a sampled version of a presumably continuous signal. We explored some implications of this sampling on spectral estimators in chapters 3 and 4.

trials at frequency f_0 if the fields possess a constant phase relation across trials at that frequency. The same relation holds for the *spike-field coherence*. However, differences arise because of the point process nature of the spike train data. These differences have profound implications with dangerous consequences. In this chapter, we explore some of these issues. For a deeper mathematical discussion and potential solutions, see [32, 33].

Mathematical Description of spike-field coherence. Let's begin with a mathematical description of the spike-field coherence. To do so, we need to introduce some notation, which is identical to that used in earlier chapters, but we include it here for completeness. A more detailed description may be found in [32].

We considered spectral estimators for a field in chapter 3 and for a point process in chapter 10. We restate the Fourier transform for a time series x,

$$X_j = \sum_{n=1}^{N} x_n \exp(-2\pi i f_j t_n), \tag{11.1}$$

where x_n is the signal at time index $t_n = n\Delta$, and the frequencies $f_j = j/T$, where $j = \{-N/2+1, -N/2+2, \ldots, N/2-1, N/2\}$. The spectral density of the time series is then

$$S_{xx,j} = \frac{2\Delta^2}{T} X_j X_j^*. \tag{11.2}$$

Here, the time series can be either a field (i.e., the LFP) or a point process (i.e., the spike train). Notice that we employ the same mathematical formula to compute the spectrum for each time series.

> For the spike train data, we first subtract the mean or expected number of spikes in each time interval and then apply the Fourier transform. In other words, the signal is the *centered increments* (see chapter 8).

Then, to estimate the coherence between two time series x and y, compute

$$\kappa_{xy,\,j} = \frac{|< S_{xy,\,j} >|}{\sqrt{< S_{xx,\,j} >}\sqrt{< S_{yy,\,j} >}}, \tag{11.3}$$

where $|< S_{xy,\,j} >|$ indicates the magnitude of the trial-averaged cross-spectrum, and $|< S_{xx,\,j} >|$ and $|< S_{yy,\,j} >|$ indicate the magnitude of the trial-averaged spectra of x and y, respectively. So far, there's nothing new here; we've just restated the standard expressions for the spectrum and coherence. To compute the spike-field coherence, we simply interpret one of the time series as a point process. To make this more apparent in the mathematical expression, we replace x in (11.3) with the symbol n as a reminder that this time series represents the number of spikes,

$$\kappa_{ny,\,j} = \frac{|< S_{ny,\,j} >|}{\sqrt{< S_{nn,\,j} >}\sqrt{< S_{yy,\,j} >}}. \tag{11.4}$$

In (11.4) the numerator is now the magnitude of the trial-averaged cross-spectrum between the field y and spikes n, and the denominator contains the trial-averaged spectrum of the spike n and the trial-averaged spectrum of the field y.[2]

Computing the Spike-Field-Coherence in MATLAB. As discussed in chapters 3, 4, and 10, many issues are involved in spectral analysis, for example, the notions of tapering. These important issues apply for the computation of spike-field coherence as well. In practice, multitaper methods are used to compute the spike-field coherence. An important advantage of the multitaper method is that we make a favorable trade-off: a reduction in variance for a loss of frequency resolution. We briefly illustrate the application of this multitaper method to compute the coherence here. To do so, we make use of the software package Chronux [2] (see section 1.24 of chapter 1).

Let's now apply the multitaper method to compute the spike-field coherence for the data of interest here. It's relatively straightforward to do so in MATLAB:

```
dt = t(2)-t(1);                 %Define the sampling interval.
Fs = 1/dt;                      %Define the sampling frequency.
TW = 3;                         %Choose time-bandwidth product of 3.
ntapers = 2*TW-1;               %Choose the no. of tapers.
%Set the parameters of the MTM,
params.Fs = Fs;                 %... sampling frequency,
params.tapers=[TW,ntapers];     %... time-bandwidth product & tapers,
params.pad = -1;                %... no zero padding.
params.trialave = 1;            %... trial average.
%Compute the MTM coherence.
[C,~,~,Syy,Snn,f]=coherencycpb(transpose(y),transpose(n),params);
```

In the third and fourth lines we define the time-bandwidth product (`TW`) and the number of tapers to apply (`ntapers`). Then, in lines 6–9, we set the parameters of the multitaper method using the variable `params`; this includes the sampling frequency (`Fs`), zero padding (here set to none), and trial averaging of the results (`params.trialave=1`). Finally, we estimate the spectra and coherence of the spike and LFP data in line 11 using the Chronux

2. We could instead write the *sample* coherence because this equation uses the observed data to estimate the theoretical coherence that we would see if we kept repeating this experiment. However, this distinction is not essential to the discussion here.

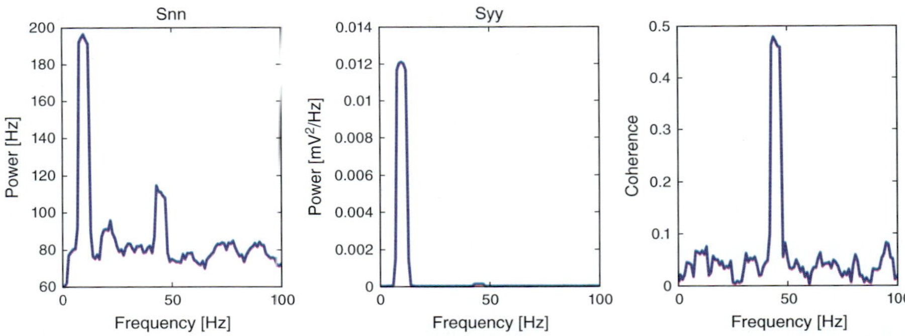

Figure 11.4
Spike spectrum (S_{nn}), field spectrum (S_{yy}), and spike-field coherence (*right*) computed using multitaper method.

function `coherencycpb`. The format of this function requires that we input the data in the form [samples × trials], and so we apply the `transpose` function to the variables `y` and `n` as inputs to `coherencycpb`. The results of these computations are plotted in figure 11.4.

Q: Consider the spike spectrum, S_{nn}, plotted in figure 11.4. What are the dominant rhythms? At frequencies beyond these dominant rhythms, the spectrum appears to fluctuate around a constant value. What is this constant value?

A: To answer the first question, we determine through visual inspection of figure 11.4 that the dominant rhythm (i.e., the frequency with the most power) occurs at 10 Hz. We also note the presence of a second peak near 45 Hz.

To answer the second question, we note that the spike spectrum asymptotes at the expected spike rate (see chapter 10). For these data, we can estimate the expected spike rate as

```
lambda = mean(sum(n,2))/(N*dt);
```

Computing this quantity in MATLAB, we find an expected spike rate of approximately 89 Hz, consistent with the high-frequency behavior of S_{nn} plotted in figure 11.4.

Q: Consider the field spectrum, S_{yy}, plotted in figure 11.4. What are the dominant rhythms? Do you observe any other interesting features in this spectrum?

A: Visual inspection of figure 11.4 reveals that the dominant rhythm occurs at 10 Hz. At first glance, no additional spectral features stand out.

These observations of the spike spectrum and field spectrum reveal that both signals exhibit rhythmic activity at 10 Hz. Therefore, a reasonable place to look for spike-field coherence is near 10 Hz, where both the spikes and the field are rhythmic. However, visual inspection of the spike-field coherence (figure 11.4) does not indicate coherence at this frequency. Instead, we find a large peak in the spike-field coherence at 45 Hz. Identifying this strong coherence at 45 Hz suggests that we reexamine the spectra. Indeed, careful inspection of the spike spectrum and field spectrum does suggest rhythmic activity at 45 Hz.

Q: Consider the field spectrum on a decibel scale (see chapter 4). What rhythms do you observe?

Q: Compare the results of your spike-field coherence analysis with the FTA plotted in figure 11.3. How does the peak in the spike-field coherence relate to interesting structure in the FTA?

The spike-field coherence results again reveal an important feature of coherence analysis. Two signals with high power at the same frequency are *not* necessarily coherent at this frequency; two signals may possess rhythmic activity at the same frequency, but these rhythms may not coordinate across trials. Conversely, two signals with low power at the same frequency may have strong coherence at that frequency; although the rhythm is weak, the two signals may still coordinate activity across trials at this frequency. These notions apply both to spike-field coherence and field-field coherence (the latter illustrated in chapter 4).

The multitaper method to compute the spike-field coherence is a powerful tool in our data analysis arsenal. There's much more to say about this approach, and interested readers are directed to [32, 34, 35].

The Impact of Firing Rate on the Spike-Field Coherence. Often, in the analysis of neural data, we compare the coherence between two pairs of signals. For example, in analysis of scalp EEG data, we might compare the coherence between voltage activity recorded at electrodes A and B with the coherence between voltage activity recorded at electrodes A and C. If we find that electrodes A and B have higher coherence at some frequency than electrodes A and C, we may conclude that the two brain regions A and B coordinate more strongly at this frequency. In this thought experiment, we are comparing the *field-field* coherence, which is not affected by the amplitude of the signals. For example, if we multiply the amplitude of signal C by a factor of 0.1, the field-field coherence does not change. To gain some intuition for this result, note that in the computation of the coherence (11.3), we divide by the spectrum of each signal. In this way, a multiplicative change in signal amplitude appears in the numerator and denominator of the coherence formula and therefore (in this case) factors out.

We might expect the same for spike-field coherence. To test this, let's manipulate the experimental data provided by our collaborator. We scale the field data by a factor of 0.1 and recompute the spike-field coherence. Scaling the field data is easy to do in MATLAB:

```
load('Ch11-spikes-LFP-1.mat')    %Reload the multiscale data,
y = 0.1*y;                       %...and scale the LFP.
```

With this change in the LFP data (y), we now recompute the spike-field coherence. We find that this multiplicative change in the amplitude of the field data does not impact the spike-field coherence (figure 11.5). This result is consistent with our intuition from field-field coherence; the height of the field does not matter. Instead, it's the consistency of the phase relation between two signals across trials that is critical for establishing the coherence.

Now, let's consider manipulating the spiking data. Right away, we notice a difference compared to the field data. In this case, a direct multiplicative change of the spiking data does not make sense. For example, consider multiplying the spike train data (n) by a factor of 0.1. Recall that the spike train data consist of two values: 0 or 1. Therefore, the new data after the scaling consist of two values: {0, 0.1} and the interpretation of the variable n no longer makes sense. What does it mean to have 0.1 spikes in a time interval? Instead, to

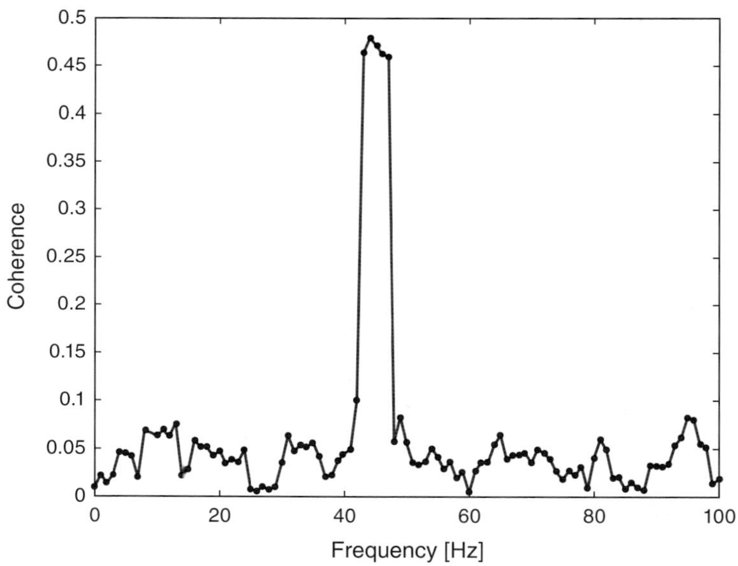

Figure 11.5

Spike-field coherence for scaled LFP data (*blue*) is identical to spike-field coherence for original, unscaled LFP data (*black dots*).

scale the spiking data, we change the average firing rate. We do so in a particular way: by removing spikes from the data, a process we refer to as *thinning*. The process of thinning is useful when comparing the spike-field coherence computed for two neurons with different firing rates. A reasonable, intuitive worry is that the firing rate of a neuron will impact the spike-field coherence. For example, we might consider that a neuron with a higher firing rate has the advantage of more opportunities to align with the field and therefore necessarily will possess a larger spike-field coherence. By thinning, we reduce the higher firing rate and establish the two neurons on an equal footing, both with the same opportunity to align with the field. The objective of the thinning procedure is to eliminate the contribution of firing rate differences to the spike-field coherence and allow direct comparison of spike-field coherence results computed for different neurons.

Let's now thin the spiking data. Here, we implement a simple procedure by randomly selecting and removing spikes from each trial of the spiking data. We assume that in selecting spikes at random to remove, we eliminate both spikes phase-locked to the field and spikes independent of the field. In this way, neither spikes coupled to the LFP nor spikes independent of the LFP receive preferential treatment in the thinning procedure. So, any relations that exist between the spikes and the field are presumably preserved, and we might expect this thinning procedure, on its own, to not affect the spike-field coherence. Let's implement this thinning procedure in MATLAB:

```
load('Ch11-spikes-LFP-1.mat')          %Reload the multiscale data.
thinning_factor = 0.5;                 %Choose a thinning factor.
for k=1:size(n,1)                      %For each trial,
 spikes = find(n(k,:)==1);             %...find the spikes,
 n_spikes = length(spikes);            %...determine no. of spikes,
 spikes=spikes(randperm(n_spikes));%...permute spikes indices,
 n_remove=floor(thinning_factor*n_spikes);
                                       %...no. of spikes to remove,
 n(k,spikes(1:1+n_remove))=0;          %... remove the spikes.
end
```

Note that within the for-loop, we first find the indices corresponding to spikes in trial `k`. We then randomly permute these indices, and select the appropriate proportion of these indices for removal. Figure 11.6 illustrates the results of this thinning procedure. In figure 11.6a we plot the thinning factor (i.e., the proportion of spikes removed from each trial) versus the empirical firing rate estimated from the data. As expected, removing more spikes decreases the firing rate.

We also plot in figure 11.6b the spike-field coherence for four different levels of thinning, one of which corresponds to the choice of a thinning factor of 0.5. We find that as the number of spikes removed increases, the peak of spike-field coherence decreases. Why? Intuition suggests that removing spikes at random (i.e., removing spikes coupled to the phase of LFP and removing spikes independent of the LFP) should preserve the spike-field

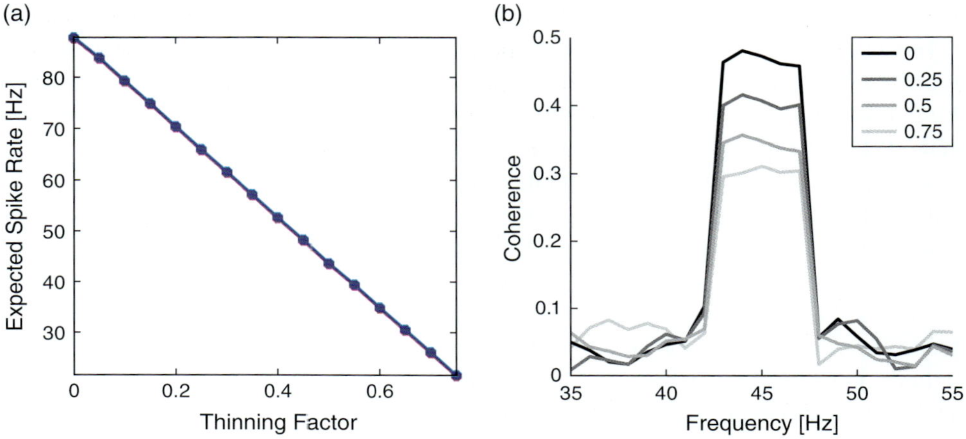

Figure 11.6

Spike-field coherence for multiscale data with original and thinned spike train. (*a*) Expected spike rate depends on the amount of thinning. (*b*) Spike-field coherence decreases as the thinning increases. Legend indicates value of `thinning_factor`.

coherence. Perhaps we were unlucky in the thinning procedure and selected to remove more phase-locked spikes than non-phase-locked spikes?

Q: Repeat the analysis with `thinning_factor = 0.5` to select another random batch of spikes to remove. How does the spike-field coherence change compared to `thinning_factor = 0.0`? Try this a couple of times, and investigate the peak spike-field coherence at 45 Hz. Is the peak in the spike-field coherence always reduced upon thinning?

Repeating the thinning procedure and selecting new instances of random spikes to remove preserves the qualitative result. The peak spike-field coherence at 45 Hz decreases. Perhaps we made a conceptual error in the thinning procedure or an error in the MATLAB code? In fact, this result is not a numerical artifact or an error in the code or a problem with the estimate; it's a property of the spike-field coherence. In [32] it's proven that the spike-field coherence depends on the firing rate. An important result from [32] is:

As the firing rate tends to zero, so does the spike-field coherence.

Therefore, we must be very careful when interpreting the spike-field coherence, especially when comparing the spike-field coherence of two neurons with different firing rates. A reduction in spike-field coherence may occur either through a reduction in association between the spikes and the field, or through a reduction in the firing rate with no change in association between the spikes and the field. This is an important and perhaps counterintuitive result of spike-field coherence. The problems at the end of this chapter further illustrate this result through simulation. In addition, some procedures exist to mitigate the dependence of spike-field coherence on the firing rate, as discussed in the next section.

> Spike-field coherence responds to overall neural spiking activity, making comparisons between two pairs of spike-field time series difficult when the average spike-rate differs in the two spike-field pairs [32].

11.2.3 Point Process Models of the Spike-Field Coherence

A variety of techniques exist to address the impact of firing rate on the spike-field coherence. We have already outlined the thinning procedure, a transformation-based technique in which the firing rates of two neurons are made equal by randomly removing spikes. Here, we focus on an additional technique that utilizes the generalized linear modeling framework. We choose this technique (described in detail in [33]) because it allows us to utilize the GLM framework applied in chapters 9 and 10. The fundamental idea of this procedure is to model the conditional intensity of the point process λ as a function of the LFP phase. More specifically, we consider the model

$$\lambda_t = e^{\beta_0 + \beta_1 \cos(\phi(t)) + \beta_2 \sin(\phi(t))}, \tag{11.5}$$

where $\phi(t)$ is the instantaneous phase of a narrowband signal in the LFP. To compute the phase, we bandpass-filter the LFP and apply the Hilbert transform, as described in our computation of the field-triggered average. Then, using the canonical log link, we fit the GLM to the spike train data (see chapters 9 and 10) to estimate the model parameters. We note that the first parameter, β_0, accounts for the overall activity of the neuron, while the other two parameters, β_1 and β_2, capture the association between the LFP phase and spiking activity. In this way, the overall firing rate and the impact of the field on the spiking activity are separately modeled, which mitigates the impact of firing rate on the measure of spike-field association.

For the analysis of spike-field association, we select a small frequency band of interest, bandpass-filter the field data, and then estimate the phase; the procedures to do so are identical to those used to compute the FTA. Building from those steps, we now focus on the procedures to estimate the phase and GLM in MATLAB:

```
phi=zeros(K,N);                %Create variable to hold phase.
for k=1:K                      %For each trial,
  Vlo=filtfilt(b,1,y(k,:));    %...apply the filter (see FTA code),
```

```
  phi(k,:)=angle(hilbert(Vlo)); %...and compute the phase.
end

n = n(:);                       %Convert spike matrix to vector.
phi = phi(:);                   %Convert phase matrix to vector.
X = [cos(phi) sin(phi)];        %Create a matrix of predictors.
Y = [n];                        %Create a vector of responses.
[b1,dev1,stats1]=glmfit(X,Y,'poisson'); %Fit the GLM.

phi0=transpose((-pi:0.01:pi));        %Define new phase interval,
X0 = [cos(phi0) sin(phi0)];           %...and predictors,
[y0 dylo dyhi]=glmval(b1,X0,'log',stats1);%...evaluate the model.
```

Note that the filter settings in the variable `b` of line 3 are defined in the MATLAB code used to compute the FTA. In fact, the first five lines are similar to the MATLAB code used to compute the FTA. The difference is that we now store the phases in a matrix (variable `phi`) that we use in the GLM procedure. To fit the GLM, we first collect the spikes and phases across trials into a vector (lines 6 and 7). Then we define the predictors (`X`, which are functions of the phase) and the response (`Y`, the spikes) and fit the model using the function `glmfit`. Finally, we evaluate the model for a chosen phase interval using the function `glmval`.

The modeling estimates are shown in figure 11.7: the FTA (identical to figure 11.3) and the estimates of the GLM (computed using the MATLAB function `glmval`). The agreement is excellent. Notice for the 9–11 Hz frequency band the lack of modulation in the estimated conditional intensity, which suggests that the probability of spiking is not affected by the phase of the LFP in the 9–11 Hz frequency range. Let's now repeat this analysis but instead bandpass-filter the LFP data for 44–46 Hz; we choose this frequency interval motivated by the spike-field coherence results (see figure 11.4). Now, for this frequency interval, we find a modulation of the estimated conditional intensity, with an increase in the probability of a spike near 0 radians. These results illustrate the close correspondence between the FTA and GLM procedures. An important advantage of the GLM approach is the ability to estimate confidence intervals. The confidence intervals in figure 11.7 are estimated in the `glmval` function and returned as outputs `dylo` and `dyhi`.

For the LFP data filtered at 44–46 Hz, let's check the significance of the parameters related to the LFP phase, β_1 and β_2, via a Wald test (see chapter 9):

```
pval1=stats1.p(2);        %Significance of parameter beta_1.
pval2=stats1.p(3);        %Significance of parameter beta_2.
```

We find that β_1 is highly significant (`pval1=1.2903e-52`) and β_2 is not significant (`pval2=0.7087`), and we conclude from (11.5) that the firing rate is highly dependent on the cosine of the LFP phase.

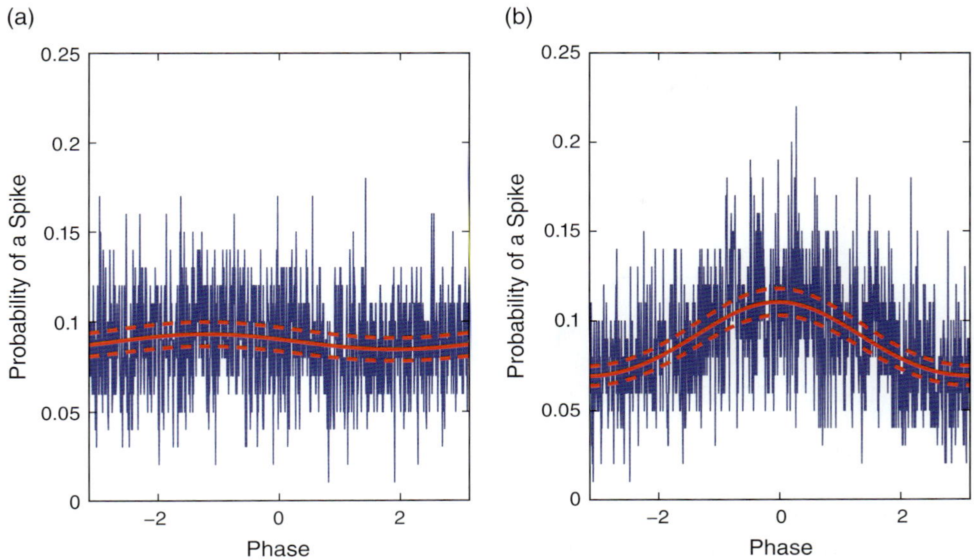

Figure 11.7
FTA (*blue*) and GLM (*red*) estimates of spike-field association for data filtered at (*a*) 9–11 Hz and (*b*) 44–46 Hz. Red dashed lines indicate 95% confidence intervals.

In chapter 9, we showed that for nested models (where one model can be made equivalent to the other by setting some parameters to specific values), under the null hypothesis that the data arise from the smaller model, the difference in the deviance between the two models should have a chi-square distribution where the number of degrees of freedom is equal to the number of extra parameters in the larger model. In this case, let's compare the model (11.5) to a model that lacks dependence on the LFP phase (i.e., a reduced model in which β_1 and β_2 are set to zero). First, we must construct and estimate this reduced model. In MATLAB,

```
X0 = ones(size(phi));                    %Define reduced predictor.
                                         %Fit reduced model.
[b0,dev0,stats0]=glmfit(X0,Y,'poisson','constant','off');
```

Here `X0` is a constant predictor, and we omit the default constant term in `glmfit`. Then, to compute the p-value for this test in MATLAB,

```
pval = 1-chi2cdf(dev0-dev1,2);           %Compare two nested GLMs.
```

Q: Why do we set the second input to the function `chi2cdf` equal to 2?

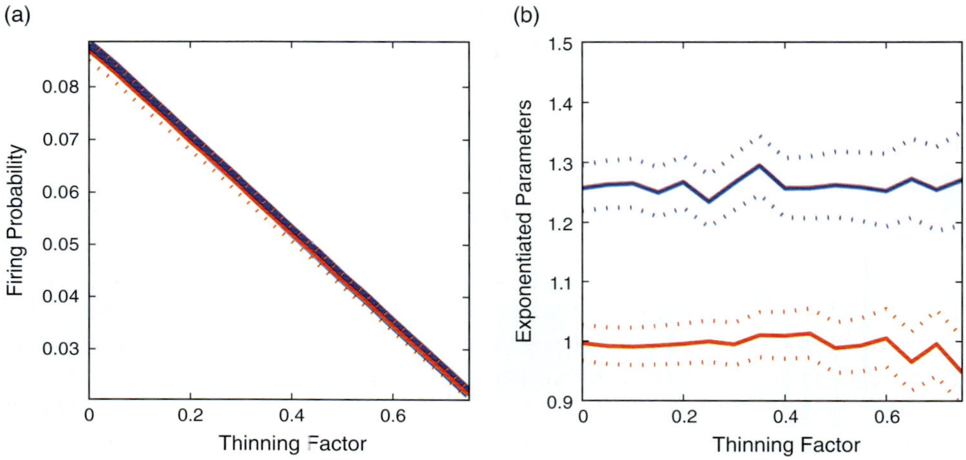

Figure 11.8

GLM framework identifies consistent spike-field association despite changing firing rates. (*a*) Direct estimate of the probability of a spike (*blue*) and GLM estimate of e^{β_0} (*red*) are consistent. (*b*) Estimates of e^{β_1} (*blue*) and e^{β_2} (*red*) do not change significantly with thinning factor. Dotted lines indicate 95% confidence intervals.

We find `pval=0` and would therefore be very unlikely to see this result if the reduced model (i.e., the model lacking dependence on the LFP phase) were correct.

Finally, let's examine how thinning the spiking data impacts the results of the GLM procedure. We choose a range of thinning factors (see figure 11.6) and repeat the analysis for LFP filtered at 44–46 Hz to recompute the FTA and estimate the GLM model for each thinning factor. Figure 11.8 shows how the estimates of the exponentiated model parameters (e^{β_0}, e^{β_1}, and e^{β_2}) vary as a function of thinning factor. As the thinning factor increases, the probability of a spike decreases. This probability can be estimated from the spike train data as

```
p = mean(sum(n,2)/(N));
```

where `n` corresponds to the thinned spike train.

Figure 11.8a plots the estimate of the probability versus e^{β_0} estimated from the GLM model. The two are in excellent agreement; as expected, as the thinning factor increases, the probability of a spike decreases. Figure 11.8b shows the exponentiated estimates of the other model parameters, e^{β_1} and e^{β_2}. Recall that these parameters represent the impact of the LFP phase on the firing rate. There are two important features to notice about these estimates. First, the thinning factor does not affect the model estimates. The changing firing rate is captured in the parameter β_0 and prevented from impacting the estimates of the spike-field association expressed in the other model parameters. Second, the exponentiated parameter e^{β_1} is significantly above 1, which indicates a significant association between the cosine phase of the LFP in the 44–46 Hz frequency band and the spiking.

Summary

In this chapter, we considered associations between data recorded from different spatial scales: the macroscale LFP and the microscale spiking. We developed methods to visualize the associations between scales, and applied tools developed in other chapters, such as the spectrum and coherence. We computed the spike-field coherence and found a strong association between the spatial scales near 45 Hz despite only the weak appearance of this rhythm in the spectra. We also considered the impact of firing rate on the spike-field coherence and illustrated that as the firing rate decreases, so does the spike-field coherence [32].

To account for the impact of firing rate on coherence we implemented a GLM, in which the firing rate depends on the LFP phase. In general, GLMs provide a powerful tool to estimate spike-field associations. The example considered here illustrates the ability of the GLM framework to estimate the influence of the LFP phase on the spiking and avoid the confounding effect of a changing firing rate. For details describing this approach and its extensions, see [33].

Problems

11.1. Analyze the properties of the spike trains in the dataset `Ch11-spikes-LFP-1.mat` using the tools developed in chapter 8. Analyze the evoked response present in the LFP using the tools developed in chapter 2. Consider these results in the context of your spike-field analysis.

11.2. Load the file `Ch11-spikes-LFP-2.mat`, available at

`http://github.com/Mark-Kramer/Case-Studies-Kramer-Eden`

into MATLAB. You will find three variables. The variable `y` corresponds to the LFP data, in units of millivolts. The variable `n` corresponds to simultaneously recorded binary spiking events. The variable `t` corresponds to the time axis, in units of seconds. Both `y` and `n` are matrices, in which each row indicates a separate trial, and each column indicates a point in time. Use these data to answer the following questions.

a. Visualize the data. What rhythms do you observe? Do you detect associations between the LFP and spikes?

b. Plot the spectrum versus frequency for these data. Are the dominant rhythms in the spectrum consistent with your visual inspection of the data?

c. Compute and display the STA and FTA. Do you find evidence for associations between the two types of data?

d. Compute and display the spike-field coherence using the methods developed in this chapter. Do you find evidence for spike-field coherence?

e. Apply the GLM model (11.5) to these data. Are the results consistent with your analysis of spike-field coherence?

f. Describe your results, as you would to a colleague or collaborator.

11.3. Load the file `Ch11-spikes-LFP-3.mat`, available at

`http://github.com/Mark-Kramer/Case-Studies-Kramer-Eden`

into MATLAB. You will find three variables. The variable `y` corresponds to the LFP data, in units of millivolts. The variable `n` corresponds to simultaneously recorded binary spiking events. The variable `t` corresponds to the time axis, in units of seconds. Both `y` and `n` are matrices, in which each row indicates a separate trial, and each column indicates a point in time. Use these data to answer the following questions.

a. Visualize the data. What rhythms do you observe? Do you detect associations between the LFP and spikes?

b. Plot the spectrum versus frequency for these data. Are the dominant rhythms in the spectrum consistent with your visual inspection of the data?

c. Compute and display the STA and FTA. Do you find evidence for associations between the two types of data?

d. Compute and display the spike-field coherence using the methods developed in this chapter. Do you find evidence for spike-field coherence?

e. Apply the GLM model (11.5) to these data. Are the results consistent with your analysis of spike-field coherence?

f. Describe your results, as you would to a colleague or collaborator.

11.4. In this question, we consider a simple example that illustrates the fundamental features of spike-field coherence. Let's consider the case in which the field is a sinusoid plus Gaussian noise, and the spike train is drawn from a Bernoulli distribution. For a Bernoulli distribution, the probability of a spike at any time is not related to previous spiking behavior. In this case, we also assume no relation between the field and point process. Therefore, we expect to find no spike-field coherence. Let's simulate some synthetic data in MATLAB, compute the spike-field coherence, and see what we find.

As a first step, let's create 100 trials of multiscale data, each trial of 1 s duration with a sampling rate of 1000 Hz. We define these parameters in MATLAB:

```
K = 100;           %Define no. of trials.
```

```
N = 1000;          %Define no. of samples per trial.
dt = 0.001;        %Define sampling interval.
```

Now, let's define the synthetic data. We create in each trial a field, which here will be a sinusoid; and a spike train, which here will be drawn from a Bernoulli distribution with a probability p of a spike in each sampling interval. In MATLAB,

```
y = zeros(K,N);             %Matrix to hold field data.
n = zeros(K,N);             %Matrix to hold spike data.
for k=1:K                   %For each trial ...
    %...define the LFP as a 10 Hz sinusoid + noise.
    y(k,:) = sin(2.0*pi*(1:N)*dt * 10)+0.1*randn(1,N);
    %...draw spikes from a Bernoulli distribution lambda=0.01
    n(k,:) = binornd(1,0.01,1,N);
end
```

In this code, the frequency of the sinusoid is set to 10 Hz, and the probability of a spike in each sampling interval is set to $p = 0.01$. These choices are reasonable yet arbitrary; as part of this simulation experiment, you are encouraged to consider the impact of different simulation choices.

With these synthetic multiscale data defined, repeat the analysis performed in this chapter. In particular, estimate

a. the spectra of the field and spiking data,

b. the STA and FTA,

c. the spike field coherence,

d. the GLM model (11.5).

Do the results of each method match your expectations? In particular, do you expect to observe spike-field coherence between these simulated data?

11.5. Let's now consider a simulation in which we expect a nonzero spike-field coherence. To produce a multiscale interaction, we must introduce a relation between the spikes and the field. We do so by making the probability of a spike in each time interval a function of the field. Let's fix the number of trials (`K=100`), the number of samples per trial (`N=1000`), and the sampling interval (`dt=0.001`) at the same values used in the first simulation example. Then let's define the spike and field data for each trial. In MATLAB,

```
f = 0.01;            % Parameter for scaling of rate.
b = 1;               % Parameter for background spiking.
y = zeros(K,N);      % Matrix to hold field data.
```

```
n = zeros(K,N);          % Matrix to hold spike data.
for k=1:K                % For each trial ...
  % ...define the LFP as a 10 Hz sinusoid + noise.
  y(k,:) = sin(2.0*pi*(1:N)*dt * 10)+0.1*randn(1,N);;
  % ...draw spikes from a Bernoulli distribution,
  p = f*(b+exp(y(k,:)));%...with probability dependent
  n(k,:) = binornd(1,p,1,N);%...LFP.
end
```

Note that the probability p of a spike in each time interval depends on three factors: (1) an overall scaling term `f`, (2) a baseline level of probability `b`, and (3) the exponentiated field `exp(y)`. We exponentiate the field y so that the probability is always positive. In this way, the probability of a spike depends on the field.

With these synthetic multiscale data defined, repeat the analysis performed in this chapter. In particular, estimate

a. the spectra of the field and spiking data,

b. the STA and FTA,

c. the spike field coherence,

d. the GLM model (11.5).

Do the results of each method match your expectations? In particular, do you expect to observe spike-field coherence between these simulated data?

References

[1] P. Wallisch, M. E. Lusignan, M. D. Benayoun, T. I. Baker, A. S. Dickey, and N. G. Hatsopoulos. *MATLAB for Neuroscientists: An Introduction to Scientific Computing in MATLAB.* San Diego: Academic Press, 2008.

[2] H. Bokil, P. Andrews, J. E. Kulkarni, S. Mehta, and P. P. Mitra. Chronux: A platform for analyzing neural signals. *Journal of Neuroscience Methods* 192(1): 146–151, 2010.

[3] H. Berger. Über das Elektrenkephalogramm des Menschen. *Archiv für Psychiatrie und Nervenkrankheiten* 87: 527–570, 1929.

[4] G. Buzsáki. *Rhythms of the Brain.* New York: Oxford University Press, 2006.

[5] P. L. Nunez and R. Srinivasan. *Electric Fields of the Brain: The Neurophysics of EEG.* 2nd ed. New York: Oxford University Press, 2005.

[6] R. E. Kass, U. T. Eden, and E. Brown. *Analysis of Neural Data.* Berlin: Springer, 2014.

[7] D. Nikolić, P. Fries, and W. Singer. Gamma oscillations: Precise temporal coordination without a metronome. *Trends in Cognitive Sciences* 17(2): 54–55, 2013.

[8] D. B. Percival and A. T. Walden. *Spectral Analysis for Physical Applications.* Cambridge: Cambridge University Press, 1993; reprinted with corrections 1998.

[9] M. B. Priestley. *Spectral Analysis and Time Series.* London: Academic Press, 1981.

[10] A. K. Engel, C. Moll, I. Fried, and G. A Ojemann. Invasive recordings from the human brain: Clinical insights and beyond. *Nature Reviews Neuroscience* 6(1): 35–47, 2005.

[11] W. H. Press, S. A. Teukolsky, W. T. Vetterling, and B. P. Flannery. *Numerical Recipes in C: The Art of Scientific Computing.* Cambridge: Cambridge University Press, 1992.

[12] E. Pereda, R. Quiroga, and J. Bhattacharya. Nonlinear multivariate analysis of neurophysiological signals. *Progress in Neurobiology* 77: 1–37, 2005.

[13] B. J. He, J. M. Zempel, A. Z. Snyder, and M. E. Raichle. The temporal structures and functional significance of scale-free brain activity. *Neuron* 66(3): 353–369, 2010.

[14] P. Bak, C. Tang, and K.Wiesenfeld. Self-organized criticality: An explanation of the *1/f* noise. *Physical Review Letters* 59(4): 381–384, 1987.

[15] G. Buzsáki, C. Anastassiou, and C. Koch. The origin of extracellular fields and currents: EEG, ECoG, LFP, and spikes. *Nature Reviews Neuroscience* 13(6): 407–420, 2012.

[16] A. Bragin, G. Jando, Z. Nádasdy, J. Hetke, K. Wise, and G. Buzsáki. Gamma (40–100 Hz) oscillation in the hippocampus of the behaving rat. *Journal of Neuroscience* 15(1): 47–60, 1995.

[17] J. Chrobak and G. Buzsáki. Gamma oscillations in the entorhinal cortex of the freely behaving rat. *Journal of Neuroscience* 18(1): 388–398, 1998.

[18] A. von Stein and J. Sarnthein. Different frequencies for different scales of cortical integration: From local gamma to long-range alpha/theta synchronization. *International Journal of Psychophysiology* 38(3): 301–313, 2000.

[19] P. Lakatos, A. Shah, K. Knuth, I. Ulbert, G. Karmos, and C. Schroeder. An oscillatory hierarchy controlling neuronal excitability and stimulus processing in the auditory cortex. *Journal of Neurophysiology* 94(3): 1904–1911, 2005.

[20] P. Lakatos, G. Karmos, A. D. Mehta, I. Ulbert, and C. E Schroeder. Entrainment of neuronal oscillations as a mechanism of attentional selection. *Science* 320(5872):110–113, 2008.

[21] R. T. Canolty and R. T. Knight. The functional role of cross-frequency coupling. *Trends in Cognitive Sciences* 14(11): 506–515, 2010.

[22] A. Tort, R. Komorowski, H. Eichenbaum, and N. Kopell. Measuring phase-amplitude coupling between neuronal oscillations of different frequencies. *Journal of Neurophysiology* 104(2): 1195–1210, 2010.

[23] M. A. Kramer and U. T. Eden. Assessment of cross-frequency coupling with confidence using generalized linear models. *Journal of Neuroscience Methods* 220(1): 64–74, 2013.

[24] D. Hearn and M. P. Baker. *Computer Graphics: C Version.* Upper Saddle River, NJ: Prentice Hall, 1996.

[25] R. Scheffer-Teixeira, H. Belchior, R. Leão, S. Ribeiro, and A. Tort. On high-frequency field oscillations (>100 Hz) and the spectral leakage of spiking activity. *Journal of Neuroscience* 33(4): 1535–1539, 2013.

[26] U. T. Eden and M. A. Kramer. Drawing inferences from Fano factor calculations. *Journal of Neuroscience Methods* 190(1): 149–152, 2010.

[27] A. L. Hodgkin and A. F. Huxley. A quantitative description of membrane current and its application to conduction and excitation in nerve. *Journal of Physiology (London)* 117(4): 500–544, 1952.

[28] S. Iyengar and Q. Liao. Modeling neural activity using the generalized inverse Gaussian distribution. *Biological Cybernetics* 77(4): 289–295, 1997.

[29] P. McCullagh and J. A. Nelder. *Generalized Linear Models.* 2nd ed. London: Chapman and Hall/CRC, 1989.

[30] H. H. Jasper and H. Andrews. Electro-encephalography. III. Normal differentiation of occipital and precentral regions in man. *Archives of Neurology and Psychiatry* 39(1): 96–115, 1938.

[31] J.-S. Brittain and P. Brown. Oscillations and the basal ganglia: Motor control and beyond. *NeuroImage* 85(pt. 2): 637–647, 2014.

[32] K. Lepage, M. Kramer, and U. T. Eden. The dependence of spike field coherence on expected intensity. *Neural Computation* 23(9): 2209–2241, 2011.

[33] K. Lepage, G. G Gregoriou, M. Kramer, M. Aoi, S. Gotts, U. T. Eden, and R. Desimone. A procedure for testing across-condition rhythmic spike-field association change. *Journal of Neuroscience Methods* 213(1): 43–62, 2013.

[34] M. Jarvis and P. Mitra. Sampling properties of the spectrum and coherency of sequences of action potentials. *Neural Computation* 13(4): 717, 2001.

[35] P. Mitra and H. Bokil. *Observed Brain Dynamics.* New York: Oxford University Press, 2008.

Index

Computational Neuroscience

Terrence J. Sejnowski and Tomaso A. Poggio, editors

Neural Nets in Electic Fish, Walter Heiligenberg, 1991

The Computational Brain, Patricia S. Churchland and Terrence J. Sejnowski, 1992

Dynamic Biological Networks: The Stomatogastic Nervous System, edited by Ronald M. Harris-Warrick, Eve Marder, Allen I. Selverston, and Maurice Moulins, 1992

The Neurobiology of Neural Networks, edited by Daniel Gardner, 1993

Large-Scale Neuronal Theories of the Brain, edited by Christof Koch and Joel L. Davis, 1994

The Theoretical Foundations of Dendritic Function: Selected Papers of Wilfrid Rall with Commentaries, edited by Idan Segev, John Rinzel, and Gordon M. Shepherd, 1995

Models of Information Processing in the Basal Ganglia, edited by James C. Houk, Joel L. Davis, and David G. Beiser, 1995

Spikes: Exploring the Neural Code, Fred Rieke, David Warland, Rob de Ruyter van Steveninck, and William Bialck, 1997

Neurons, Networks, and Motor Behavior, edited by Paul S. Stein, Sten Grillner, Allen I. Selverston, and Douglas G. Stuart, 1997

Methods in Neuronal Modeling: From Ions to Networks, second edition, edited by Christof Koch and Idan Segev, 1998

Fundamentals of Neural Netowrk Modeling: Neuropsychology and Cognitive Neuroscience, edited by Randolph W. Parks, Daniel S. Levine, and Debra L. Long, 1998

Neural Codes and Distributed Representations: Foundations of Neural Computation, edited by Laurence Abbott and Terrence J. Sejnowski, 1999

Unsupervised Learning: Foundations of Neural Computation, edited by Geoffrey Hinton and Terrence J. Sejnowski, 1999

Fast Oscillations in Cortical Circuits, Roger D. Traub, John G. R. Jeffreys, and Miles A. Whittington, 1999

Computational Vision: Information Processing in Perception and Visual Behavior, Hanspeter A. Mallot, 2000

Graphical Models: Foundations of Neural Computation, edited by Michael I. Jordan and Terrence J. Sejnowski, 2001

Self-Organizing Map Formation: Foundations of Neural Computation, edited by Klaus Obermayer and Terrence J. Sejnowski, 2001

Neural Engineering: Computation, Representation, and Dynamics in Neurobiological Systems, Chris Eliasmith and Charles H. Anderson, 2003

The Computational Neurobiology of Reaching and Pointing, Reza Shadmehr and Steven P. Wise, 2005

Dynamical Systems in Neuroscience, Eugene M. Izhikevich, 2006

Bayesian Brian: Probabilistic Approaches to Neural Coding, edited by Kenji Doya, Shin Ishii, Alexandre Pouget, and Rajesh P. N. Rao, 2007

Computational Modeling Methods for Neuroscientists, edited by Erik De Schutter, 2009

Neural Control Engineering, Steven J. Schiff, 2011

Understanding Visual Population Codes: Toward a Common Multivariate Framework for Cell Recording and Functional Imaging, edited by Nikolaus Kriegeskorte and Gabriel Kreiman, 2011

Biological Learning and Control: How the Brain Builds Representations, Predicts Events, and Makes Decisions, Reza Shadmehr and Sandro Mussa-Ivaldi, 2012

Principles of Brain Dynamics: Global State Interactions, edited by Mikhail Rabinovich, Karl J. Friston, and Pablo Varona, 2012

Brain Computation as Hierarchical Abstraction, Dana H. Ballard, 2015

Visual Cortex and Deep Networks: Learning Invariant Representations, Tomaso A. Poggio and Fabio Anselmi, 2016

Case Studies in Neural Data Analysis: A Guide for the Practicing Neuroscientist, Mark A. Kramer, and Uri T. Eden, 2016

From Neuron to Cognition via Computational Neuroscience, edited by Michael A. Arbib and James J. Bonajuto, 2016